International Conference and Youth Scientific School on Materials and Technologies of New Generations in Modern Materials Science 2016

IOP Conference Series: Materials Science and Engineering
Volume 156

Tomsk, Russia
9 – 11 June 2016

Editor:

Sergey Panin

ISBN: 978-1-5108-3228-2
ISSN: 1757-8981

Printed from e-media with permission by:

Curran Associates, Inc.
57 Morehouse Lane
Red Hook, NY 12571

Some format issues inherent in the e-media version may also appear in this print version.

Printed by Curran Associates, Inc. (2016)

For permission requests, please contact the Institute of Physics
at the address below.

Institute of Physics
Dirac House, Temple Back
Bristol BS1 6BE UK

Phone: 44 1 17 929 7481
Fax: 44 1 17 920 0979

techtracking@iop.org

Additional copies of this publication are available from:

Curran Associates, Inc.
57 Morehouse Lane
Red Hook, NY 12571 USA
Phone: 845-758-0400
Fax: 845-758-2633
Email: curran@proceedings.com
Web: www.proceedings.com

Table of contents

Volume 156

International Conference and Youth Scientific School on "Materials and Technologies of New Generations in Modern Materials Science"

9–11 June 2016, Tomsk, Russian Federation

Accepted papers received: 25 October 2016
Published online: 8 November 2016

Preface

Papers

Preface

The International Conference and youth scientific school "Materials and technologies of new generations in modern materials science" (MT2016) was held at the Institute of High Technology Physics at Tomsk Polytechnic University, June 9-11, 2016, Tomsk, Russia. The conference is devoted to the 120th anniversary of Tomsk Polytechnic University and the 60th anniversary of the Department of Materials Science in Mechanical Engineering.

The conference goal was discussing the current issues in materials science, problems and methods of their solutions for creating new materials with targeted properties, development and manufacturing of construction materials, surface hardening and protective coatings, advanced materials and technologies. New approaches and ideas presented in the conference proceedings can find their practical application in enhancing reliability of machine parts and equipment in chemical, energy, building and other industries.

The mains topics of the conference were as follows:

- *Problems of strength, ductility and fatigue life of advanced structural materials*
- *Surface hardening and protective coatings*
- *Silicate and refractory nonmetallic materials*
- *Photonic Materials and technologies*

The Organizing Committee would like to thank Institute of Strength Physics and Materials Science SB RAS (Tomsk, Russia), Russian Material Science Center (Tomsk, Russia), National Research Tomsk State University (Tomsk, Russia) and Novosibirsk State Technical University (Novosibirsk, Russia) for their participation in the conference organization.

We also appreciate the financial support from the Russian Foundation for Basic Research.

The Organizing Committee of MT2016

Prof. Sergey Panin

Institute of Strength Physics and Materials Science SB RAS, Tomsk, Russia

Tomsk Polytechnic University, Tomsk, Russia

Peer review statement

All papers published in this volume of *IOP Conference Series: Materials Science and Engineering* have been peer reviewed through processes administered by the proceedings Editors. Reviews were conducted by expert referees to the professional and scientific standards expected of a proceedings journal published by IOP Publishing.

The empirical definition of total emissivity of modern super-thin liquid composite thermal insulators

M V Anisimov[1], V S Rekunov[1], M N Babuta[2], D V Lychagin[3], U N Kuznetsova[1], Nguyen Thi Hong Bach Lien[2], E V Ivanova[1, 4] and Z T Taalaybekov[1]

[1] Tomsk State University of Architecture and Building, 2, Solyanaya square, Tomsk, 634003, Russia

[2] National Research Tomsk Polytechnic University, 30 Lenina ave., Tomsk 634050, Russia

[3] Tomsk State University, Tomsk, 634050, Russia

[4] Institute of Monitoring of Climatic and Ecological Systems, SB RAS, Tomsk, 634055, Russia

E-mail: teploproekt@mail2000.ru

Abstract. Modern world trends in the field of energy and mineral resources preservation policy involves the need for a more cost-efficient use of the Earth's natural resources, including in the field of construction industry. Using insulation modern materials would largely solve this problem. The acceptability appraisal of various advanced heat-insulating blankets is a crucial task, which requires experimental verification of total emissivity empirical definition of modern super-thin liquid composite thermal insulators and their real value definition. Method of investigation is as follows: an empirical definition of blankets emissivity using the proposed laboratory equipment, which comprises a system of "gray" bodies, thermocouple probe and a source of continuous heat flux. Total emissivity of modern super-thin liquid composite thermal insulators is experimentally determined. It amounted $\varepsilon = 0.89$ for sample # 1, and $\varepsilon = 0.87$ for sample # 2 at a temperature of 35-65 °C. It was found that the actual emissivity of the samples was higher than it had been declared.

1. Introduction

The current global trends in policy of energy saving and resource economy demand different material-intensive productions (including construction industry) to use resources in efficient way. The use of modern heat insulating materials allows to solve the problem of sustainable use of the Earth's natural resources.

The ultrathin liquid composite insulating coatings (hereinafter insulating paints) of new generation have recently appeared on the market. They are presented by mascoat, tsmceramic, thermalcoat, Isollat, Astratek, Alfatek, Teplokott, Corund, etc. designed for thermal insulation of building faces and building structures as well as heat transfer pipes.

---

[1] Maksim V Anisimov
E-mail address: teploproekt@mail2000.ru

The manufacturers of these paints assure they possess exceptional heat insulating properties. One layer of such paint with thickness of 1 to 3 mm spread on engineering pipes can successfully replace well-known mineral wool heat insulators with thickness of few centimeters [1, 2]. At the same time, some authors proved in their works [3, 4] that the real values of the thermal conductivity and emissivity factor of these materials differ greatly from published data of manufactures. Studies undertaken in 2015 [5, 6] had shown that the stated thermal conductivity ratio of two selected liquid insulating paints of known brands was not confirmed.

Also manufacturers often provide data on special heat-insulating properties of paints. They claim the capacity covered with the specified coatings takes on the properties of "thermos".

In the present context the urgent objective is to determine experimentally the total emissivity as a key indicator of the body's ability to reflect heat radiation for modern liquid insulating coatings. In this connection, we decided to carry out the experiment on determination of the total emissivity of some liquid insulating coatings to find their true value.

2. Review of existing methods

While planning an experiment to determine the total emissivity of liquid insulating coatings, the review of existing methods on the stated research topic was carried out [7-10]. We also considered the works of foreign [11-18] and Russian authors [19, 20], who conducted research on similar topics. The analysis of the existing methods of experimental determination of the total emissivity of various materials allows to make the following conclusions:

1. Typically, the stated methods are complicated and require a large amount of expensive equipment, or the presence of vacuum in the system.

2. A number of methods are not designed to conduct research ε for ultrathin liquid composite insulating coatings.

3. In some methods heat condition can cause internal deformation or heat damage of insulating paints, as they are not designed to work at temperatures above 200 °C. Furthermore, the change of heat condition can result in changes of coatings' thermal properties.

4. There are restrictions on the samples geometry.

5. The amount of heat consumed for test samples heating requires accurate records.

3. Theory of measurement

Based on the results of the analysis of the existing research methods, we decided to develop our own algorithm for calculation of ε for the stated insulating coatings. It would allow to perform measurement with satisfactory accuracy taking into account the physical properties of the samples.

Stefan-Boltzmann law proposes the expression for the calculation of gray body self-radiation flux density (emissivity):

$$E = \varepsilon \cdot E_0 = \varepsilon \cdot \sigma_0 \cdot T^4 = \varepsilon \cdot c_0 \cdot \left(\frac{T}{100}\right)^2 = c \cdot \left(\frac{T}{100}\right)^4 \quad (1)$$

where $c = \varepsilon \cdot c_0$ is the coefficient of gray body emissivity, W/(m^2·K^4); T – temperature of body, K.

In cases where we observe a radiation heat transfer in closed system consisting of two gray bodies separated by diathermancy environment (Figure 1), the resulting heat radiation flux should be calculated as follows:

$$Q_{w,1} = \varepsilon_c \cdot \sigma_0 \left(T_2^4 - T_1^4\right)\varphi_{21} \cdot F_2 \qquad Q_{w,1} = c_c \left[\left(\frac{T_2}{100}\right)^4 - \left(\frac{T_1}{100}\right)^4\right] \cdot \varphi_{21} F_2$$
$$\text{or} \tag{2}$$
$$Q_{w,2} = \varepsilon_c \cdot \sigma_0 \left(T_1^4 - T_2^4\right)\varphi_{12} \cdot F_1 \qquad Q_{w,2} = c_c \left[\left(\frac{T_1}{100}\right)^4 - \left(\frac{T_2}{100}\right)^4\right] \cdot \varphi_{12} \cdot F_1$$

where ε_c is the reduced degree of blackness in the system of two bodies; T_1 and T_2, F_1 and F_2 are absolute temperature and heat transfer area of the first and second body; φ_{12} and φ_{21} are radiation angular coefficients, which show the part of hemispherical radiation energy falling from one body to another.

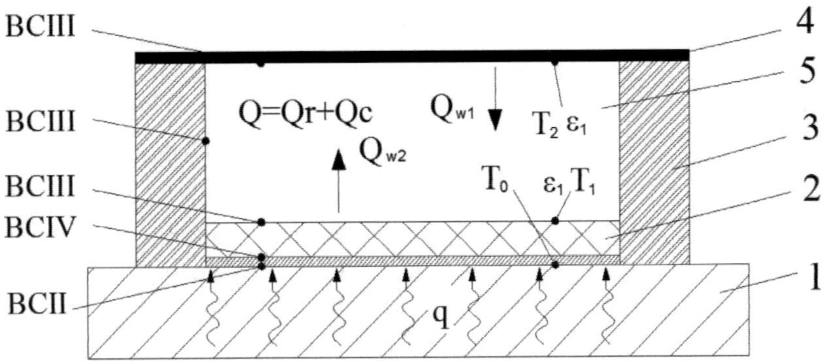

Figure 1. The proposed measurement scheme: 1 is the source of steady heat flow; 2 is the layer coating of the test applied on a copper plate with a known thickness and conductivity coefficient thermal conductivity; 3 is the insulator (polystyrene); 4 is the layer of material with a certain degree of blackness; 5 is the diathermancy environment (air)

For a closed system (figure 1) from the law of conservation of energy it follows the equality:

$$Q_{w,2} = -Q_{w,1}. \tag{4}$$

Taking into account that the heat source (1, figure 1) is in steady condition, we can calculate heat transfer rate in the present system by the law. The boundary conditions of the 2nd kind (BCII) between a layer of copper plate coated with insulated paint layer are formed.

$$q_u = \frac{\Delta Tu}{\frac{d_u}{\lambda_{effu}} - 2R_L}, \tag{5}$$

where d_u is the thickness of the sample during the test, m, ΔT_u is the temperature difference on the surface of copper plate and sample, °C ($\Delta T_u = T_0 - T_1$); R_L is the thermal resistance in copper plate coated with sample paint layer (m²·°C)/W. Such change is correct and does not contradict the theory of thermal process study [20].

As the air layer is absent at the border of the copper plate and coated paint, we can accept the boundary conditions of the 4th kind (BC IV). We assume that all heat goes into the paint layer.

We can accept boundary conditions of the 3rd kind (BC III) at the boundaries of the coating surfaces of reference material and heat insulator. In this case, the heat flow from the coating to the reference sample is the sum of heat transferred by convection and radiation heat presented as $Q_{w,2} = Q_R + Q_C$. As the body system (figure 1) is closed, we consider there is no convection due to the "side" horizontal movements of air. In free convection the convective coefficient value (to calculate the appropriate amount of heat *Qc*) can be calculated by Nusselt number *Nu*.

Furthermore, we consider that the entire heat flow is transferred from the sample 2 to the reference material 4 (figure 1) and areas of the stated objects are equal.

Based on our assumptions, the reduced degree of blackness in the system of bodies shall be calculated by the law

$$\varepsilon_{пр} = \frac{q}{\sigma_0 (T_1^4 - T_2^4)} \quad (6)$$

Consequently, the total emissivity of insulating paint shall be calculated by the expression

$$\varepsilon_{пр} = \frac{1}{\frac{1}{\varepsilon_1} + \left(\frac{1}{\varepsilon_2} - 1\right)} \text{ from here } \varepsilon_1 = \frac{1}{\frac{1}{\varepsilon_{пр}} - \frac{1}{\varepsilon_2} + 1}. \quad (7)$$

where ε_1 and ε_2 are the total emissivity of the first and the second bodies.

The present method demands the heat conductivity coefficient of the layer of tested insulating paint to perform the experiment.

4. Description of the experiment

We considered a list of the most well-known manufacturers of liquid insulating coatings to perform the experiment.

We selected two samples of liquid insulators for our study. These are «Izolatt» and «Teplomett» (hereinafter Sample #1 and Sample #2).

We determined in advance the heat conductivity coefficient values of samples of insulating coatings [5, 6]. The heat conductivity of copper plate with thickness $\delta = 0.5$ mm, equals to $\lambda = 384$ W / (m·°C). The ambient room temperature during the experiment was equal to $t = 24$ °C, the relative humidity is equal to $\varphi = 48\%$.

Check measurements were made to regulate meter readings during its "warming up" and heat flow transfer into steady condition (heat flow flattening). However, we recorded the dynamics of meter readings of 3 thermocouple sensors for half an hour with 5 minutes' intervals of measurement.

Calibration works were performed to test the adequacy of the developed device measuring total emissivity of insulating paints. The radiometer sensor was placed on the surface of tested coating to show the heat flow. The discrepancy does not exceed 5%.

The obtained data suggest that the error of this method is within tolerable limits. It proves the chosen scheme is correct.

5. Analysis of the results of the experiment

Sample #1 and sample #2 were tested in various temperature conditions at various heat flow to analyze the dynamics of the total emissivity, depending on the temperature of the paint sample. The obtained results are shown in figure 2. As can be seen from figure 2, the average value of the blackness degree increases as paint sample temperature also increases. This observation corresponds satisfactorily with similar results presented in [19, 20].

The experiment shows that the sample #1 has average emissivity $\varepsilon = 0.89$, the sample #2 has average emissivity $\varepsilon = 0.87$ within the temperature range from 35÷65 °C. A further increase of temperature is inappropriate, since test samples were used for painting pipelines of heating systems where the temperature conditions are limited to 100 °C. The obtained results correspond satisfactorily with experimental data obtained by other authors [6].

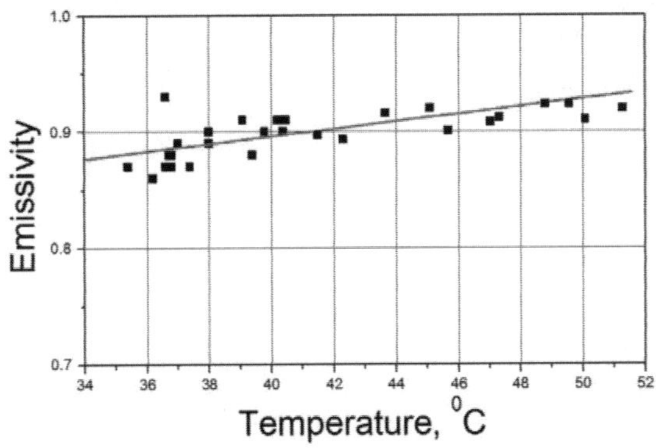

a)

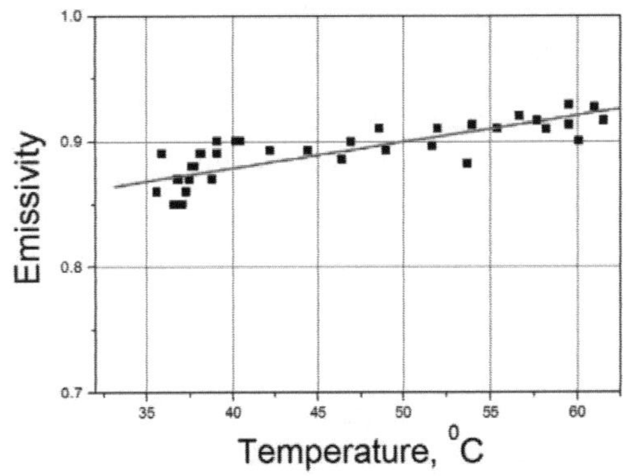

b)

Figure 2. Experimental results on determination of paint sample emissivity: a) sample #1; b) sample #2

6. Conclusion

As a result, we experimentally determined the total emissivity (ε) of ultrathin liquid composite insulating coatings. It makes ε = 0.89 for the "Sample #1» and ε = 0.87 for the "Sample #2» within the temperature range from 35÷65 °C. Comparing the results with the available data we can conclude that the degree of blackness of the tested materials is comparable by its properties to asbestos paper (ε = 0.93) or gypsum (ε = 0.9). Also the obtained emissivity is very close to white enamel varnish spread on metal sheet (ε = 0.906). At the same time, the statements of manufactures of insulating paints, assuring they possess high reflectance, are overdone.

The authors of the present article assume the discrepancy between experimental and stated characteristics is possible due to the fact, that as the selected paint samples are filled with gas but not bubble microspheres. Some manufacturers indicate this difference in microspheres, because their thermal properties are not the same.

Despite this, such liquid insulating coatings are of great interest to builders. They allow to warm objects of complex geometric shapes (valves, complex nodes, etc.), that in some cases makes them irreplaceable.

Acknowledgments
We express our gratitude to the department of "Heat supply", TSUAB (Tomsk) for providing measuring equipment, as well as to construction companies of Tomsk for providing sample materials.

References

[1] Maneshev I O, Pravnik Y I, Sadykov R A, Safin I A and Eremin S A 2014 Experimental determination of thermal conductivity and thermal insulation efficiency of ultrathin coatings *News KazGASU* **1 23** 135–142
[2] Barbolina M S and Pavlov M B 2016 Determination of liquid thermal insulation thermal conductivity coefficient of non-stationary thermal regime *Technical science* **3 39** 26–32.
[3] Shirinian V T 2007 Hike liquid ceramic "super heat-insulating" coating on the Russian heat networks *Heating News* **9** 46-51
[4] Dekusha L B, Grishchenko T G, Mazurenko L I, Vorobev A G and Shapovalov B I 2006 On the real physical properties and possibilities of "insulating" paints *Industrial Heat Engineering* **28 5** 93-96
[5] Anisimov M V and Rekunov V S 2015 Experimental determination of thermal conductivity of ultrathin liquid composite thermal barrier coatings *News Tomsk Polytechnic University* **326 9** 15-20

[6] Anisimov M V, Rekunov V S, Babuta M N and Nguyen Thi Hong Bach Lien 2016 The empirical evaluation of thermal conduction coefficient of some liquid composite heat insulating materials. *IOP Conf. Series: Materials Science and Engineering* **116 012014** doi:10.1088/1757-899X/116/1/012014).
[7] Patent RF No. 2510491 5.
[8] Patent RF No. 2132549 10.
[9] Patent RF No. 1766172 7.
[10] Patent RF No. 770333 15.
[11] Salvaggio C and Miller D P 2004 Temporal variations in the apparent emissivity of various materials *Proceedings of SPIE* **5425** doi: 10.1117/12.546321.
[12] Fronapfel E L and Stolz B J 2006 Emissivity Measurements of Common Construction Materials *InfraMation* **12 4** 22–30.
[13] Nerry F, Labed J and Stoll M P 1988 Emissivity signatures in the thermal infrared band for remote sensing: calibration procedure and method of measurement *Applied Optics* **27** 758–764.
[14] Xiaodong He, Yibin Li, Lidong Wang, Yue Sun and Sam Zhang 2009 High emissivity coatings for high temperature application: Progress and prospect *Thin Solid Films* **1 9** 5120-5129.
[15] Neuer G and Jaroma-Weiland G 1998 Spectral and Total Emissivity of High-Temperature Materials *International Journal of Thermophysics* **19 3** 122-136
[16] Watson K 2003 Two-temperature method for measuring emissivity *U.S. Geological Survey, Denver, USA* doi:10.1016/0034-4257(92)90095-2.
[17] Le Yuan, Xiaolong Weng and Longjiang Deng 2013 Influence of binder viscosity on the control of infrared emissivity in low emissivity coating *Infrared Physics & Technology* **56** 25–29.
[18] Chang-Da Wen and Issam Mudawar 2004 Emissivity characteristics of roughened aluminum alloy surfaces and assessment of multispectral radiation thermometry (MRT) emissivity models *International Journal of Heat and Mass Transfer* **47 17–18** 3591–3605.
[19] Proshkin S S 2015 Measurement of the emissivity of polymer composite materials *News MAX* **3** 73-78.
[20] Dombrovsky L A 2005 Modeling of thermal radiation of polymeric coating containing hollow microsphere *Thermal physics high temperatures* **43 1** 1–11.

IOP Conf. Series: Materials Science and Engineering **156** (2016) 012002 doi:10.1088/1757-899X/156/1/012002

Research and optimization of the technological process of manufacturing a GTE blades using computer-aided design

R A Vdovin[1, 2] and V G Smelov[1]
[1]Samara National Research University,
Moskovskoe sh. 34, Samara, 443086 Russia

[2]E-mail: vdovin.ssau@gmail.com

Abstract. This paper discusses an approach to prediction of the generation of pouring defects using the example of a technological process of manufacturing GTE blades. A computer-based modeling system for foundry processes, ProCast, was used as a tool for this research. Possible reasons of the formation of defects were considered and the recommendations for their elimination were provided.

1. Introduction

GTE blades are the most stressed and critical parts of a gas turbine engine. They are manufactured using special heat-resistant and highly durable steels, titanium- and refractory alloys featuring stringent requirements to precision and quality of the surface layer. Therefore the manufacturing processes for the production GTE blades must be efficient and must ensure the required precision parameters [1, 2].

The cost of manufacture of each blade is several thousand Russian rubles, therefore a set of full-scale experiments in order to determine the optimum parameters of the entire manufacturing process may turn out to be cost prohibitive.

Thus, there emerges a need to perform computer modeling of the existing manufacturing process (MP) of casting with the purpose of selecting the optimum process parameters.

The advantages to perform a stage of computer modeling are obvious: design optimization, reduction of processing costs, improved product quality, reduced operational expenses, etc. The mathematical modeling of foundry processes is based on heat and mass transfer equation: Heat flow equations, Navier-Stokes, diffusion, kinetic equations of crystalline transformations, etc. [3].

2. Selection of the modeling tool

There is a wide range of special software focused on foundry processes. The following systems are most widely used in the world: Magmasoft (Germany), WinCast (Germany), ProCast (France), QuickCast (France), PAM-Cast (France), CalcoSoft (France), Flow3D (USA), PowerCast (USA), SolidCast (USA), CastCAE (Finland), LVMFlow (Russia, Izhevsk), Poligon (Russia, St. Petersburg), FlowVision (Russia, Moscow), and Castflow (Australia).

Unfortunately, only a few of these software solutions proved to be efficient in the manufacturing process, i.e. Magmasoft, ProCast, SoldCast, as well as two Russian software products: Poligon and LVMFlow.

The method of finite elements is the best solution for computer modeling of the MP for casting blades of GTE. This is because this method enables to most accurately describe the geometrically complex model of a blade airfoil. Therefore, at this stage is it expedient to exclude the software products based on the methods of finite differences and control volumes. We also excluded software suites which failed to perform as powerful tools for the modeling of foundry processes aimed to produce parts of a gas turbine engine [4].

As a result of their comparison we identified three powerful CAE software suites for modeling of the foundry processes: Poligon, SolidCast, and ProCast. It is expedient to further compare these software suites based on their individual functionality.

A literature-based review of the practical experience of using the computer modeling software enables to make conclusions regarding functionality of each system [5-10]:

✓ Poligon software product is recommended for use together with another foundry software as a verification of critical castings;

✓ SolidCast software product is entry-level foundry software. It enables to model main tasks to be tacked by the existing technology. It is recommended for companies manufacturing a wide range of ***not very complex*** castings;

✓ ProCast software product has unlimited number of axes in the system, a number of units and elements in the lattice, a calculator of thermodynamic properties of materials, and all of its results may be analyzed and demonstrated in real time mode (thanks to its multicore processor). It is best suited for tackling complex engineering problems in the area of materials science and resistance of materials as it is able to reliably and accurately model the task of a strain-stress state of a casting and predicts the formation of crystallite structure in a casting.

Based on the above analysis of CAE systems of computer modeling we concluded that ProCast software is the most optimal solution for optimization MP of a GTE blades using a computer-based experiment.

3. Description of the modeling method

Below is a good case of identification and analysis of foundry defects in a casting of a GTE blades using the process of computer modeling in ProCAST software suite. As of today, the company experts have developed a technology for casting blades. However, this technology is far from ideal: after pouring-ins, the foundrymen end up with 60% of quality castings and 40% of rejects due to microporosity in the upper and lower flanges of blades and microporosity in blade airfoils. The aim of this work was to research and optimize the existing technology of manufacture of GTE blades with the purpose of increassing the quality of part blanks.

The computer modeling methodology reviewed consists of four stages [11]:

Stage One: Building a 3D geometry in a CAD system and preparation of an estimated finite-element lattice. In order to build the geometry one can use any CAD software product because the conversion to ProCAST is performed via an intermediary format, for instance, Parasolid or Step. In our case UGS NX 8.5 (figure 1) was used to build the geometry of a casting and mold.

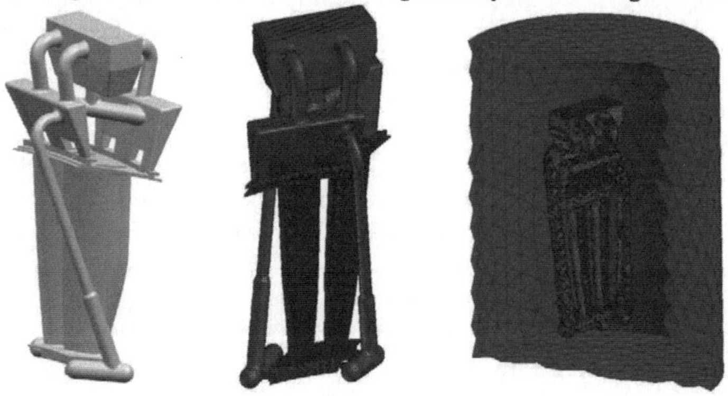

a *b* *c*

Figure 1. General view of geometry of mold piece and form:

IOP Conf. Series: Materials Science and Engineering **156** (2016) 012002 doi:10.1088/1757-899X/156/1/012002

a – 3D casting model; *b* – estimated lattice of the casting; *c* – estimated lattice of the casting and virtual mold

Stage Two: Selection of materials, borderline conditions and initial conditions of the process, and calculation parameter. We should emphasize the relevance of a thermodynamic database of ProCast which solves the problem of searching for properties of domestic alloys or selection of their foreign analogs. Using the thermodynamic database it is possible to calculate thermophysical- and mechanical properties of an alloy using its chemical composition. Calculation of properties was performed for the alloys based on Fe, Al, Cu, Mg, Ni, and Ti using the main allying components. The resultant properties have variable values in the required temperature interval which ensures high level of accuracy of the calculation.

Stage Three: Launching the task for calculation. We should note that all the modules work simultaneous, and calculation of hydrodynamics is performed concurrently – a mold is filled with an alloy, crystallization and cooldown of a casting, formation of stresses and deformation of the casting. This scenario describes the foundry processes in the most comprehensive manner.

Stage Four: Review and analysis of the calculation results. It is possible to show the results of modeling in the postprocessor both during and after the calculation.

4. Analysis of the modeling results

In the process of observation of filling the alloy in the mold one can determine the type of filling and evaluate efficiency of the gating system determine the areas of turbulent flow, the areas of high velocity of flow, possible areas of erosion of the mold or rod, and the areas of formation of air inclusions which may contribute to the formation of gas inclusions in the casting.

Based on the type of hydraulics of the mold filling (figure 2) it is possible to say that the pouring-in initial velocity was selected in the optimum way (approximately 20 seconds) because the level of metal in the pouring basin is constantly maintained, and there is no excessive splashing and inhibition effect.

The dynamics of filling the mold with metal testifies to the need of changing the metal feeding, namely the gating system. Since the side gates are not operational it is necessary to make a forced-feed riser to feed the metal from below, so that the pouring is initially performed only from the bottom, and then to ensure that a certain level of metal is fed into the risers, heating them. This is how it is possible to generate directional crystallization (bottom-up) and to ensure smoother pouring-in of a blade hollow, i.e. to make a riser with a two level feeder to the lower- and upper sections. In this case feeding of the lower flange of the blade can be improved.

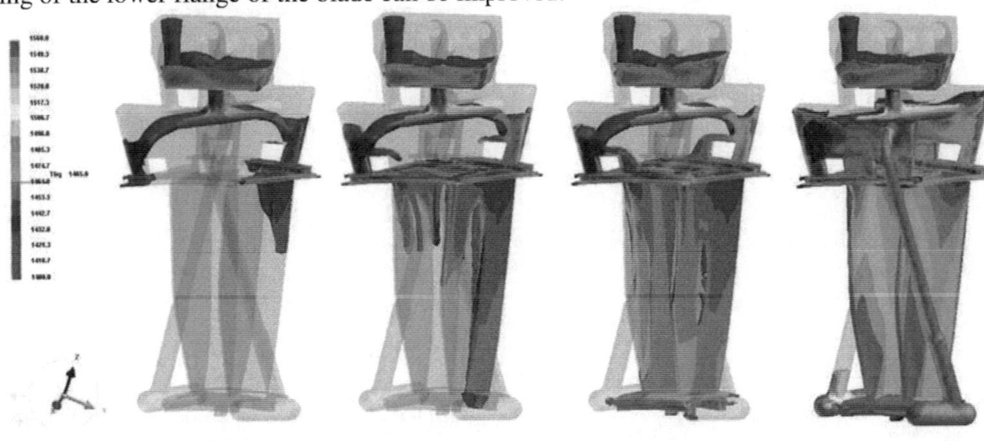

Figure 2. Pattern of filling of the casting with liquid metal.

It is important to pay attention to the pattern of filling the mold with metal (figure 3): the metal is divided into two flows inside the blade body. During pouring-in (outdoors condition) in this setting gas bubbles may appear and, consequently, this may result in the formation of an oxide film. In our case when pouring-in in a vacuum it is possible to predict the emergence of cast seams.

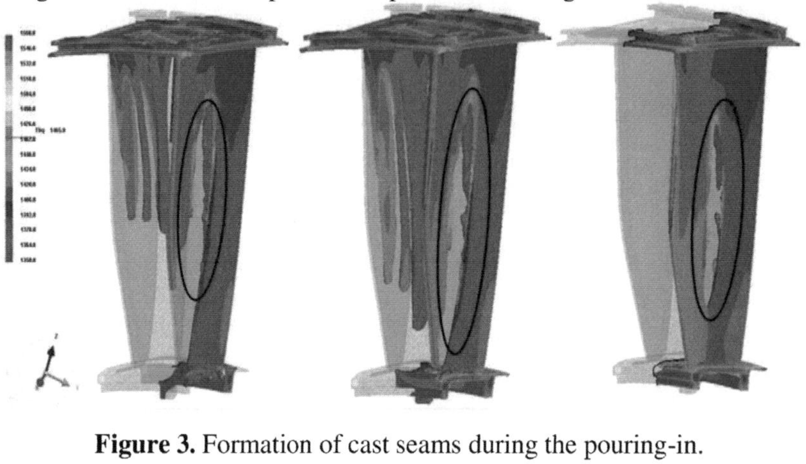

Figure 3. Formation of cast seams during the pouring-in.

Figure 4a shows the pattern of distribution of the filling velocity. Based on this velocity one can determine that velocity only increases in the blade which is due to the change in the flow section when the flow diminishes. The average velocity pressure is around 1 m/second which is not a critical value that is able to wash away the electric corundum sinterskin.

Figure 4b shows the metal flow vectors. The flow is slowed (shown in black) hence the metal is cooled. This phenomenon takes place, most likely, due to a contact with the rod in the central area.

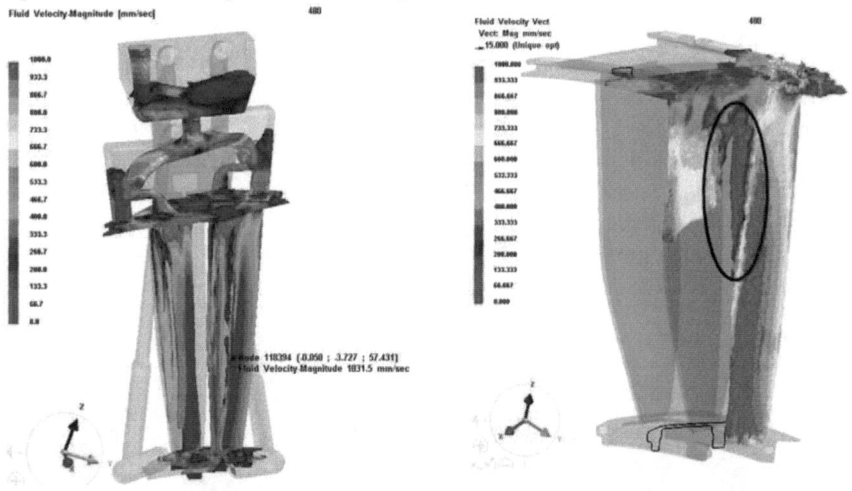

a *b*

Figure 4. Pattern of distribution of the metals dynamic pressure:
a – metal flow velocity; *b* – velocity vectors

The crystallization process begins in the lower flange of blades and in the side gates (figure 5a). The lower section of the casting – the thinnest spot – is crystallized right after the pouring-in. Crystallization of the casting takes place very quickly (approximately 18.5 seconds) as all the casting elements have thin walls. This proves that the pouring-in time was chosen correctly. The blade is thin, it is crystallized quickly, and the metal in the side gates is cold, so when the casting is not yet fully crystallized the side gates stop operating (figure 5b). Thus, two heat units (spots) appear in the lower flange of the blade (figure 5c), and this is where shrinkage and porosities may be predicted.

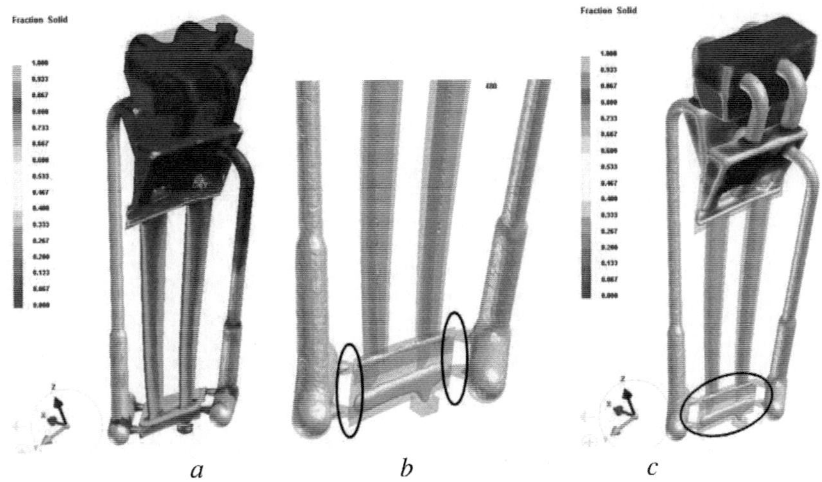

a *b* *c*

Figure 5. Crystallization of the casting of GTE blades:
a – at the initial moment; *b* – inefficiency of side risers; *c* – formation of hot spots.

Crystallization of a central section of the blade airfoil takes place quite dynamically (figure 6a). Micro-porosity is possible for this type of crystallization which is due to the phenomenon of dimensional crystallization. The blade airfoil does not crystallize evenly and the so-called 'peaks' take place – an elongated section of the liquid metal which has an unfavorable impact on the crystallization process because such zones are typified by their lack of crystallization (figure 6b). The upper risers work quite well, however, in the area of the upper flange of blades it is possible to predict minor porosities and internal shrinkage (figure 6c).

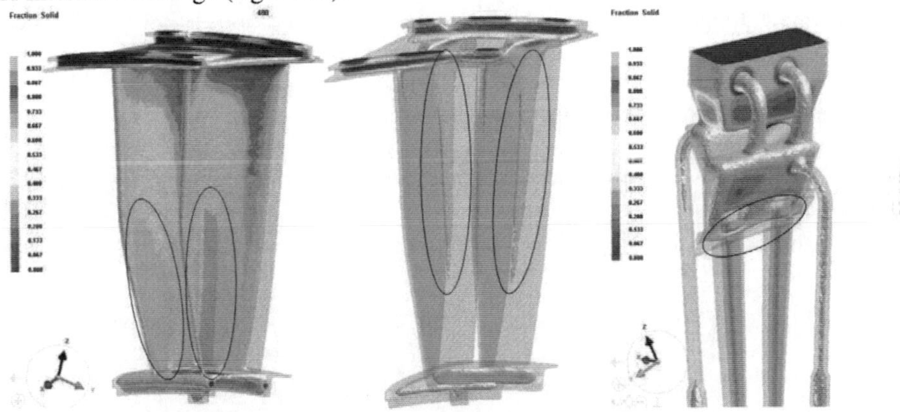

Figure 6. Control for the prediction of defects in an airfoil of blades.

Hot spots (figure 7a) indicate the crystallization time of zones in the casting; the heat zones which are the last to crystallize and where shrinkage is possible. These are the areas where internal shrinkage takes place as a result of slowed crystallization – slower than in the adjacent areas of the casting [6].

Porosity in the casting (figure 7*b*) appears in the blade airfoil, as well as in the area of upper- and lower flanges. This is due to the fact that the lower feeders and the side risers failed to compensate for the metal shrinkage.

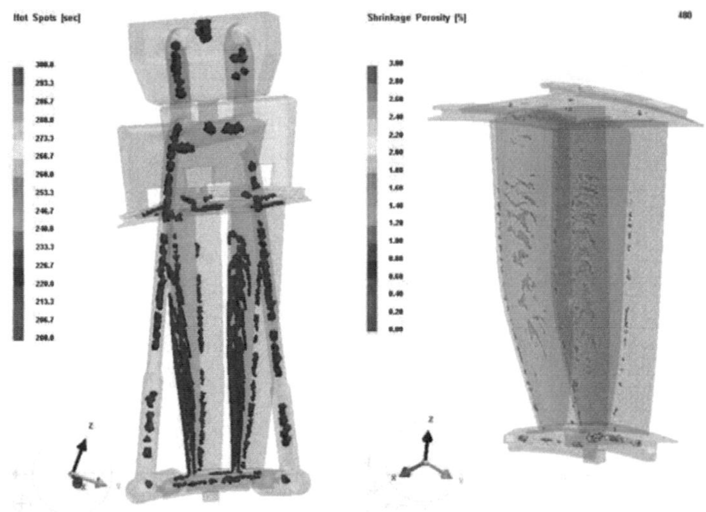

a *b*

Figure 7. Pattern of observation of porosity at macro- and micro levels:
a – formation of hot spots; *b* – formation of porosity.

5. Conclusion

The research of MP for the manufacture of GTE blades using elements of computer modeling enabled us to better understand the mechanism of formation of a casting, the structure and reasons for the formation of pouring defects, based on which recommendations were made as to ways to eliminate the defects. In order to optimize the MP for casting a blades it was proposed to slightly modify the design of gating system, i.e. to make a forced-feed riser to feed the metal from below which would avoid cast seams, formation of hot spots and porosities in the lower fen and to achieve a more even filling of the blade airfoil. The optimized parameters of the manufacturing process were tested in the ProCast computer modeling medium. The results of this modeling demonstrated lack of certain pouring defects in the casting. Sample castings of the blades were then manufactured using the adjusted MP parameters. These sample castings showed no defects in the form of cast seams and porosities in the upper- and lower fens (figure 8).

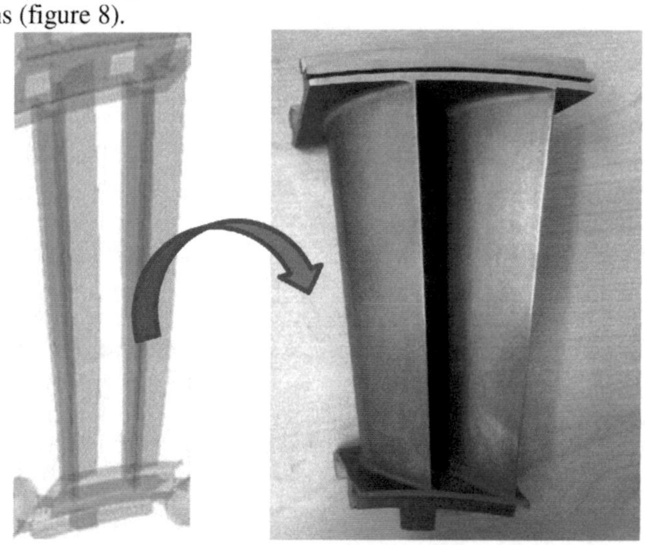

Figure 8 Lack of pouring defects in the casting of a block of nozzle unit.

IOP Conf. Series: Materials Science and Engineering **156** (2016) 012002 doi:10.1088/1757-899X/156/1/012002

References

[1] Dai H J, Souza N D and Dong H B 2011 Grain selection in spiral selectors during investment casting of single-crystal turbine blades: Part 1. Experimental investigation *Metallurgical and materials transactions.* A **42** 3430-38

[2] Rzyankina E Szeliga D Mahomed N and Nowotnik A 2013 Investigation of the effect of solidification velocity on the quality of single crystal turbine blades *Applied Mechanics and Materials.* **372** 54-61

[3] Vdovin R A and Smelov V G 2014 Elaboration of a casting defects prediction technique via use of computer-aided design systems *IJET.* **6** 2269-75

[4] Vdovin R A 2013 Multiproduct manufacture process improvement *Izvestiya of Samara Scientific Center of Russian Academy of Sciences.* **15** 612-619

[5] Kuchin P S and Maltsev H 2008 Mathematical model operation of processes of molding of metals and alloys *Molding production.* **10** 37-39

[6] Bikmeev A T, Ivanov S V and Perepelitsa S I 2012 Computer model operation of process of cooling of a hollow blade of GTE after superplastic molding *Vestnik UGATU.* **16** 38-42

[7] Szeliga D, Kubaiak K and Jarczyk G 2013 The influence of the radiation baffle on predicted temperature gradient in single crystal CMSX-4 castings *International Journal of Metalcasting.* **7** 17-23

[8] Vdovin R A, Smelov V G, Bolotov M A and Pronichev N D 2016 Paths of improving the technological process of manufacturing of GTE turbine blades *IOP Conf. Series; Materials Science and Engineering.* **142** 1-8

[9] Kermanpur A, Eskandari M, Pumohamad H, Soltani M A and Shaten R 2010 Influence of mould design on the solidification of heavy forging ingots of low alloy steels by numerical simulation *Materials and Design.* **31** 1096-1104

[10] Tarasevich N I, Korniyets I V, Tarasevich I N and Dudchenko A V 2010 Comparative analysis of systems of computer model operation of metallurgical and foundry processes *Metal and molding of Ukraine.* **5** 20-25

[11] Vdovin R A 2012 3D virtual simulation and casting process optimization for "Swirler of II duct" part of GTE with use of computer technologies *Vestnik of Samara State Aerospace University.* **3** 115-120

IOP Conf. Series: Materials Science and Engineering **156** (2016) 012003 doi:10.1088/1757-899X/156/1/012003

Research of annealing mode for high accuracy stamped parts production from titanium alloy 83Ti-5Al-5Cr-5Mo after tooling

A V Balaykin, E A Nosova and N V Galkina
Samara University, 34, Moskovskoye Shosse, Samara, Russia

E-mail: nata12_92@mail.ru

Abstract. The aim of the work is to solve question of accuracy increase in tolled and annealed parts made from forged rod of titanium alloy. Plate pieces were cut from cross-section, annealed at 800°C during 1, 2, 3, 4 and 5 hours. The criterion combining minimum bending radius and spring back angle was found. This criterion shows the maximum values after tooling and annealing for 3 hours.

1. Introduction

Alloy 83TI-5AL-5CR-5MO is used for manufacturing of high-load parts and constructions, which work for long periods of time at the temperatures of up to 350 - 400 °C (loaded components of airframes, wings, forged products, control system components, fastening pieces, e.g. eye bolts) [1]. These products are complex and require the obtaining high-accuracy of dimensions. Such products are manufactured from ingots by means of forging, stamping and finishing up of final dimensions with the use of tooling [2]. In the process of manufacturing of such products different operations of heat treatment are applied, including stress- relief annealing and mechanical hardening.

State Standard19807-91 [3] sets forth the grades of titanium alloys, intended for production of semi-finished products (sheets, bands, strips, rods, tubes, forged products and stamping billets) by means of plastic straining. Billets in the form of rods are used for manufacturing of parts for the aviation industry. According to the State Standard 26492-85 [4] the rods are manufactured without heat treatment (hot-rolled). In turn, the billets undergo the heat treatment (annealing or quenching and ageing). After annealing the material becomes highly ductile and weak. Cutting modes are chosen on the basis of the properties of the material. These properties, in their turn, are determined by the heat treatment mode. The process of mechanical treatment is accompanied by heating in the cutting area. This heating results in the structural changes of the material [5]. Such changes lead to the change of geometry of finished products. That is why additional calibration operations (fitting) may be required after tooling. Flat elements of parts are subjected to fitting by means of bending. For implementation of bending and fitting modes it is necessary to know the most appropriate heat treatment (annealing) modes, that ensure manufacturing of high accuracy products.

Finding out the cutting and fitting modes is based on the range of properties (maximum strength, yield strength, impact strength, minimum radius of bend, spring-back ratio, etc.) [6, 7]. These properties may change depending on the modes of deformational processing and heat treatment, demonstrating different tendencies: as a rule, after annealing viscosity increases, but strength, hardness and rigidity decrease [8, 9]. Consequently, there is no decisive criterion for choosing the temperature or duration of annealing. The temperature of annealing is chosen depending on the grade of material orthe alloyingdiagram. The duration of annealing depends on the size of the product, temperature and speed of heating [10]. Heating and structural transformations processes proceed slowly in massive elements of products and quickly- in thin elements. This different speed of transformations leads to inhomogeneity of the cross-sectional structure of the product [11].

In order to determine the most practical annealing modes of products made of the alloy 83TI-5AL-5CR-5MO it is necessary to find out the criterion, that comprises technological properties and that would definitely determine the best combination of properties. The above-mentioned criterion can be useful aftertooling and annealingbefore fitting.

2. Materials and methods

Samples with dimensions 2x10x30 mm were cut from the rod made of the alloy 83TI-5AL-5CR-5MO by the electro erosion cutting, which ensures the stability of structure and properties of the initialbillet. In accordance to the State Standard 90013-, the alloy has the following chemical composition: 78.485…86.6 mass.% Ti (base element), 4.4…5.9 mass.% Al, 0.5…1.5 mass.% Fe, up to 0.1 mass.% C, up to 0.15 mass.% Si, 0.5…2 mass.% Cr, 4…5.5 mass.% Mo, 4…5.5 mass.% V, less than 0.05 mass.% N, less than 0.3 mass .% Zr, less than 0.2 mass.% O, less than 0.015 mass.% H.

After cutting the samples were subjected to bending in the tool die with the apex angle of 90 ^{O} and spherical radius of 10 mm. After bending the samples were annealed at the temperature of 800 ^{O}C for the period of 1, 2, 3, 4 and 5 hours.

Geometry changes of the samples were determined on the basis of the change of the angle between the flanges of the billets after annealing. The measurements were carried out with the help of the software product COMPASS-3D. The scanned samples were used for measuring the expansion angles before and after annealing.

Then the samples were again subjected to deformation/strain until crack appearing. The radius at the billet apex was measured in order to determine the geometry change (The measurements were carried out with the help of the software product COMPASS-2D). The measurement principle is shown in figure 1.

Minimum radius of bend was calculated according to the formulae (1).

$$r_{\min} = \frac{r_{mes}}{t}, \qquad (1)$$

where r_{mes} - radius measured on the sample, mm; t - thickness of the sample, in our case $t = 2$ mm.

The spring-back ratio was calculated according to the formulae (2).

$$\gamma = \frac{\gamma_{ann} - \gamma_{init}}{\gamma_{init}}, \qquad (2)$$

where γ_{ann}- flanges expansion angle, measured on the sample after annealing, grad, γ_{init}- die apex angle, grad, in our case $\gamma_{init} = 90°$.

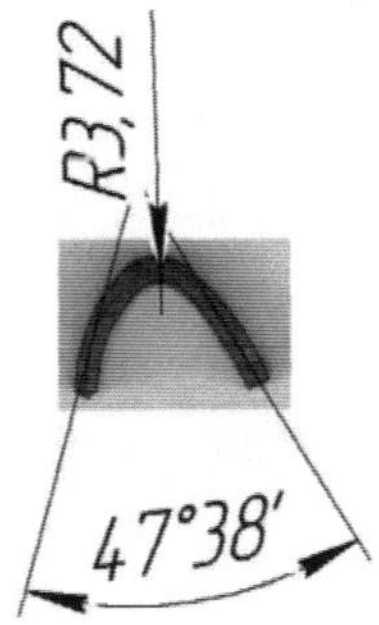

Figure 1. Radius of bend measurement scheme

3. Results

The found dependence r_{min} from the annealing duration is shown in the figure 2.

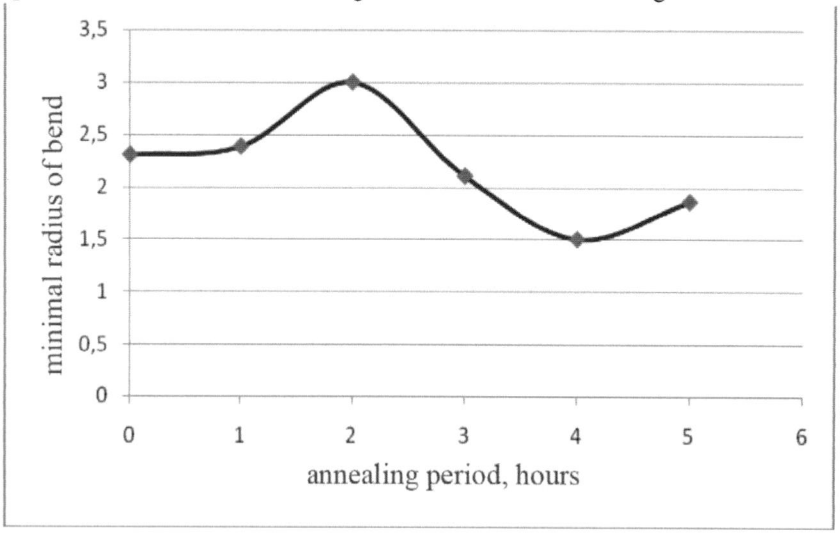

Figure 2. Dependence r_{min} from the annealing duration

The diagram shows the least value of the radius of bend hat equals 1.5, was obtained after annealing for the period of 4 hours. The minimal radius of bend (r_{min}) may be obtained without formation of a crack. Consequently, the lesser this indicator, the more effective is the conducting of bending operations, and the more exact are the obtained spherical radii of reinforcing ribs. That means that in order to conduct fitting it is necessary to perform preliminary annealing of the billet for the period of 4 hours.

Figure 3 shows the dependence of the spring-back ratio from the annealing duration. It can be seen that the diagram is non-monotonic and there are 2 minimums and 1 maximumvalue. Minimums are observed in respect of 2 and 4 hours of annealing (0.18 and 0.14 respectively). In this case it is impossible to definitely determine the optimal duration of annealing, because in case annealing is performed for the period of 2 hours, the high value of the minimal radius of bend is observed, and it equals 3 (figure 2).

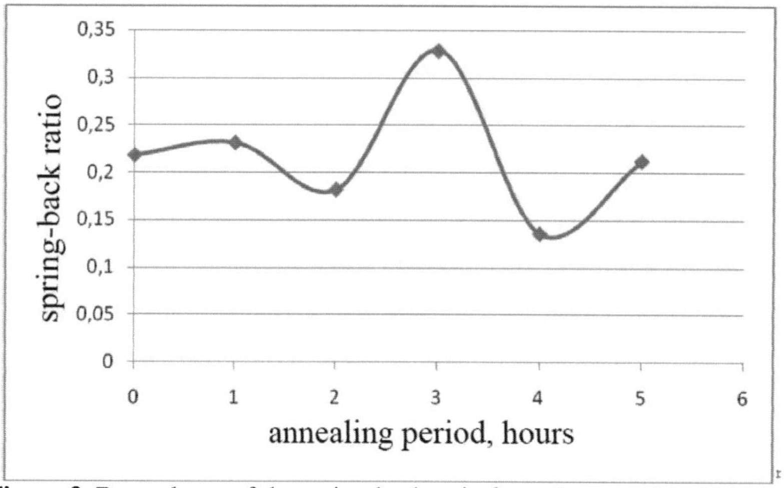

Figure 3. Dependence of the spring-back ratio from the annealingduration

In order to find out the optimal annealingduration it is necessary to take into account both indicators that demonstrate the bestfitness and suitability of the material for bending operations at their minimum values. For this purpose, the composite indicator of fitness of pressed/stamped and annealed products for fitting by means of bending was developed. This indicator is represented in the formulae (3).

$$k_{bend} = \frac{1}{r_{\min} \cdot \gamma}, \tag{3}$$

where $r_{\min}$ - minimum radius of bend, γ - spring-back ratio.

While both indicators are non-dimensional, k_{bend} will also be non-dimensional. Analyzing the structure of the formulae, the lesser the denominator, i.e. the values of $r_{\min}$ and γ, the higher is the indicator k_{bend}. That is why maximum values of k_{bend} suggest the best fitness of finished products to bending operations within fitting. Dependence of k_{bend} from the annealing duration is shown in the figure 4.

The diagram shows that the composite indicator, combining the spring-back ratio and minimum radius of bend reaches its maximum value in case of exposure to annealing for the period of 4 hours.

Microstructural analysisof the deformation zone shows, that annealing during 1 hour doesn't lead to any significantchange in the size of the grain. And in case of exposure to annealing during 2 hours some equiaxial grains were observed on the background of elongated grains. This indicates the beginning of primary recrystallization. Annealing for the period of 3 hours leads to obtaining the most fine-grained structure that may cause the increase of the spring-back ratio. Annealing for the period of 4 hours leads to a non-material grain coarsening, but the grain structure retains its homogeneity. After annealing for the period of 5 hours, the structure is characterized with the presence of some coarse equiaxial grains on the background of small grains that suggests the beginning of secondary recrystallization.

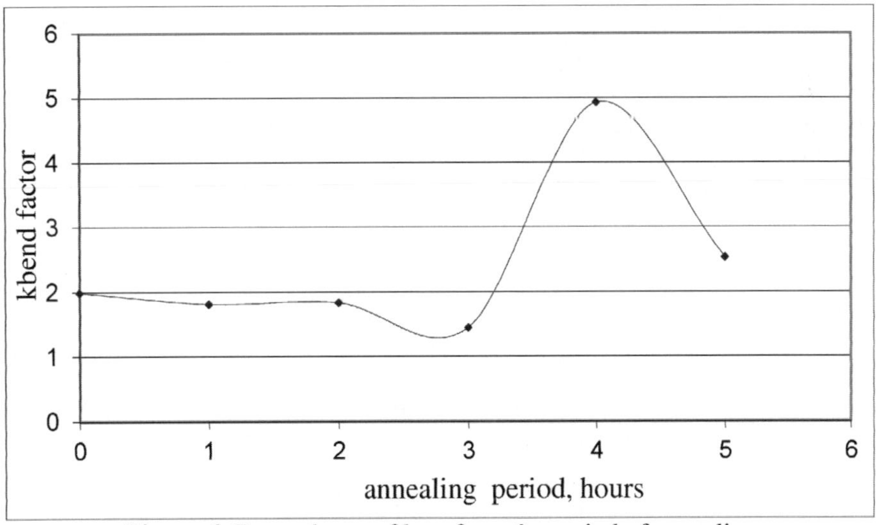

Figure 4. Dependence of k_{bend} from the period of annealing

4. Conclusions

1. Dependence of minimum radius of bend from the period of annealing shows that the most favorable mode of annealing of titanium alloy 83TI-5AL-5CR-5MO +parts is heating to the temperature of 800 ^{o}C with the period of exposure to annealing amounting to 4 hours aftertooling before fitting by bending. In these conditions the minimum radius of bend with the value of 1.5 is reached.

2. Dependence of spring-back ratio from the period of annealing reaches minimum values in case of two variants of annealing: 0.18 in case of 2 hours and 0.14 in case of 4 hours.

3. The developed criterion shows, that the most favorable combination of both indicators is observed in case annealing takes place for the period of 4 hours.

Acknowledgements
These studies were conducted on the equipment of CAM technology common use center (RFMEFI59314X0003).This work was supported by the Ministry of education and science of the Russian Federation in the framework of the implementation of the Program "Research and development on priority directions of scientific-technological complex of Russia for 2014-2020".

References

[1] Brennan F P, Ngiam S S and Lee C W 2008 An Experimental and Analytical study of Fatigue Crack Shape Control by Cold Working *Eng. Fracture Mech.* **75** 355 363
[2] Bobrov V F 1975 *Basics of the theory of cutting metals* (Moscow) p 344
[3] State Standard 19807-91 «Titanium and titanium wrought alloys. Grades»
[4] State Standard 26492-85 «Rolled rods from titanium and titanium alloys. Specifications»
[5] Afonin A V and Nosova E A 2015 Experimental research of the temperature profile in the cutting area within the framework of milling of the alloy 83Ti-5Al-5Cr-5Mo *Eng. Mag.: Science and Innovations* 5 7
[6] Catalogue of cutting instruments Sandvik http://www.sandvik.coromant.com/ru-ru/products/pages/tools.aspx
[7] Smelov V G, Sotov AV and Agapovichev A V 2016 Recovery technology features of aerospace parts by layering synthesis *Key Eng. Mat.* **684** 316 322
[8] Vdovin R A and Smelov V G 2014 Elaboration of a casting defects prediction technique via use of computer-aided design systems *Int. J Eng. Tech.* **6** 69 75
[9] Krivoukhov V A and Chubarov A D 1970 *Processing of titanium alloys by means of cutting* (Moscow: Machine engineering) p 180
[10] Iljin A A, Kolachev B A and Polkin I S 2009 *Titanium alloys. Composition, structure, properties* (Moscow: VILS-MATI) p 520
[11] Kolachev B A, Elagin V I and Livanov V A 1999 *Metallurgy and heat treatment of non-ferrous metals and alloys* (Moscow: MISIS) p 416

Experimental study of stress-strain state of adhesive joints steel / carbon fiber under tension with a bend by digital image correlation

D G Kopanitsa[1], A M Ustinov[1], A A Kondratyuk[2], Y A Abzaev[1], A I Potekaev[3], A A Klopotov[1,3], G D Kopanitsa[2]

[1]Tomsk State University of Architecture and Building, Tomsk, Russia
[2]National Research Tomsk Polytechnic University, Tomsk, Russia
[3]National Research Tomsk State University, Tomsk, Russia

E-mail: artemustinov@bk.ru

Abstract. This paper presents a study of characteristics of an evolution of deformation fields in surface layers of medium-carbon low-alloy specimens under compression. The experiments were performed on the "Universal Testing Machine 4500" using a digital stereoscopic image processing system Vic-3D. A transition between stages is reflected as deformation redistribution on the near-surface layers.

1. Introduction

It is impossible to perform a strength analysis of important composite constructional elements without a knowledge how the composite destructs. Dissipative processes lead finally to the development of cracks during plastic deformation. This requires investigation of a behavior of steel-carbon fiber adhesive joints on all the deformation phases from the initial till the out-of-limit phase.

An adhesive joint loses its strength continuously while a load increases. On the plastic phase of deformation of a loaded composite junction, one can observe a local loss of shear strength. Flows of deformation defects originate on mesoconcentrators of stress and move in the direction of the maximum shear stress. Each flow causes highly localized simultaneous shifts and turns of the material [1, 2]. To better understand these processes and reveal the mechanics of a deformation destruction it is necessary to study the evolution the deformation defects flows in the adhesive joints.

Non-uniformity of deformation's development can occur not only due to the direction of an external load but also because of a structural heterogeneity of a composite that originates on the formation phase. In case of layered composite metal-glue-carbon fiber we have different characteristics and different micro and macro structure of each layer. This leads to the different deformation processes in each layer causing a destruction of the layered composite.

During a long operation of adhesive joints there originate multiple defects that change the characteristics of the material. In this new defect state the joint still has a remaining bearing strength. This means that the mechanical characteristics are defined by the interaction of regions with different defects in different structural elements of the joint.

Composite materials are difficult to study at every deformation phase [3]. It is shown that an adhesive joint under a tensile deformation has localized deformation in various locations on a surface of a specimen. With a growth of the deformation, one can observe an evolution of deformations from a chaotic to the regular wave deformation with a specific wavelength. This makes important the studies of the behavior of adhesive junctions at supercritical deformation phase.

The object of this study is an angle with a simulated defect strengthen by carbon fiber and a resin binder (figure 1).

The goal of the research is to study *in situ* evolution of the distribution of local deformation regions in the near surface layers of composite fabrics and main metal angle under a tensile load.

2. Materials and methods

2.1. Adhesive joint metal/carbon fiber

We have prepared specimens of adhesive joints. Figure 1.a shows the adhesive specimens to be tested under an eccentric tensile deformation.

To ensure full utilization of the applied CFRP material, surface preparation of the steel must be undertaken to enhance the formation of chemical bonds between the adherend and the adhesive. This requires a chemically active surface that is free from contaminants. Most surface treatment involves cleaning, followed by removal of weak layers and then re-cleaning [4].

The most effective means of achieving a high-energy steel surface is by grit blasting [5, 6].

We used a steel roller stock angle as bases for the specimens. The mechanical characteristics of the steel were studied in the previous experiments.

The second component of the specimens was a unidirectional carbon fabric FibARM Tape-200/300. The mechanical characteristics of the lamellae are presented in table 1.

A resin compound FibARM Resin 530+ was used as a binder for the external reinforcement. The mechanical characteristics of the glue are presented in Table 1.

Table 1. Mechanical characteristics of the materials

Mechanical characteristics	Material		
	Steel	Carbon fabricFibARM Tape-200/300	GlueFibARM Resin 530+
elastic modulus E, GPa	212	~165	-
yield point, MPa	231	-	-
ultimate strength, MПa	335.4	-	-
aspect ratio, %	24.65	-	-
tensile strength, MPa	-	~ 2800	-
shear strength of Glue specimens (7 days at 23°C), MPa	-	-	~ 13

2.2. Experimental methods

The tensile tests were performed on the «INSTRON 3386» machine with a maximum tensile load of 100 kN (10.19 kgF). The deformation was performed at a constant speed of 0.3 mm/min. A digital optical system VIC-3D [4, 5] was used to investigate the evolution of the deformed state

To measure deformations on the surface of the specimen the cameras were installed opposite to the inner fold of the angle to ensure that the filmed surface will fit into the intersection of depth-of-fields of the cameras (figure 1, c). The point of intersection of optical axes matched the surface of the specimen to provide a high quality image. The value of the angle between optical axes was 30 degree. The calibration of the cameras resulted in the definition of the variable associated with positioning and specific parameters of the cameras: focal distance and sensor's midpoint (a point on the sensor that corresponds to the center of a lens), distances and angles between cameras X, Y, Z.

As the result we have obtained data about the change in positions of the lenses in 3D space. Displacement fields in our case is a projection of shifts along the OX axis (lateral deformation), OY axis (longitudinal deformation) and along OZ axis (lateral deformation) of local surface regions [6].

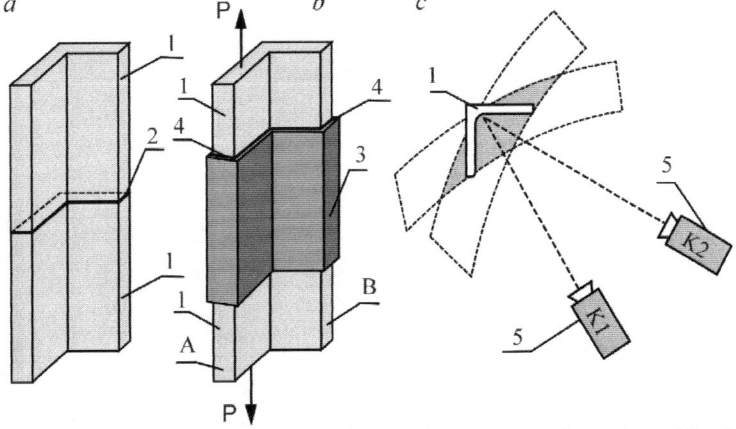

Figure 1. Experimental design for an eccentric tension: a –steel angle with simulated defect; *b*–adhesive junction metal/glue/carbon fiber and the load direction; *c*–digital cameras positioning. 1 – metal angle; 2 – simulated defect; 3 – carbon fabric; 4 – binder; 5 – digital cameras K1 and K2; A –loaded plate of the angle; B –not loaded plate of the angle

3. Results and discussion

The data about the deformation in the near to surface layers was measured using a VIC-3D system. This data was used to build a deformation diagram (figure 2) that shows the dependency of overall deformation on the value of a load. Besides that the VIC-3D system allowed to build the deformation patterns that represent a detailed evolution of the distribution of isolines of relative deformations during the load process (figure 3).

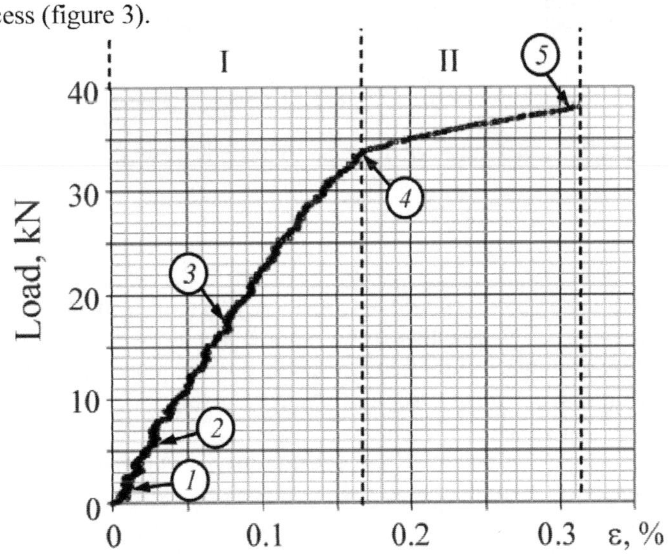

Figure 2. Loading diagram for the layered composite (metal-glue-carbon fiber). The digits show the regions of the diagrmation that correspond to the deformational patterns in the figure 3

We can distinguish 2 phases on the deformation diagram (figure 2). The first phase is an elastic deformation. The surface deformation patterns in this phase can be characterized by a chaotic distribution of local deformation regions with a distinguished overall deformed state. In this phase we can observe tension of the loaded plate of the angle (Plate A in the figure 1) and a compression on the not loaded plate (plate B in the figure 1). In the joint zone, we could observe origination and considerable increase of local deformation zones under a 8.2 kN with 0.038. This phase ends at 34 kN load and overall relative deformation value of 0.173%.

A transition to the elastic-plastic phase 2 is accompanied by the decrease of a strain-hardening coefficient $\theta = \partial P/\partial \varepsilon$ (figure 2) and a change in the positioning of structural elements (figure 3).We can observe origination of the first plastic deformation lines in the loaded plate of the angle in the region of the junction (Figure3, pattern 4). This phase ends at 38.2 kN load and overall relative deformation value of 0.323%.

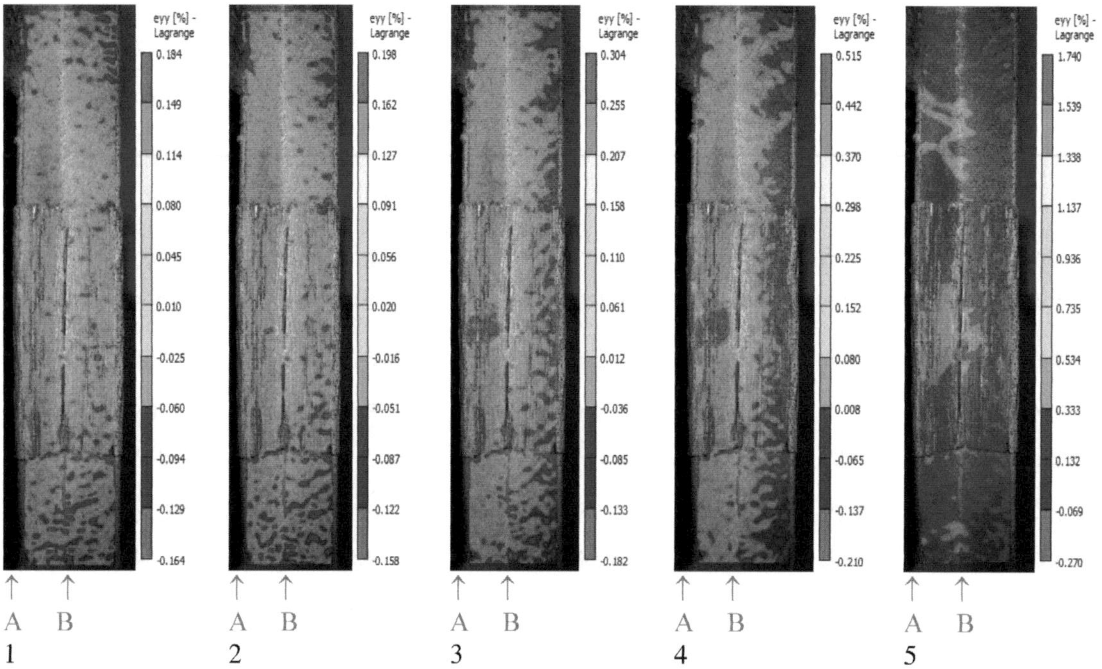

Figure 3. Patterns of vertical relative deformation of the inner surfaces of a steel angle and carbon fabric plates on the different deformation phases. A – loaded side of the angle; B – not loaded side of the angle. 1, 2, 3, 4, 5 — points in the figure 2

4. Conclusion

1. The results of the study showed a 2 phase deformation process on the strain-hardening diagrams for the steel angles joint by adhesion with carbon fabric plates under a tensile-turn load.
2. With the increase of loads and transition from the phase I to the phase II we can observe a decrease up to 8.5 times of strain-hardening coefficient θ. This is accompanied by the change in the way the structural elements are positioned.
3. On the phase II deformations of the both carbon fabric plates (loaded and not loaded) are almost similar. The local deformation regions that were observed on the phase I flow together to form a single area along the whole plate surface. This illustrates the change in the mechanics of deformation. However, a region of localized deformation on the loaded plate remains in the joint area, which originated by the end of the phase I.

Acknowledgement

The research was supported by the Tomsk State University Academic D.I. Mendeleev Fund Program in 2015-2016.

References

[1] Panin V E 1998 Modern problems of plasticity of solid bodies *Russian Physics Journal* **1** 7-34 (in Russian)

[2] Marushak P O, Panin S V, Student A Z et al 2013 *Deformation of destruction of thermoresistant steels* (Tomsk: TPU) p 236 (in Russian)

[3] Ustinov A M, Kopanitsa D G, Koshko B O et al 2015 *Development and localization of plastic deformations on a surface of metal specimens with layered composites:metal / glue / carbon fiber* (Tomsk: TPU) pp. 21-24 (in Russian)

[4] Mays G C and Hutchinson A R 1992 *Adhesives in Civil Engineering* (New York: Cambridge University Press) p 333

[5] Hollaway L C and Cadei J 2002 Progress in the technique of upgrading metallic structures with advanced polymer composites *Progress in Structural Engineering Materials* **4** **2**, 131-148

[6] Hutchinson A R 1987 Surface pretreatment – the key to durability, *Proceedings of the International Conference on Structural Faults & Repair*, University of London, July 1987, pp. 235-244

[7] Tretyakova T V 2014 VIC 3D software for studying non elastic deformations Mechanics of continua 7 pp. 162-171 (in Russian)

[8] Tretyakova T V, Tretyakov M P, VildemanV E 2011 *Estimating the precision using video system for deformation fields Reports of PGTU. Mechanics* (Perm: Perm. GTU) **2** 92-100 (in Russian)

[9] Ustinov A M, Kopanitsa D G, Potekaev A I et al 2015 Distribution of local deformations on the near-surface layers and scale effects of low alloy steel specimens *AIP Conference Proceedings* **1683** 020233

IOP Conf. Series: Materials Science and Engineering **156** (2016) 012005 doi:10.1088/1757-899X/156/1/012005

Modal testing circuit board assembly of an electronic apparatus by laser vibrometry

V A Krasnoveikin[1], I Yu Smolin[1,2,3], N V Druzhinin[1], E A Kolubaev[1] and D A Derusova[3]

[1] Institute of Strength Physics and Materials Science SB RAS, 2/4, pr. Akademicheskii, Tomsk, 634021, Russia

[2] Tomsk State University, 36 Lenin Ave.,Tomsk, 634050, Russia

[3] Tomsk Polytechnic University, 30 Lenin Ave.,Tomsk, 634050, Russia

E-mail: volodia74ms@yandex.ru

Abstract. The operating capacity and service life of printed circuit boards in various electronic equipment and devices depends on their ability to resist vibroacoustic loads, including vibration and acoustic noises. In this paper, non-contact laser vibrometry has been applied to perform the modal analysis of a circuit board assembly in order to identify its vulnerable spots and to find solutions to protect the assembly from external vibroacoustic loads. A broadband periodic chirp signal was used to excite vibration, which enabled a rapid generation of results. The paper provides data on eigenfrequencies, vibration velocity fields, and vibration displacement profiles. Frequency ranges have been determined in which eigenfrequencies with the highest vibration amplification lie. The obtained data can be used to develop a quality control technique for printed circuit boards and to optimize their construction as early as the design stage.

1. Introduction

Modern electronics equipment during service and transportation is often subjected to dynamic loads, such as vibrations, shocks and overloads, acoustic noises. These loads can disturb the normal operation mode of equipment, initiate damage of its components, and lead to failure. In many cases, failure of the components and parts of electronic circuit boards widely used in various industries is induced by vibroacoustic loads from various operating mechanisms and devices. Vibroacoustic frequencies can coincide with vibration eigenfrequencies of circuit board components, causing resonance and subsequent failure. For example, aerospace equipment experiences vibratory loads while moving in the atmosphere with high velocities and overcoming environmental resistance. Additionally, jet noises and airflow pulsations give rise to variable pressures of 170–190 dB in aircraft skin at the vibration frequency 0–5 kHz. As a result, the vibration amplitude level can reach 20g and higher, which leads to resonance phenomena in electronic circuit boards [1–3]. Resonance makes the overloads acting on the electronic components mounted on circuit boards to increase drastically [3, 4].

Vibration-induced failure can be avoided by determining eigenfrequencies and eigenvibrations of components and by taking them into account in further design [1, 5, 6]. Data on vibration modes, vibration decrement and other modal characteristics are also necessary for the verification of finite element models used in the development of electronic systems [7] and in the computer design of new materials [8]. With knowledge of eigenvibrations and vibration velocities, it is possible to calculate the dynamic stress field distribution in bench testing as well as under real operation conditions of electronic equipment. These data can be obtained using laser vibrometry methods [9].

Laser vibrometry is a modern, qualitatively new technique of measuring mechanical vibration parameters of objects. Unique physical characteristics of laser methods largely contribute to their advantages. These are the possibility of distant, non-contact vibration measurement and the absence of effect on the resonance properties of objects; the possibility of vibration measurement in different points of an object in human hazardous environments (chemically aggressive, high temperature, radiation, etc.). Today laser vibrometry is extensively applied in dynamic testing of various printed circuit boards. Quality control approaches are developed for fault inspection of electronic equipment and its components and for optimization of circuit board arrangement [3, 6, 7, 9-11].

This paper reports the experimental data obtained using a three-axis scanning laser vibrometer, namely, eigenvibrations and eigenfrequencies of a circuit board assembly.

2. Equipment

A three-axis scanning laser Doppler vibrometer was used to measure the vibration velocity characteristics and to perform modal and harmonic analysis.

A scanning laser vibrometer PSV-500 (Polytec Scanning Vibrometer) measures three orthogonal components of vibration velocity based on the principle of laser interferometry (a laser beam reflected from the moving surface interferes with a reference beam). Since the measurement technique is non-contact, the modal characteristics of objects excited by vibration remain unchanged in the experiment and the obtained vibration modes are undistorted, which cannot be achieved in measurements with the use of mounted detectors.

Modal tests require that the boundary conditions, as a rule these are the conditions of workpiece mounting, must be as close to real conditions as possible. Another requirement is an adequate and physically justified choice of the exciting signal and the exciter for the vibrational loading of the workpiece. In this paper, the exciter was a TMS 2075E vibration shaker driven by the internal generator of the vibrometer. A broadband periodic chirp signal was used. Frequency spectral analysis was performed within the range from 0 to 4 kHz.

3. Modal testing of a printed circuit board assembly of an electronic apparatus

The performance of modal tests and the processing of obtained vibration velocity fields of electronic equipment units and components include the following tasks:

- to perform modal testing of electronic equipment units and components;
- to determine the principal modes of circuit board components and parts, and of the whole circuit board;
- to provide efficient operation of electronic equipment systems and components through the modification of the electronic equipment construction and its optimization based on modal test data.

Let us consider modal test performance by the example of a printed circuit board assembly.

The tested board was fixed by hex pillars and screws to a solid steel plate in position 1 and position 2 (Figure 1a, b). The steel plate was screwed to the vibration shaker.

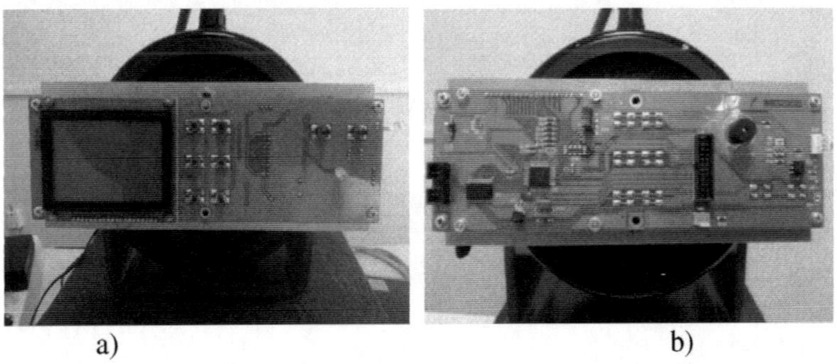

a) b)

Figure 1. View of the board screwed to the shaker in position 1 (a) and 2 (b)

The first few vibration modes are always of particular interest. Subsequent modes are shown to illustrate the capabilities of the apparatus.

The results of laser Doppler vibrometry has shown that different vibration modes (see Figures 3-10) are observed simultaneously on the both sides of the circuit board fixed in position 1 and 2 (Figure 1a, b). It was revealed that different circuit board regions experience maximum displacement in different vibration modes, with particular board components being in these regions. These circuit board components also lie in the regions of maximum vibration velocities, and arising sign-alternating loads can destroy solder joints or component mountings due to fatigue damage accumulation. For example, testing in position 1 within the resonance frequency range 550–765 Hz (mode 3–6) showed that the vibration velocity fields and their profile in the left part of the board are significantly different from those in the right part; there is an antinode at one of the fixed edges (left) of the liquid crystal screen for which the detected vibration velocity was 0.2–2 mm/s. The right edge of the fixed screen had the maximum displacement on modes 13–16 (2399, 2606, 2713 and 2936 Hz, respectively), accelerating up to vibration velocities of 5–50 μm/s. It should be noted that the both screen edges are in resonance on modes 13 and 14. In so doing, the acoustic radiator (speaker) on the back side of the board (position 2) which is near the connector and vibrates with a velocity of 5–20 μm/s enters into resonance (modes 17–19; 2625, 2860 and 3103 Hz). Modes 19, 20 and 22 demonstrate that there is an antinode in the entire circuit board region behind the liquid crystal screen (3103, 3236 and 3760 Hz, respectively) vibrating with a velocity of 10–50 μm/s.

Thus, the screen mountings (vibration velocity up to 2 mm/s) and the speaker (vibration velocity up to 20 μm/s) are most vulnerable to vibration. Consequently, if the apparatus operates in the aforementioned working frequency range, the circuit board construction must be changed (Figures 2a, b). Attention must also be paid to the arrangement of other components: their mountings must not be in regions with considerably different vibration velocities, especially where the vibration velocity changes its sign or direction (see Figure 3-10).

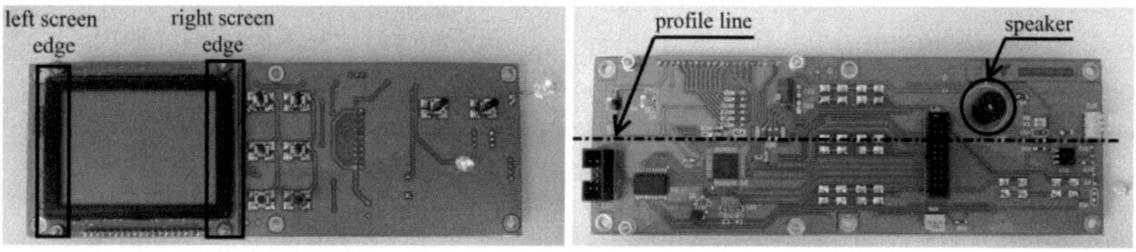

a) b)

Figure 2. Fixed screen edges entering into resonance (a) and speaker entering into resonance (b)

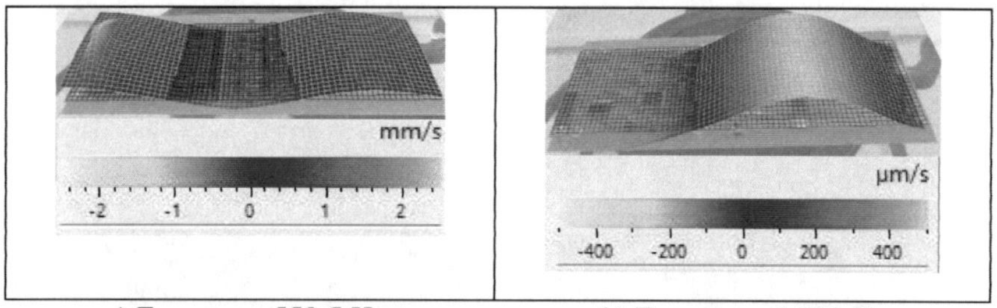

a) Frequency 550.5 Hz b) Frequency 347.3 Hz

Figure 3. View of the mode 3 of board in position 1 (a) and 2 (b)

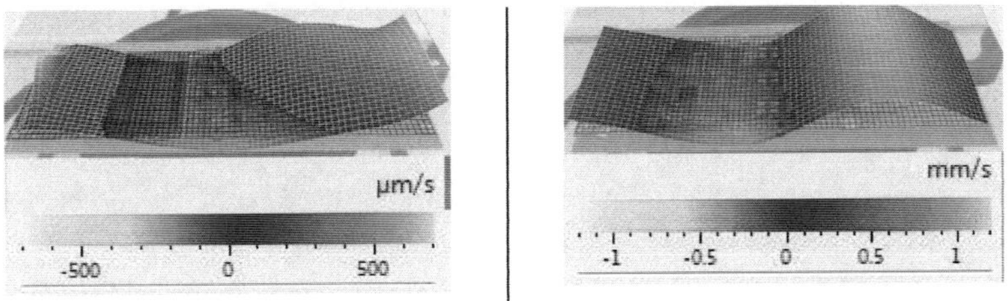

a) Frequency 620.6 Hz b) Frequency 431.9 Hz

Figure 4. View of the mode 4 of board in position 1 (a) and 2 (b)

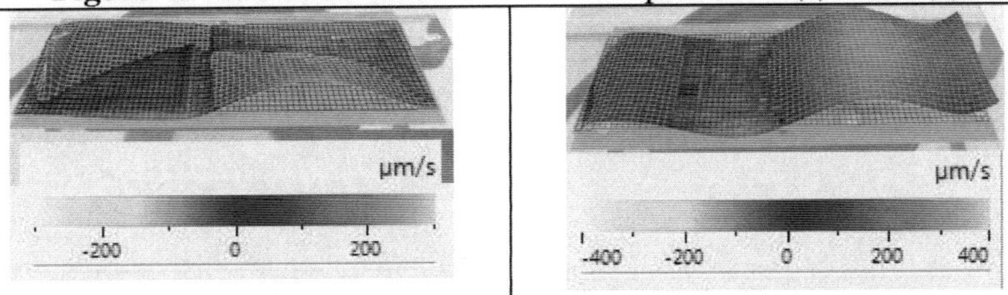

a) Frequency 731.8 Hz b) Frequency 570.3 Hz

Figure 5. View of the mode 5 of board in position 1 (a) and 2 (b)

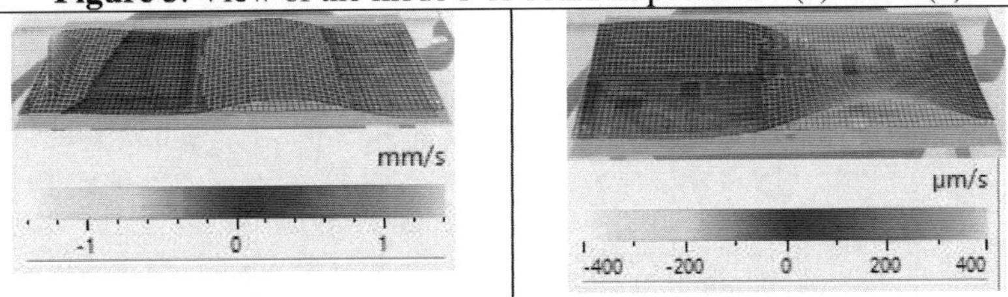

a) Frequency 765 Hz b) Frequency 695.5 Hz

Figure 6. View of the mode 6 of board in position 1 (a) and 2 (b)

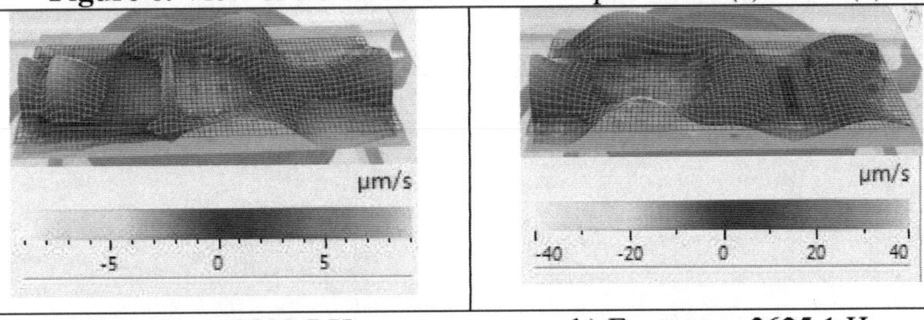

a) Frequency 2399.7 Hz b) Frequency 2625.1 Hz

Figure 7. View of the mode 13 and 17 of board in position 1 (a) and 2 (b)

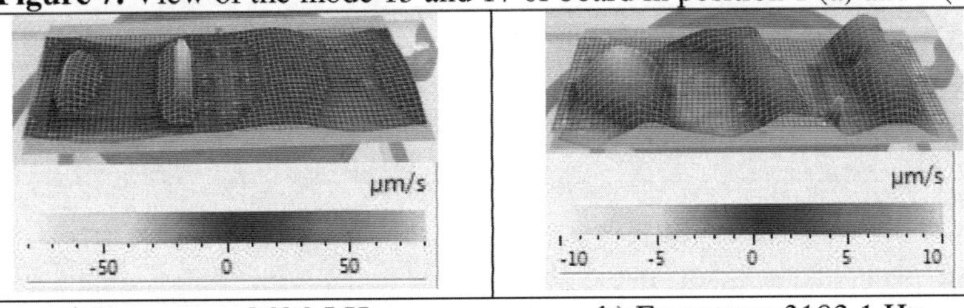

a) Frequency 2606.5 Hz b) Frequency 3103.1 Hz

Figure 8. View of the mode 14 and 19 of board in position 1 (a) and 2 (b)

IOP Conf. Series: Materials Science and Engineering **156** (2016) 012005 doi:10.1088/1757-899X/156/1/012005

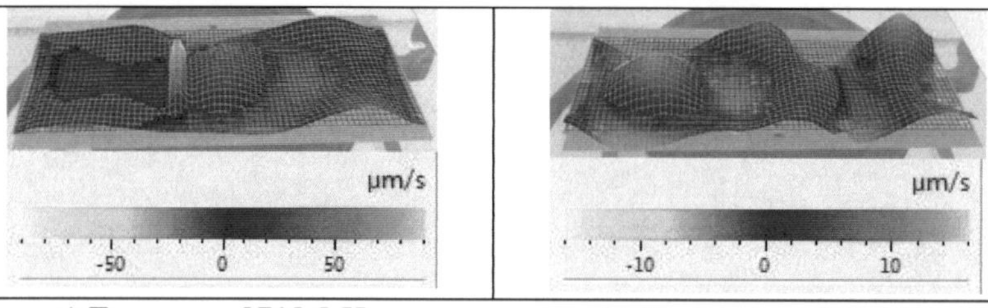

a) Frequency 2713.8 Hz b) Frequency 3236.9 Hz

Figure 9. View of the mode 15 and 20 of board in position 1 (a) and 2 (b)

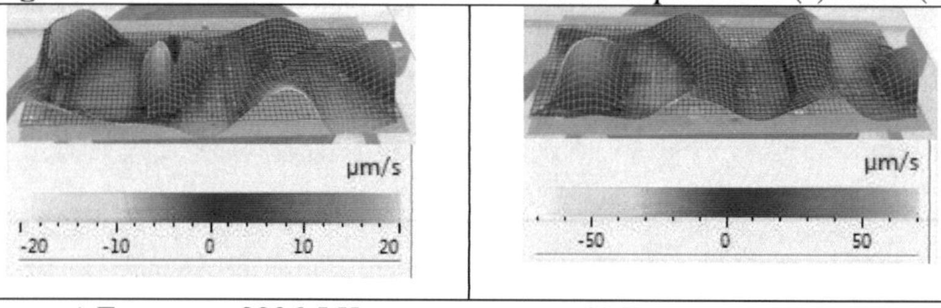

a) Frequency 2936.5 Hz b) Frequency 3760 Hz

Figure 10. View of the mode 16 and 22 of board in position 1 (a) and 2 (b)

Vibration displacement amplitude profiles were obtained for vibration modes 1–6 (see Figures 11-16). The highest amplitude reaches 2.5–2.7 µm for the first two modes. The highest amplitude for the first mode is achieved in position 1, and for the second—in position 2. Amplitudes for other vibration modes do not exceed 0.7 µm. These data were obtained at an average shaker speed of 0.13 mm/s. It was found for the first vibration mode that the maximum vibration velocity on the circuit board at the given shaker speed is 1.16 mm/s. Correspondingly, the vibration amplitude is amplified by a factor of 9 on this mode (frequency 131.9 Hz), which is significant. For other vibration modes a 3-4-fold vibration amplification was detected.

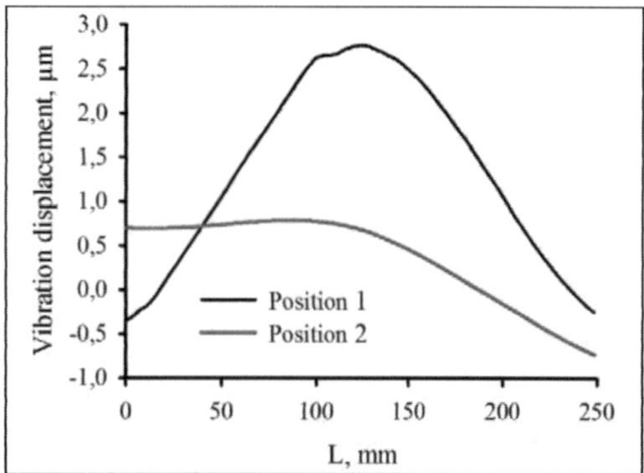

Figure 11. Profiles of natural mode 1 vibration displacement (position 1 – 131.9 Hz, position 2 – 109.4 Hz)

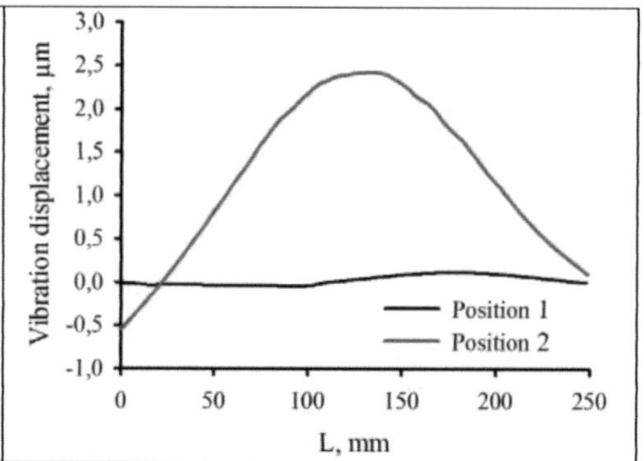

Figure 12. Profiles of natural mode 2 vibration displacement (position 1 – 409.4 Hz, position 2 – 131.9 Hz)

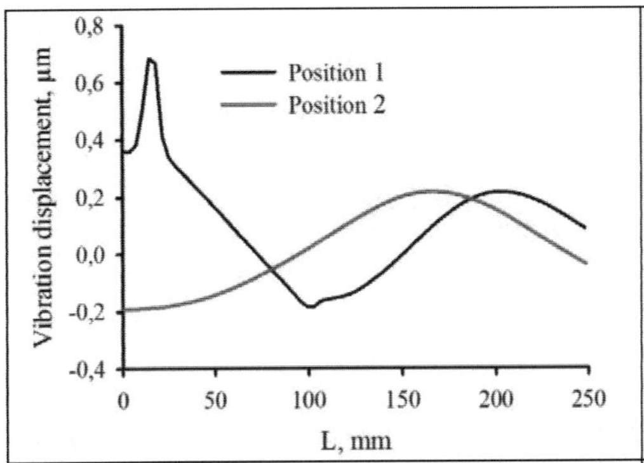

Figure 13. Profiles of natural mode 3 vibration displacement (position 1 – 550.5 Hz, position 2 – 347.3 Hz)

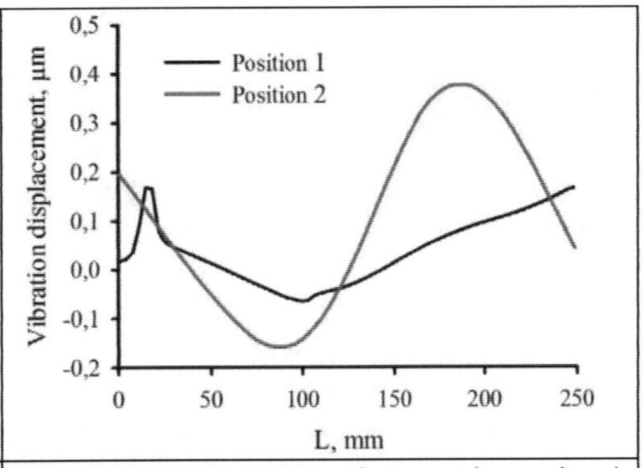

Figure 14. Profiles of natural mode 4 vibration displacement (position 1 – 620.6 Hz, position 2 – 431.9 Hz)

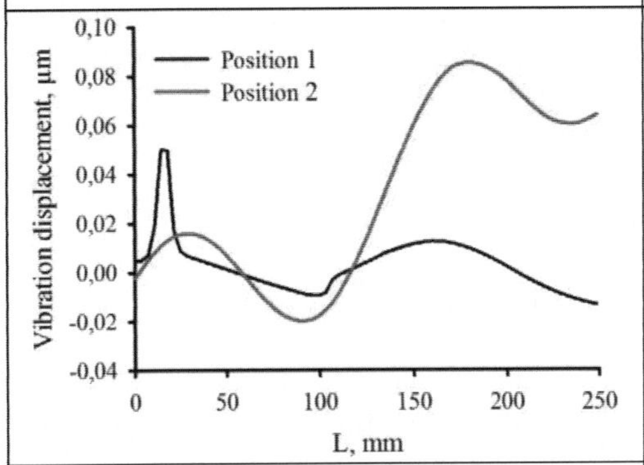

Figure 15. Profiles of natural mode 5 vibration displacement (position 1 – 731.8 Hz, position 2 – 570.3 Hz)

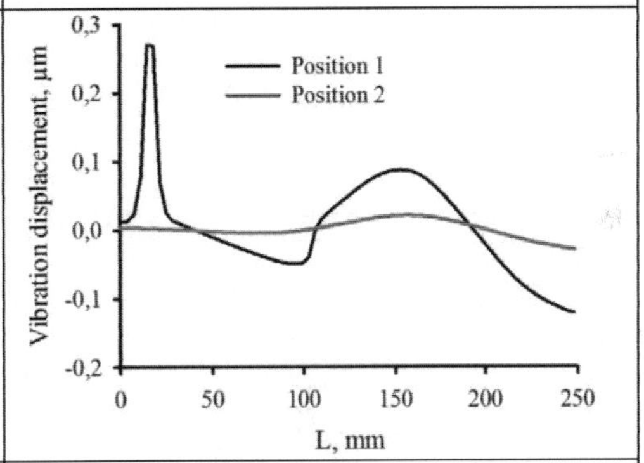

Figure 16. Profiles of natural mode 6 vibration displacement (position 1 – 765 Hz, position 2 – 695.5 Hz)

Data on the amplification factor were obtained for the frequency range 0–4 kHz (Figure 17 a, b). As one can see, frequencies of up to 1 kHz are most dangerous; vibration amplitudes within this range increase many-fold. The eigenvibrations whose profiles exhibit abrupt changes in vibration displacements also fall within this range (see Figures 11-16). In the remaining frequency range from 1 kHz to 4 кHz no many-fold amplitude increase is observed.

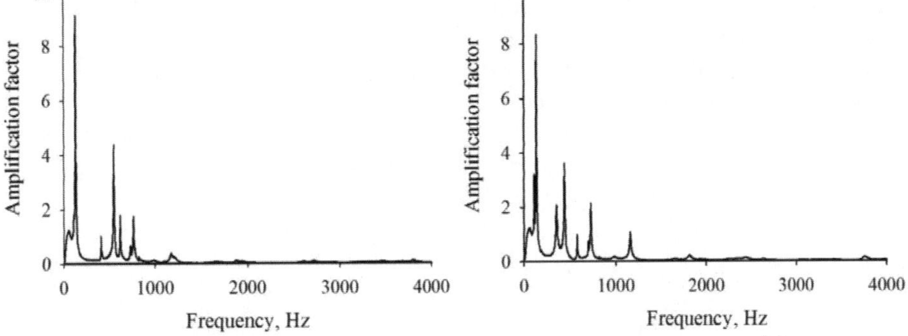

a) b)

Figure 17. Vibration amplification factor for the circuit board fixed in position 1 (a) and 2 (b)

4. Conclusion

The eigenvibrations obtained in circuit board scanning on both sides are generally similar. The difference in the two experiments is mostly due to the contribution of individual components mounted on different sides of the board, such as a liquid crystal screen and a speaker. The chosen circuit board design and assembly provide strongly different vibration velocity amplitudes from 5 μm/s to 2 mm/s at the frequencies 550–760 Hz (left screen edge), 2399–2936 Hz (right screen edge) and 2625–3103 Hz (board region behind the screen). The most dangerous factors are the eigenvibrations that correspond to abrupt amplitude changes in profiles and a many-fold vibration amplitude increase with respect to the shaker vibrations. Such eigenvibrations are observed in the range up to 1 kHz. In order to improve the device reliability, it is recommended either to avoid vibrations with the mentioned frequency ranges or to modify the circuit board design and component arrangement so that to shift dangerous frequencies beyond the working frequency range. No critical amplification of vibrations is observed in the range 1–4 kHz.

References

[1] Agletti G S 2002 A lighter enclosure for electronics for space applications *Proceedings of the Institution of Mechanical Engineers, Part G: Journal of Aerospace Engineering* **216** 131-142

[2] Kim Y K, Hwang D S 2015 PBGA packaging reliability assessments under random vibrations for space applications *Microelectronics Reliability* **55** 172-179

[3] Sakri M I, Mohanram P V 2014 Experimental investigations on board level electronic packages subjected to sinusoidal vibration loads *International Journal of Current Engineering and Technology.* Special Issue **2** 427-431

[4] Amy R A, Aglietti G S, Richardson G 2009 Reliability analysis of electronic equipment subjected to shock and vibration - A review *Shock and Vibration* **16** 45-59

[5] Meier K, Roellig M, Schiessl A, Wolter K-J 2011 Life time prediction for lead-free solder joints under vibration loads 2011 *12th Int. Conf. on Thermal, Mechanical and Multi-Physics Simulation and Experiments in Microelectronics and Microsystems* (Linz: IEEE) pp 1-8

[6] Tang W, Ren J, Feng G, Xu L 2007 Study on vibration analysis for printed circuit board of an electronic apparatus *Proceedings of the 2007 IEEE International Conference on Mechatronics and Automation* (Harbin: IEEE) pp 855-860

[7] Lowe R D, Dodson J C, Foley J R, Mougeotte C S, Geissler D W, Cordes J A 2014 Constitutive response of electronics materials 2013 *Proceedings of the 2013 Annual Conference on Experimental and Applied Mechanics* (Lombard, United States: IL) pp 57-74

[8] Psakhie S G, Popov L V, Shilko E V, Smolin A Yu and Dmitriev A I 2009 Spectral analysis of the behavior and properties of solid surface layers. Nanotribospectroscopy *Physical Mesomechanics* **12** 221-234

[9] Huang Z, Zang C, Friswell M I 2014 Parameter identification of a printed circuit board structure using model updating and scanning laser vibrometer measurements *26th International Conference on Noise and Vibration Engineering, ISMA 2014, Including the 5th International Conference on Uncertainty in Structural Dynamics, USD 2014* (Leuven, Belgium: KU Leuven) pp 2671-2682

[10] Vanlanduit S, Cauberghe B, Guillaume P, Verboven P 2003 Development of an autonomous scanning laser doppler vibrometer vibration tracking method for inline quality control *Proceedings of SPIE - The International Society for Optical Engineering* (Munich: SPIE, 2003) (Optical Measurement Systems for Industrial Inspection III vol 5144) pp 737-744

[11] Vanlanduit S, Guillaume P, Schoukens J 2004 Robust data reduction of high spatial resolution optical vibration measurements *Journal of Sound and Vibration* **274** 369-384

Subminiature eddy current transducers for studying metal–dielectric junctions

S Dmitriev[1], A Katasonov[1], V Malikov[1], A Sagalakov[1], M Davydchenko[1], L Shevtsova[2], A Ishkov[3]
[1] Altay State University, 61, Lenina ave., Barnaul, 656049, Russia
[2] Department of Materials Science, Novosibirsk State Technical University, 20, Karla Marksa ave., Novosibirsk, 630073, Russia
[3]Altay State Agrarian University, 98, Krasnoarmeyskiy ave., Barnaul, 656049, Russia

E-mail: osys@me.com

Abstract. Based on an eddy current transducer (ECT), a probe has been designed to research metal–dielectric structures. The measurement procedure allowing one to detect defects in laminate composites with a high accuracy is described. The transducer was tested on the layered structure consisting of paper and aluminum layers with a thickness of 100 μm each in which the model defect was placed. The dependences of the ECT signal on the defect in this structure are given.

1. Introduction

A subminiature eddy-current transducer [1-6] has been designed to monitor physical parameters when studying the properties of metal–dielectric junctions. The material and its distribution over the surface and in the thickness of the studied object. The developed measuring system allows effective investigate the metal-insulator transitions in miniature metal-polymer composite objects. Similar composites may include several metal layers separated by thin dielectric layers of polymer. The typical defects of such materials are, for example, the disturbance of layers' continuity and the formation of link between the layers.

The analysis of a recent investigations points to the miniaturization tendency for eddy-current transducers. Transducers with a size of 5 × 5 mm and a 0.15 mm diameter of the wire have been designed [7]. However, they do not provide the required penetration depth and localization of the magnetic field that are necessary for local measurements in different nonuniform media. Ferrite magnetic field concentrators are often used to increase the area of the magnetic field. A similar design provides an advantage that is related to the absence of the scatter of eddy currents [8]. In addition, a 2.5 mm penetration depth is attained.

Experimental methods based on two eddy-current transducers that operate in the differential mode have been developed [9]. Similar connection circuits significantly decrease parasitic noise that arises during high-speed scanning in real time. The analysis of a recent investigations points to the miniaturization tendency for eddy-current transducers.

L. Barbato et al. [10] scanned two aluminum plates with a model flaw in the centre and tested cracks between the plates. The diameter of the measuring winding was 7 mm. The scanning was performed at 1 and 5 kHz. In this case, the penetration depth of eddy currents into the studied plates at the above-mentioned frequencies was 3.82 and 1.71 mm.

In the past there were several attempts to use highly sophisticated magnetic field sensors like SQUID and Fluxgate sensors for sensitive low frequency eddy current testing to detect deep defects in metal parts [11–12]. Although very good results could be achieved such testing systems can hardly be used in real industrial applications because of their complexity, prices and insufficient robustness [13].
In this connection, the challenge is to design subminiature eddy-current transducers that provide a penetration depth of up to 5 mm and an area of 2500 μm^2. Since the eddy-current inspection method is insensitive to non-conducting paint layers, it can also be used for diagnosing parts with paint coats.

2. The design of a subminiature eddy-current transducer and measuring system

The design of the measuring system includes two differentially connected subminiature transducers, which provide a large area of the magnetic field.
The tested parameter is the electric conductivity of the material and its distribution over the studied object. Exciting subminiature transmitter winding consists of 10 turns, and its diameter is 0.13-0.12 mm. Measuring winding consists of 130 turns and its diameter is 0.05-0.08 mm. In order to minimize the influence of the exciting windings on the received signal, compensating winding is included into the circuit. The compensating winding is connected to the measuring winding so that the voltage of the exciting windings can be subtracted. The compensating winding consists of 20 turns. Copper wire 5 μm thick is used to the wind the turns.
The windings are wound on the pyramidal core of the ferrite 2000 HM3 with relative magnetic permeability (μ_{max}=500) or (if necessary, a higher localization of the magnetic field) of the annealed by a special technique 81NMA alloy. The core is a pyramid in the shape of a tetrahedron 1 mm in height, with a side of the bases 0.2 mm. The measuring winding is located on the tip of the pyramid, which improves the magnetic field localization.

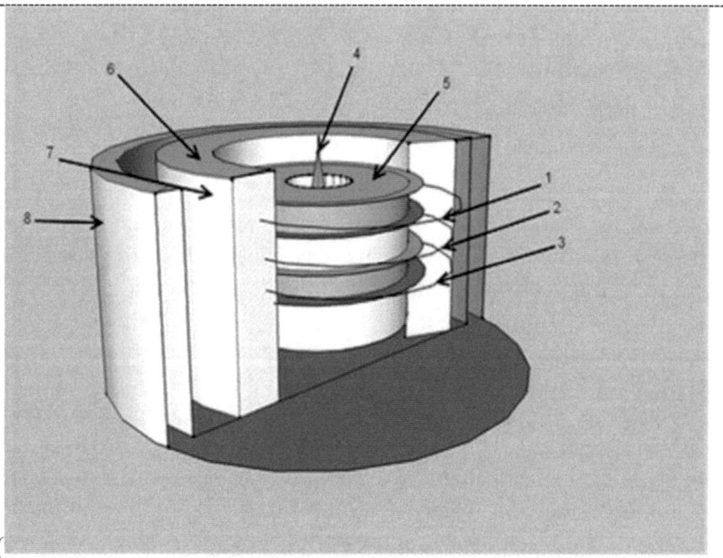

Figure 1. The scheme of the eddy-current transducer (ECT).

The scheme of the subminiature eddy-current transducer (ECT) is shown in Figure1. The eddy-current transducer is a transformer with measuring (1), exciting (2), and compensation (3) windings and a magnetic circuit 4, which is located inside the cylindrical platform 5 with tracks that are cut on the external side for windings. The platform is impregnated with a compound 6 at a temperature of 200°C to prevent the disintegration of the windings when the ferrite screen 7, which is intended for the localization of the electromagnetic field on the tested object, is put in place. From the outside the

transducer is contained in a corundum washer **8**, which protects the core 4 from contacting the tested object.
The characteristics of the designed transducers allow one to efficiently localize the magnetic field within 2500 μm^2 and provide penetration of the magnetic field into the studied object at a depth of up to 1.4 mm.
The eddy-current transducer is connected to a set of the designed amplifiers and band-pass filters and is controlled by the sound map of a personal computer with special software, which applies a voltage to the generator winding of the transducer.
Digital signal from the virtual oscillator is fed to analog-to-digital converter (ADC) input of sound card, after which the analog signal through the power amplifier is fed to the exciting winding transducers. Passing through the excitation windings, sine wave signal creates electromagnetic field which induces voltage in the measuring windings of eddy-current transducer. The voltage depends on the parameters of the object under control. The transducers are included counter-currently, whereby the resultant signal represents the difference between two values of voltage. This signal is fed to a series of amplifiers and bandpass filters, and then supplied to the microphone input of the sound card and then through the preamplifier - to the input of an analog-to-digital converter (ADC) of sound card. The analog signal is converted to the digital one and transferred to software processing and control unit. The processing and control unit records the level of the digital signal in conventional units corresponding to the difference between the voltages on the measuring windings.

3. Experimental results

To demonstrate the proposed method operability, the structure of alternating aluminum foil of 100 µm thick and paper of 100 µm thick has been used. As a defect model between the layers a hollow parallelepiped with a wall of 300 µm thick has been placed. The defect was at a distance of 600 µm from the sensor in the depth of the layered structure.

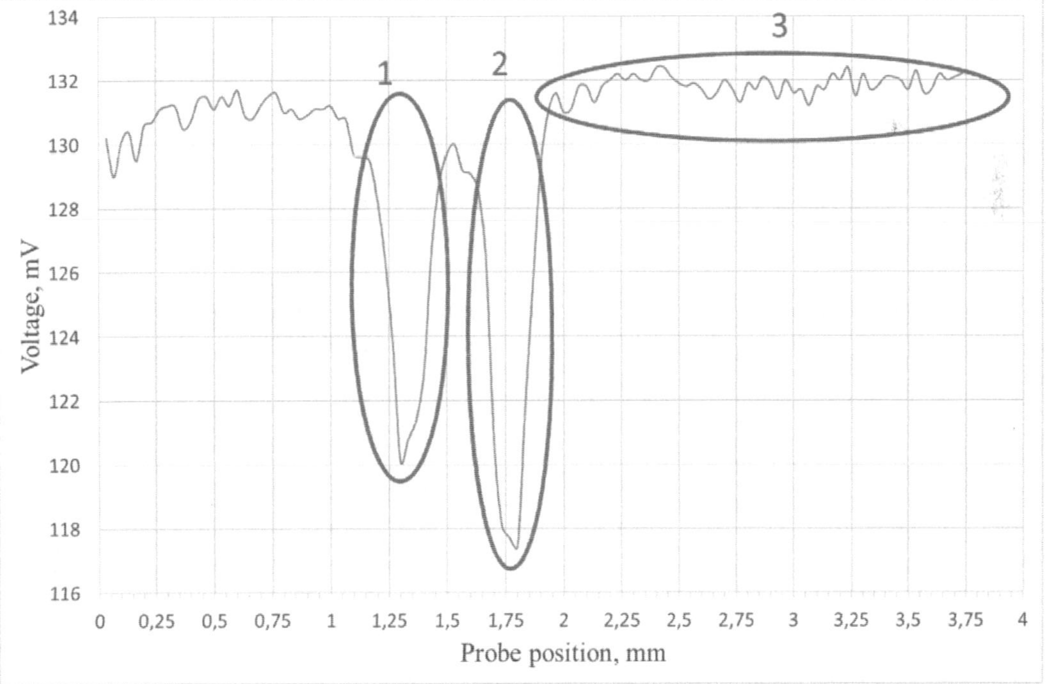

Figure 2. Plot of signal observed when the probe moves along the layered medium with a defect. The signal frequency is 1000 Hz: (1, 2) walls of the parallelepiped, (3) defectless part of the sample.

Figure 2 shows plot of signal when the probe moves above the layered medium inside which a defect is located. The signal level from the measuring winding characterizes conductivity values on the survey section. For basic operating frequency of 1000 Hz the voltage level included in the measuring winding was 130 ± 2 mV. Areas 1 and 2 on the graph where in voltage level drop is up to 115 mV correspond to the walls of the parallelepiped. This change in the signal amplitude is equal to 11% of the signal level, corresponding to the defectless section of the sample. At the same time the fluctuations of signal amplitude at the defectless section does not exceed 4mV, representing 3% of the signal corresponding to the defectless region of the sample.
From the graph shown in Figures 3,4 clearly visible change in amplitude when testing the transducer at other frequencies. These changes are caused by the increase in frequency at condition less penetration depth of the eddy currents in the depth of the layered structure and increase in the influence of different small cracks on the surface of the layered structure, as well as external influences.

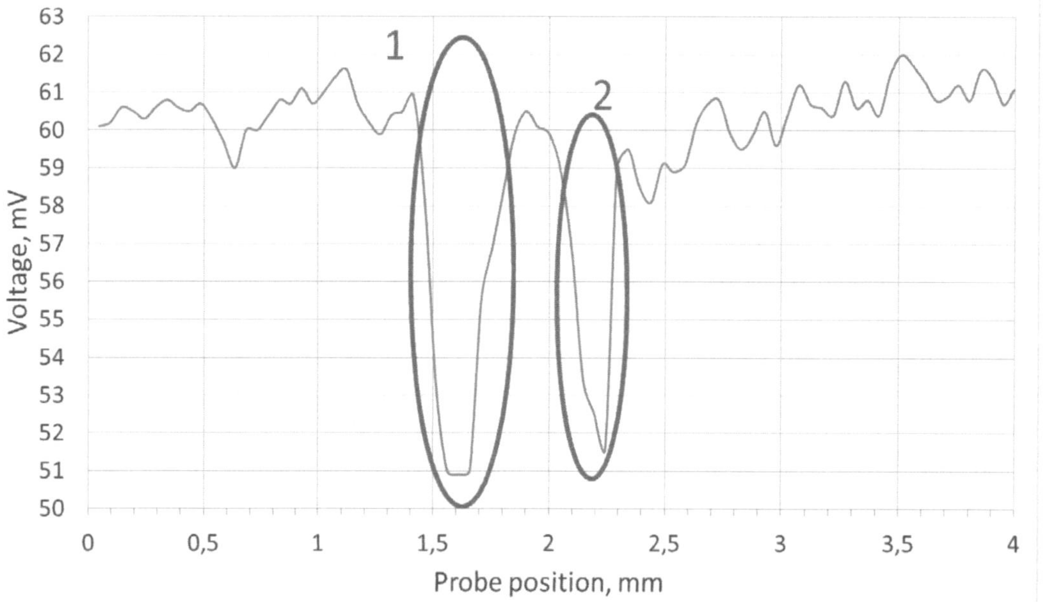

Figure 3. Plot of signal observed when the transducer moves along the layered medium with a defect. The frequency of the transducer is 6000 Hz.

At a frequency of measurement equal to 6000 Hz (Figure 3), the defect model is still clearly visible, however the amplitude of oscillation on a defectless portion of the sample, already exceed 7% of the signal level corresponding to the defectless portion of the sample.
In laboratory and industrial measurements, such oscillation amplitude can be caused by external influences, unconditioned the presence of defects or small defects on the surface of the metal layer. In the study of the test object with unknown defects, similar changes amplitude may mistakenly be interpreted as a defect.

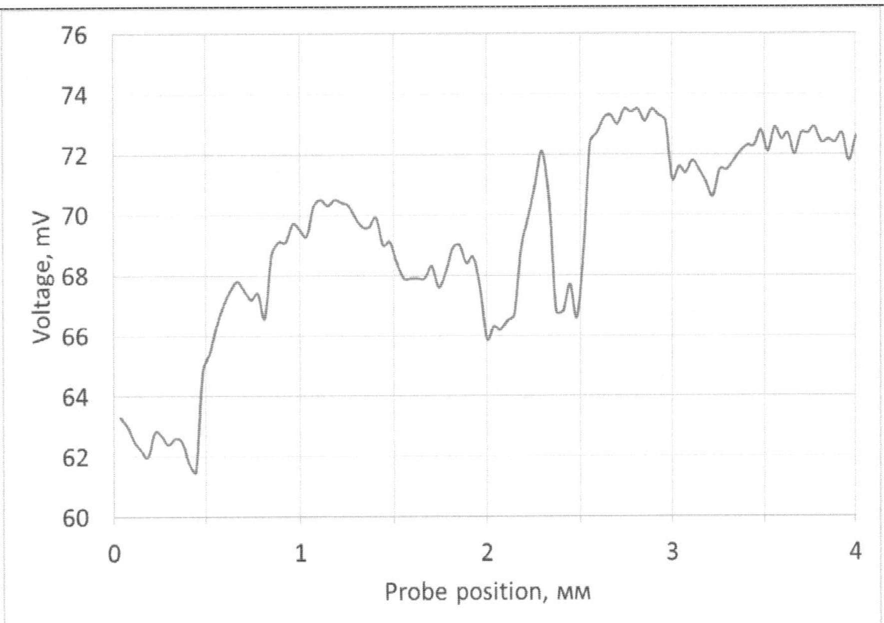

Figure 4. Plot of signal when the probe moves along the layered medium with a defect. The frequency of the transducer is 7000 Hz.

When operating frequency of the device is beyond the given limits the results of the measurements will be distorted by amplitude fluctuations caused by microcracks on the surface of the sample or by the decreasing of field localization inside the layered structure. Figure 4 shows the measurement results of the sample at a frequency of 7000 Hz. As seen from the graph, the amplitude changes, caused, in this case, by the microcracks on the sample surface, are much higher than the amplitude changes, caused by the defect directly.

4. Conclusion

Right up to the depth of the defect, equal to 1400 μm, there is a clear dependence of the response of transducer, on the position of the transducer over the defect.

By fixing change in amplitude response in the converter caused by a defect, it is possible to change the frequency of the current in the exciting winding so that the eddy-currents are concentrated in the layers of composite placed above the defect.

The solution of the inverse problem allows to determine the depth of the defect. After calibration, the sensor for typical defects, can use the eddy-current transducer for the diagnosis of composite laminates with a thickness of 1 to 1400 μm.

References

[1] Malikov V N, Dmitriev S F, Sagalakov A M, and Ishkov A V 2014 *Instruments and experimental techniques* **6** 751-4

[2] Polyakov V.V., Dmitriev S.F., Ishkov A V, Kolubaev E A and Malikov V N 2014 *Advanced Materials Research* **880** 105-8

[3] Dmitriev S F, Katasonov A O, Malikov V N, and Sagalakov A M, 2014 *Russian Journal of Nondestructive Testing* **1** 32-7

[4] Malikov V N, Davydchenko M A, Dmitriev S F and Sagalakov A M 2015 *Proc. of Int. Conf. on Mechanical Engineering, Automation and Control Systems 2015, Tomsk*

[5] Dmitriev S F, Malikov V N, Sagalakov A M, Shevtsova L I and Katasonov A O 2016 *Journal of Physics: Conference Series* **643**
[6] Dmitriev S F, Katasonov A O, Malikov V N and Sagalakov A M 2016 *Russian Engineering Research* **8** 626-9
[7] Prance R J, Clark T D, and Prance H 2000 *Sens. Actuators* **85** 361-4.
[8] Prance R J, Clark T D, and Prance H 2003 *Rev. Sci. Instrum* **74** 3735-9
[9] Rocha T J., Ramos H G, Ribeiro A L, and Pasadas D J 2015 *Measurement* **228** 55-61
[10] Barbato L, Poulakis N, Tamburrino A, and Theodoulidis,T 2015 *IEEE Transactions on Magnetics* **51**
[11] Gabor V, Antal G 2002 *Journal of Electrical Engineering* **53**
[12] Kreutzbruck M, Allweins K and Heiden C 2000 *Proc. of DACH Jahrestagung 2000,* Innsbruck
[13] Hesse O, Pankratyev S 2005 *Measurement science review* **5** 1-8

IOP Conf. Series: Materials Science and Engineering **156** (2016) 012007 doi:10.1088/1757-899X/156/1/012007

Experimental studies of ultrasonic stabilization of rings of rolling bearings

A V Korolev[1], A A Korolev[1], A F Balaev[1], O.V. Zakharov[1], O Y Davidenko[1], E V Mukhina[1], A S Yakovishin[1], S A Savran[1]
[1]Yuri Gagarin State Technical University, Saratov, Russia,

E-mail: bandrey.sc@gmail.com

Abstract. The article describes the results of experimental studies on ultrasonic stabilization of the bearing rings. The mean and root-mean-square values of non-roundness of the rings after ultrasonic and thermal treatment were compared. Comparative analysis showed that the root-mean-square deviation and the mean value of non-roundness after ultrasonic stabilization was lower than that after thermal stabilization.

1. Introduction

The ability of bearings to retain their geometric accuracy for a long time mainly determines their reliability and service life of the machine as a whole. Therefore, ensuring the stability of the geometric accuracy of the bearing rings is a task of primary importance.

In the process of manufacturing the bearings are subjected to repeated physical, mechanical, chemical and thermal effects. As a result of this the internal stresses are accumulated in the bearing material. Sometime after the production a relaxation of residual stresses occurs that results in deformation of the original shape of the product. Further, some bearings after the manufacturer's quality control are discarded at the input control of the enterprise customer. The major hazard for machine reliability is related to the bearings, which have passed the input control and still have residual stresses.

The operation of these bearings is characterized by repeated dynamic loads when the relaxation of unrelieved residual stresses occurs. During the relaxation of residual stresses, the size of the surfaces, contact area, magnitude of interference, and friction force change. This leads to a deterioration of the operating conditions and further reduction of the service life of bearings. Residual stress relaxation mechanism is described in numerous studies [1-5], according to them the value of relieved stresses is not linearly dependent on the time of operation. The operation can be roughly divided into three periods: the initial, main, and final. Stress relaxation is most intense in the initial and final periods of operation of the product. At the initial stage of operation more intense stress relaxation is caused by the running-in when the system goes into equilibrium-stressed state. During the subsequent period, the main relaxation occurs relatively slowly over a long period of time, which determines the service life of the product. In the final period of operation, the strain magnitude caused by residual stresses relaxation reaches a critical value and leads to the limit state of rolling bearings and the completion of the service period.

The service life of a rolling bearing can be increased due to the artificially accelerated relaxation of residual stresses of the first period prior to the operation of bearings. The application of the operation of residual stresses relaxation is expected to stabilize the geometric parameters of the bearings over time and substantially increase reliability of the machine as a whole.

2. Comparative analysis of experimental data

Currently, there are many different methods of residual stress relaxation. According to the physical principle of operation these methods can be divided into thermal, chemical and thermal, mechanical, electromagnetic [6-9]. The analysis of the existing methods enabled to choose an ultrasonic method of residual stresses relaxation as it was considered the most economical and productive [10]. To evaluate the effectiveness of this method a process of stress relaxation of bearing rings in a liquid medium was developed. For the implementation of this process an automated equipment that cleans and removes the residual stresses of bearing rings was designed and manufactured. For the treatment of the outer rings of bearings of a combined construction the equipment had special technological means. The cycle of ring treatment using the equipment was divided into four stages. In the first stage the sequential automatic feed of rings from the feeder on the conveyor into the liquid medium with anti-corrosion and detergent additives was performed. In the second stage the rings were exposed to the ultrasonic effect of piezoceramic emitters by transmitting vibrations in the liquid medium. In the third stage the rings were directed to the drying zone. Here the moisture from the rings was removed by a mechanical vibrator and a fan. In the fourth stage the rings were directed from the drying zone to the collector bunk.

In order to confirm the effectiveness of the technology experimental research was performed on the manufactured equipment.

Two groups of bearing rings 2112-1006135-01 containing 30 pieces each were randomly selected as the object of the study. Both groups were subjected to the same process operations. Bearing rings 2112-1006135-01 contained metal final machined parts and their outer surfaces had polyamide flanges melted on them. In the course of flange welding the heating caused the deformation and change in the original geometric parameters of the rings. This lead to changes in operating conditions of the bearings and reduction of their service life.

Non-roundness was selected as a controlled geometrical parameter of the rings. It was measured using the equipment Talyrond R-105 in normal conditions. Non-roundness of each ring was determined prior to the flange casting, immediately after the casting, 3 hours after stabilizing treatment, and 15 days after the treatment.

In order to conduct a comparative analysis, the first group of the rings was subjected to ultrasonic stabilizing treatment on the equipment described above according to the authors' technology. The second group was subjected to the conventional thermal tempering at 150 °C for 3 hours in an oven under production conditions in JSC "EPK Saratov" according to the existing stabilizing technology.

According to the results of statistical processing of the measurement the mean (Figure 1) and root-mean-square (Figure 2) values of non-roundness before and after stabilizing treatment were obtained. The statistical difference between the indices in the two groups of bearing rings was assessed using Student's t-test with a confidence level of 0.95.

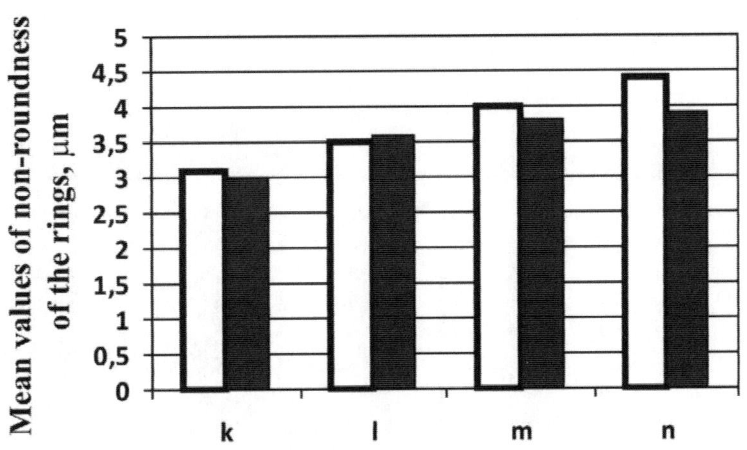

Figure 1. Mean values of non-roundness of the rings after thermal stabilization (white) and after ultrasonic stabilization (black): k - before the flange casting; l - after the flange casting; m - after stabilizing treatment; n - 15 days after stabilizing treatment

Figures 1 and 2 show that the difference between the mean and root-mean-square deviations of non-roundness of the rings before and after pouring of flanges in both groups was insignificant. Immediately after the flange casting mean non-roundness in both groups increased by 14-15% (Figure 1) and the root-mean-square deviations of non-roundness increased by 12-14% (Figure 2). 3 hours after thermal tempering the mean value of non-roundness increased by 13% (Figure 1) and the root-mean-square value of non-roundness by 29% (Figure 2). After the ultrasonic stabilization the mean roundness value increased by 5% only (Figure 1) and the root-mean-square value of non-roundness by 3% (Figure 2). After 15 days the value of the root-mean-square deviation of roundness after thermal and ultrasonic treatment did not change significantly, i.e. within 3%. 15 days after thermal treatment the mean non-roundness increased by 10%, whereas after the ultrasonic treatment it increased by 3% only.

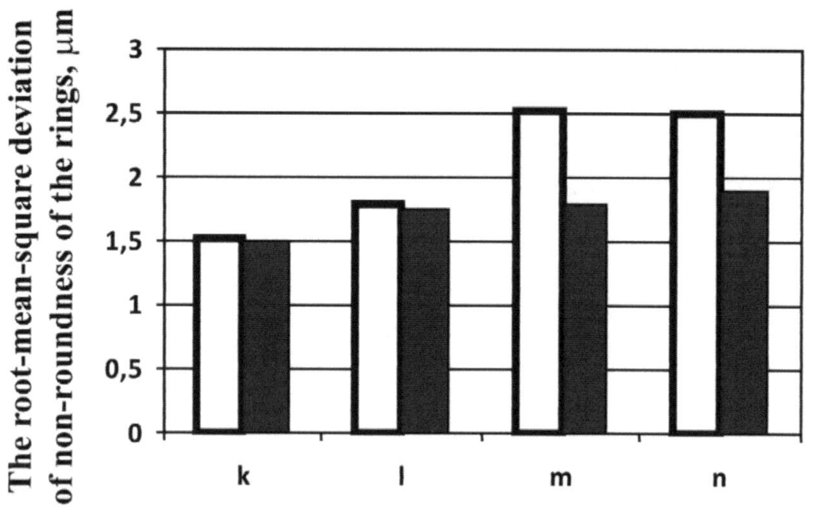

Figure 2. The root-mean-square deviation of non-roundness of the rings after thermal stabilization (white) and ultrasonic stabilization (black): k - before the flange casting; l - after the flange casting; m - after stabilizing treatment; n - 15 days after stabilizing treatment

Comparison of the measurement results after two kinds of treatment showed that the ultrasonic treatment allowed to achieve lower and more stable changes of the geometrical parameters of rings compared to thermal treatment. Additionally, it should be noted that the full cycle of the ultrasonic treatment of one part at a continuous conveyor feed equals 0.6 seconds. The full cycle of thermal stabilization together with the auxiliary time, loading and unloading from the furnace of maximum amount of 10000 parts is 7.6 hours. The mean thermal treatment cycle for one part is 2.74 seconds. Thus the ultrasonic stabilization is 4.5 times more productive than thermal tempering.

3. Conclusion

After the comparative analysis of the experimental results it was found that the ultrasonic stabilization method can significantly reduce the effect of residual stresses on the geometric parameters of rings compared to the thermal tempering. Taking into consideration the higher performance and lower power consumption the ultrasonic method can be regarded as a rational replacement for thermal tempering.

Acknowledgments

This paper has been developed under the state order of the Ministry of Education and Science of the Russian Federation (No. 9.896.2014/K).

References

[1] Balayev A F Korolev A V Kochetkov A V Sklyarova A I Zakharov O V 2016 Energy Approach-Based Simulation of Structural Materials High-Cycle Fatigue *IOP Conf. Series: Materials Science and Engineering* **116 1**

[2] Vardanjani M J Ghayour M Homami R M 2016 Analysis of the Vibrational Stress Relief for Reducing the Residual Stresses Caused by Machining *Experimental Techniques* **40 2** 705–713

[3] Yang Y J 2009 Understanding of Vibration Stress Relief with Computation Modeling *Journal of Materials Engineering and Performance* **18 7** 856–862
[4] Jurcius A Valiulis A V Cernasejus O Kurzydlowski K J Jaskiewicz A Lech-Grega M 2010 Influence of Vibratory Stress Relief on Residual Stresses in Weldments and Mechanical Properties of Structural Steel Joint *Journal of Vibroengineering* **12** 133–141
[5] Edwards P Ramulu M J 2015 Surface Residual Stresses in Ti-6Al-4V Friction Stir Welds: Pre- and Post-Thermal Stress Relief *Journal of Materials Engineering and Performance* **24 9** 3263–3270
[6] Aniyi J A Bello-Ochende F L Adeyemi M B 1996 Effects of pressure, die, and stress-relief temperatures on the residual stresses and mechanical properties of squeeze-cast aluminum rods *Journal of Materials Engineering and Performance* **5 3** 399–404
[7] Korolev A V Korolev A A Zhuravlyov M M 2015 The Mechanism of Hardening and Internal Stress Stabilization in the Process of Laser Treatment *Journal of Russian Laser Research* **36 4** 356–363
[8] Dong Y Shao W Jiang J Zhang B Zhen L 2015 Minimization of Residual Stress in an Al-Cu Alloy Forged Plate by Different Heat Treatments *Journal of Materials Engineering and Performance* **24 6** 2256–2265
[9] Barney E 2003 Klamecki Residual stress reduction by pulsed magnetic treatment *Journal of Materials Processing Technology* **141 3** 385–394
[10] Korolev A V Chistyakov A M Krivego V A Moiseev V G 1999 Vibroaging method of parts Patent 2140842 RU **8**

Numerical study of composite bulkhead partition strength with in-situ X-ray monitoring

A N Anoshkin, V Yu Zuiko, V M Osokin and P V Pisarev
Perm National Research Polytechnic University, Perm, 614990, Russian Federation

E-mail: pisarev85@live.ru

Abstract. The aim of this work is the stress-strain analysis of bulkhead partition made of carbon fiber reinforced plastic (CFRP) at mechanical testing. The problem was solved in a general 3-D statement for an anisotropic elastic body. The strength prediction of the construction was done by stress analysis in each ply of composite laminate. The numerical simulation was obtained with ANSYS Workbench software. The detailed ply-by-ply stress analysis with special attention to areas of plies curvature was done. The results of numerical simulation were compared with the test data for verification of the proposed numerical algorithm.

1. Introduction

Nowadays composite materials are increasingly being used for design of high-loaded parts and components of aircraft machinery and space-rocket industry. Generally, implementation is done as a replacement of the metallic component with its composite analog. Such implementation involves structure interchangeability in attaching points, moreover the structure must be unified and standardized, but not lose its strength.

For this it is necessary to design accurate numerical models of a composite part, allowing detailed research of stress strain in the most loaded areas with complex reinforcement geometry. As a rule, these areas include joints, technological junctions and various surface attachments in complex geometry parts. These areas can be considered *a priori* a weak link in the structure, as they are mostly prone to interlayer fracture due to low contact and interlayer resistance of polymer composite materials. Moreover, the probability of process defects is increased in these areas.

The object of current study is the composite bulkhead partition (Figure 1) of aircraft construction consisting of several plies of composite material. When in service, the bulkhead is exposed to compound structural loading, which may cause interlayer cracks and stiffness decline.

Figure 1. A view of bulkhead partition.

In the context of this work the boundary value problem for a laminate composite with different spatial orientation of layers and contact conditions is set up. A detailed structure ply-by-ply stress-strain analysis with the use of refined finite element mesh was done and the strength of bulkhead section was evaluated.

2. Calculation of bulkhead partition stress-strain

Geometrical model of the bulkhead and mechanical testing scheme is presented in Figure 2. The bottom part of the bulkhead is attached to the base by bolting, the top shelf is in contact with the test rig; the width of the contact area is 6 mm.

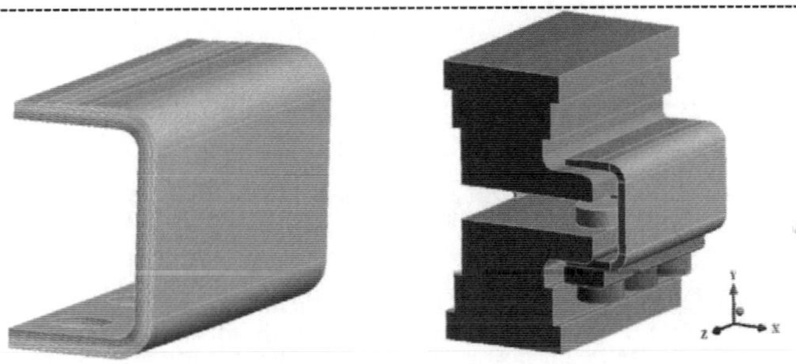

a b

Figure 2. Geometrical model of the bulkhead (a); test rig with the bulkhead (b).

The mathematical formulation of the problem corresponded to the theory of anisotropic elasticity. The variational formulation of this statement is to find minimizer of the Lagrange functional with additional terms in the form of geometric Cauchy relations [1].

Strength evaluation of the bulkhead partition made of CFRP carried out by the criteria of ultimate stress. Strength of orthotropic material corresponds to satisfy the inequality system in each point of the structure:

$$S_{11}^{-} \le \sigma_{11} \le S_{11}^{+},\ S_{22}^{-} \le \sigma_{22} \le S_{22}^{+},\ S_{33}^{-} \le \sigma_{33} \le S_{33}^{+},\ S_{12} \le \sigma_{12},\ S_{13} \le \sigma_{13},\ S_{23} \le \sigma_{23}, \quad (1)$$

where S_{ii}^{+}, S_{ii}^{-}, S_{ij} – the static strength of the material in tension, compression and shear, respectively, in the local coordinate system of the layer.

CFRP factor of safety for various components of stress state was estimated by the equation:

$$n_{ij} = \min_{\mathbf{r} \in V^{(1)}} \left(\frac{S_{ij}}{\sigma_{ij}(\mathbf{r}) k} \right), \quad (2)$$

where k– coefficient of static strength reduction due to the various environmental factors.

Numerical simulation of mechanical testing carried out by finite element system with the use of ANSYS Workbench software. The bulkhead partition under consideration consists of 24 anisotropic plies of epoxy carbon fiber on the basis of Porcher 3692 fabric.

Technical elastic constants of the carbon fiber used in the numerical simulation was adopted both as the results of standard sample (for E11, E22, G12) testing, and on estimates of the research works [2, 3] (Table 1).

Table 1. Elastic properties of the material used in the calculation

Material	E_{11} (GPa)	E_{22} (GPa)	E_{33} (GPa)	G_{12} (GPa)	G_{13} (GPa)	G_{23} (GPa)	ν_{21}	ν_{13}	ν_{32}
CFRP	63.9	63.9	20.0	19.5	2.7	2.7	0.04	0.3	0.3

The multilayer bulkhead model completely illustrated all 24 structure layers. The constant load of 1.5kN was set at the low surface of the upper bulkhead rim. Surface area of loading corresponds to the contact area of bulkhead and the beam in laboratory tests at the appropriate load. The kinematic boundary conditions were set as restrain of displacement in all directions in the bolting of low bulkhead rim.

Contact between the bulkhead layers was considered to be ideal. Figure 3 illustrates the scheme of the boundary conditions.

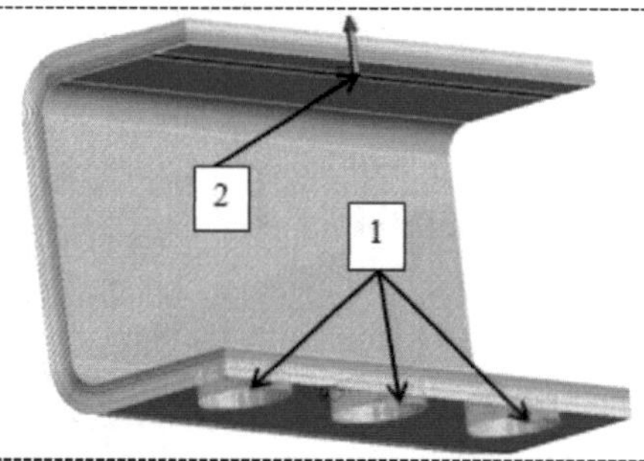

Figure 3. The scheme of boundary conditions: 1-bolting; 2 – load 1.5kH at the attachment of bulkhead and beam.

The computational mesh with prismatic type cells was generated to ensure better convergence of the solution and reduce errors in results. Within the construction and local mesh refinement sharp differences in geometric dimensions of the neighboring elements more than two times were not allowed. The maximum size of the element for each ply was 1mm, the minimum - 0.1mm. The total number of finite elements was about two million. Figure 4 shows a view of the finite element mesh.

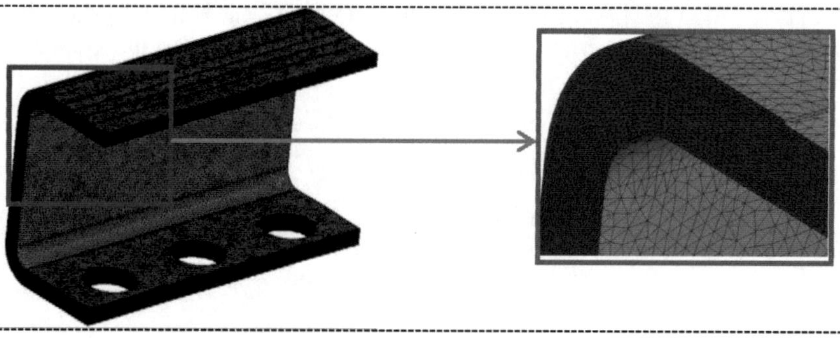

Figure 4. View of finite element mesh.

Stress distribution analysis in the warp and weft directions indicates that they reach the highest value around the inflexion zone. The maximum stress values arise on the surface of the upper layer. The maximum normal tensile stress along the warp is σ11 = 168.69 MPa, along the woof is σ22 = 42.55 MPa, and the maximum compressive stress is σ11 = -116.28 MPa, σ22 = -33.171 MPa. According to the loading nature the normal stress σ33 is the most dangerous and destructive for the given structure.

Normal stresses fields σ33 are shown in Figure 5. The maximum values of normal stress σ33 are observed in the central plies of the bulkhead in the upper and lower bending areas. Compressive stress σ33 = -22.72 MPa and tensile stress σ33 = 28.917 MPa.

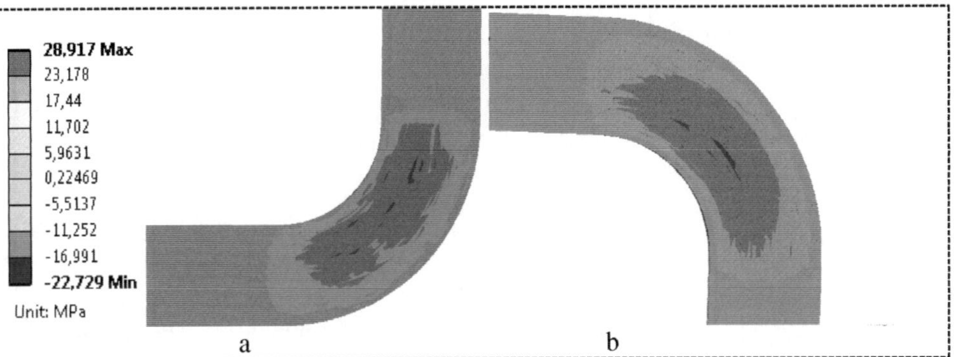

Figure 5. Interlayer normal stresses in the regions: a- lower bending, b- upper bending.

Maximum values of shear stresses τ_{12} occur between the first and second layer at the upper inflexion point. The maximum value of the interlayer shear stresses τ_{13}; τ_{23} amounted 12.8MPa and 3.38 MPa, respectively. The maximum value τ_{12} of shear stresses in the reinforcement plane reached 22.9 MPa.

Evaluation of the static strength of bulkhead partition conducted by the maximum stresses criteria (1). Table 2 demonstrates the static strength of the carbon fiber values. The presented data have been obtained experimentally, except of strength properties across S33 layer, which estimates have been taken from research paper [4].

The coefficient k in the equation (2), considering the material static strength reduction due to the various environmental factors, was assumed to be 1.2, that corresponds to the requirements imposed at similar structures.

Table 2. CFRP static strength properties

Material	S_{11}^{+}, MPa	S_{11}^{-}, MPa	S_{22}^{+}, MPa	S_{22}^{-}, MPa	S_{33}^{+}, MPa	S_{33}^{-}, MPa	S_{12}, MPa	S_{13}, MPa	S_{23}, MPa
CFRP	809	804	809	804	56.4	128	150	77	77

The conducted evaluation gives the safety factor of carbon fiber plastic static strength in the bulkhead partition for stresses in the layers as follows:

- the factor of safety n_{11} is 4.8 with tensile stresses alongside fiber warp;
- the factor of safety n_{22} is 19.0 with tensile stresses alongside fiber woof;
- the factor of safety n_{33} is 1.6 on interlaminar normal tensile stresses ;
- the factor of safety n_{13} and n_{23} is 6 and 22.8 respectively on interlaminar shearing;
-the factor of safety n_{12} for shear stresses in reinforcement plane is 6.6.

Thus, the normal interlaminar stresses are the most dangerous in the structure of bulkhead from composite materials. These stresses determine the static strength of carbon fiber plastic bulkhead which factor of safety is 1.62. Structure failure is expected to be delamination type in the central plies of the structure, mainly in the region of bend.

Taking into account that the limit of static strength S_{33}^{+} for considered carbon fiber plastic is 56.4 MPa [4], the structure failure delamination type is expected to take place with a load of 2.93 kN.

3. Comparison of numerical modeling results and mechanical testing data

Stress-strain diagram analysis (Figure 6) obtained during the mechanical testing indicated that load, corresponding to the load bearing capacity of the structure makes up 3 kN. Further loading leads to partial delamination in the laminated bulkhead structure reducing stiffness and bearing capacity. After multiple delamination further laminate deformation of the bulkhead takes place at a relatively constant load.

X-Ray testing was conducted at three strain levels, characterizing the main stages of damage development (Figure 6). Monitoring was performed with one projection. The source and the X-ray receiver located that way to detect interlaminar cracks reliably [5]. Herewith, the exposure time was fifty seconds at an anode voltage of 40 kV and anode current of 40 mA. In order to fix the crack opening, the applied load at the time of measuring remained constant. X-ray images were obtained and processed within monitoring.

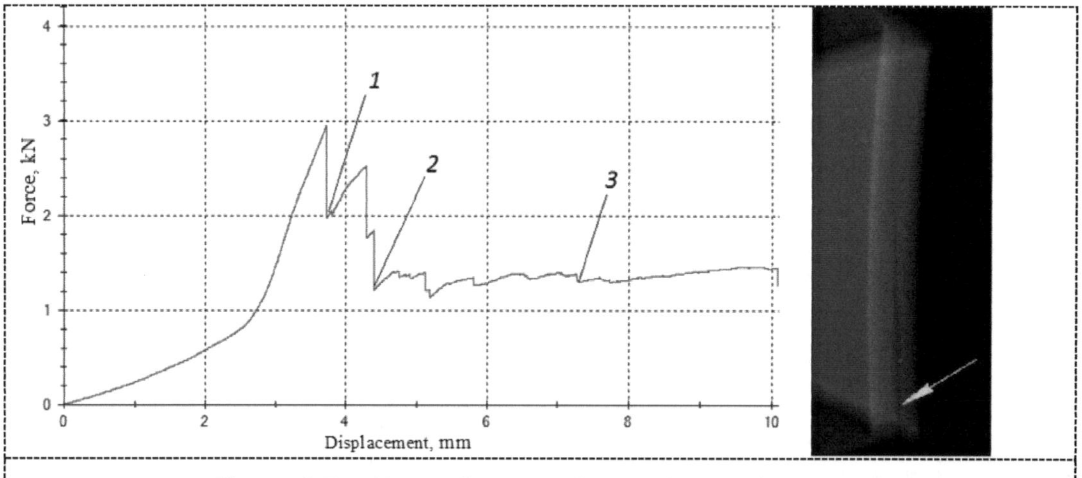

Figure 6. Tensile test diagram and X-Ray image of bulkhead (deformation 3.8 mm, load 3 kN).

Comparison of mathematical simulation and mechanical testing data revealed the difference between calculated and experimental values of the ultimate load limit in about 1%. High stresses regions in the central bend segments of layers coincide with areas of crack formation identified by radiography testing.

4. Conclusion

Within the conducted research the technique of stress-strain analysis and strength prediction for composite bulkhead partition was developed. The study was done in a general 3-D statement for anisotropic elastic body. Stress analysis in the layers of bulkhead revealed that interlaminar normal stresses are the most dangerous for structural strength. The structural failure is expected to start with delamination in bend regions in the central laminar of the structure.

Approbation of mechanical testing technique including microfocus x-ray is presented. This technique allows the precise determining of intelaminar cracking under different loads of structure.

Numerical results compared to experimental data obtained from mechanical testing of bulkhead section show that the difference between estimated and experimental values of the ultimate load is not more than 1%.

Acknowledgement

The study was performed in Perm National Research Polytechnic University with support of the Russian Science Foundation (project No. 15-19-00259).

References

[1] Abovsky N P, Andreev N P and Deruga A P 1978 *Theory of elastic body and shells variation principles* (Moscow: Science) p 287

[2] Lukina N F, Dementjeva L A and Kutsevich K E 2014 *Adhesive prepregs based on tissue Porcher – perspective materials for parts and units out of polymeric composite materials* (Moscow: Viam works No.6)

[3] Anoshkin A N and Tashkinov A A 1998 *Prediction of carrying capacity of composite flanges for the aircraft engine casing parts* (Perm: PSTU) p 101

[4] Anoshkin A N, Zuiko V Yu, Shipunov G S and Tretyakov A A 2014 *Technologies and problems of composite materials mechanics for production of outlet guide vane for aircraft jet engine* (Perm: PNRPU Mechanics No.4) pp 5-44

[5] Brault R, Germaneau A, Dupré J C, Doumalin P, Mistou S and Fazzini M 2013 *In-situ* Analysis of Laminated Composite Materials by X-ray Micro-Computed Tomography and Digital Volume Correlation *Experimental Mechanics* **53 7** 1143–1151

Mechanical state assessment using lamb wave technique in static tensile tests

M V Burkov[1,2], R T Shah[1], A V Eremin[1,2], A V Byakov[1,2], S V Panin[1,2]
[1]Institute of High Technology Physics, National Research Tomsk Polytechnic University, 634050, Lenina ave 30, Tomsk, Russia
[2]ISPMS SB RAS, 634021, Akademicheskiy ave 2/4, Tomsk, Russia

E-mail: burkovispms@mail.ru

Abstract. The paper deals with the investigation of Lamb wave ultrasonic technique for damage (or mechanical state) evaluation of AA7068T3 specimens in the course of tensile testing. Two piezoelectric transducers (PZT), one of which is used as an actuator and the other as sensor, were adhesively bonded on the specimen surface using epoxy. Two frequencies of testing signals (60 kHz and 350 kHz) were used. The set of static tensile tests were performed. The recorded signals were processed to calculate the informative parameters in order to evaluate the changes in stress-strain state of the specimens and their microstructure.

1. Introduction

The modern non-destructive testing (NDT) techniques are used in manufacturing and in-service inspection to ensure product integrity and reliability, as well as to control of manufacturing processes in order to reduce the production costs while maintaining high quality. During construction, NDT is used to monitor the uniformity of materials, welded joints, etc., while in-service NDT inspections are used to confirm that the products in use continue to have the integrity necessary to provide the sufficient reliability [1].

The emerging trend in the field of inspection during operation is to increase the NDT performance allowing saving time as well as fund. Thus over the past several decades, a significant research efforts have been focused on Structural Health Monitoring (SHM). The SHM concept is based on the use of in-situ non-destructive sensing and analysis of structural characteristics including the structural response in order to detect changes that may indicate damage or degradation. The output of SHM process is periodically updated informing about the ability of the structure to continue performing its intended function in the light of the inevitable aging and damage accumulation caused by the operational environments. Under an extreme event, such as an earthquake or unanticipated blast loading, SHM is used for rapid condition screening. This screening is intended to provide, in near real-time, reliable information about system performance during such extreme events and the subsequent integrity of the system [2, 3].

All active SHM operations require the development of complex algorithms and software for data processing and continuation of operation decision-making [4, 5]. The basis for the software is a deformation, damage accumulation and fracture mechanics of different materials. To test designed system and software the joint consideration of experimental data and computer modelling is required. There are a lot of papers dealing with research of ultrasonic systems utilizing Lamb wave principle and algorithms for damage detection. However, in most of them the PZTs without substrate are used.

Such PZTs will be inappropriate for cyclic loading during operation due to brittleness and low ultimate stress of piezoceramics. The aim of the present work is to assess the applicability of PZTs with steel substrate for damage evaluation as well as to investigate the response of the actuator-sensor pairs on different stress-strain state of the loaded aluminum specimens. The experimental results of ultrasonic examination using Lamb waves of AA7068T3 specimens tested with static tension are presented in the paper.

2. Materials and technique

The investigated ultrasonic technique was applied for evaluation of AA7068 specimens' mechanical state during static uniaxial tensile testing. The specimens were cut in the dogbone shape (figure 1) from the 4 mm sheet. The specimens were split on two groups: the first was used for testing in the initial state; the specimens from the second group were cut across gage length and then welded using gas tungsten arc welding. The AW1E12G-190EFL1Z piezoelectric discs used as transducers were adhesively bonded to the surface using 3M Scotch Weld DP105.

The ultrasonic technique used during tensile testing is described in details in [6]. There was a 100 signals averaging used to increase S/N ratio. Then the informative parameters (Maximum Envelope [7], Normalized Correlation Coefficient NCC and Variance μ_2) were calculated in order to associate the stress-strain state evolution with characteristic changes of propagated ultrasonic signal.

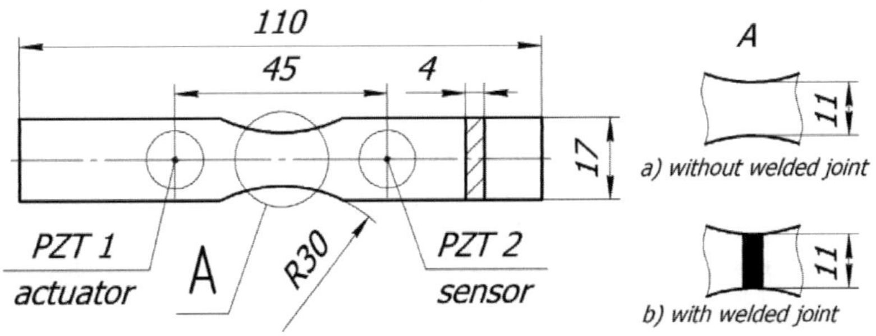

Figure 1. The sketch of the specimen with the PZTs on the specimen surface.

The static tensile testing was carried out using electromechanical testing machine Instron 5582 with the strain rate of 0.6 mm/min. The loading was stopped in defined points for capturing of ultrasonic data (the specimen was fixed in grips and subjected to tensile load). During the elastic region the step of 1 kN by load was used, when the plasticity started the step was changed to 0.2 % of elongation. So the dependence of the recoded signal amplitude on the stress-strain state was studied. The main goal of the static tests is to investigate the response of the actuator-sensor pairs during loading and to ensure the possibility of PZTs and epoxy adhesive application for further fatigue evaluation tests of the AA7068 specimens.

During each tensile test the strain and stress values were captured using gages of the Instron testing machine while the specimen was imaged using the Vic-3D digital image correlation (DIC) system. DIC system allowed us to obtain strain values calculated from the surface of the deformed specimen. This data can be used as the additional source of strain information during analysis of results.

3. Results and discussion

The preliminary study was carried out in order to choose the appropriate frequency for further testing: the actuator-sensor response of the specimen fixed in grips was characterized by the dependence (figure 2) in terms of maximum envelope of the signal versus ultrasound frequency. The frequency was varied in the range from 10 to 400 kHz with the step of 1 kHz. It is easily seen from the figure 2 that there are three main peaks of the maximum envelope corresponding to the frequencies of 60 kHz, 150 kHz and 350 kHz. The dispersion curves for aluminum plates obtained experimentally and theoretically [8-10] allow calculating the group velocities and the wavelengths of A_0 and S_0 modes for the specimens with thickness of 4 mm. Thus the first peak at 60 kHz corresponds to the high amplitude of A_0 mode while the amplitude of S_0 is negligible because it has long wavelength at this frequency. The third peak (350 kHz) corresponds to the S_0 mode. The second peak at ~150 kHz is also associated with the mix of mainly A_0 and S_0. Actually the association of these peaks with different modes of Lamb waves is calculated with an error about 5-15 % in comparison with the theoretical values. This discrepancy is related to the different boundary conditions of the specimen which is fixed in grips as well as due to its small size where side reflections can greatly change the signal sensed by the receiver.

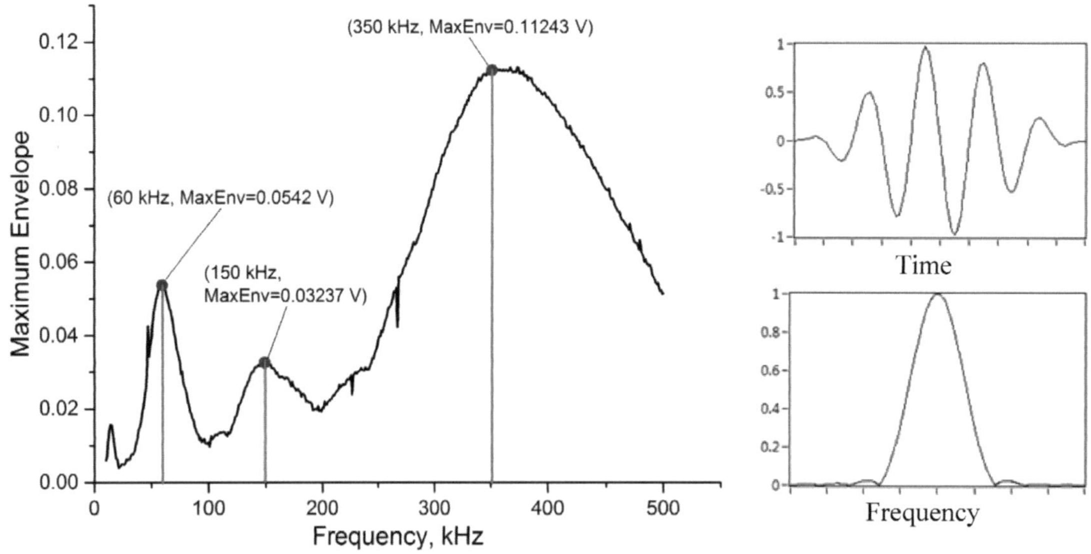

Figure 2. The dependence of the sensed signal amplitude on the frequency. The examples of the signal in time and frequency domains are shown.

This work deals with the investigation of the possibility of ultrasonic technique to characterize the changes in the specimen dimensions due to the deformation as well as microstructural changes of the material in the highly stressed gage length. Due to higher magnitude the first and the third peaks corresponding to the actuating frequency of 60 and 350 kHz were used during tensile testing. Thus the A_0 mode of Lamb waves was generated using low frequency scanning signal while S_0 mode with higher group velocity was generated at high frequency.

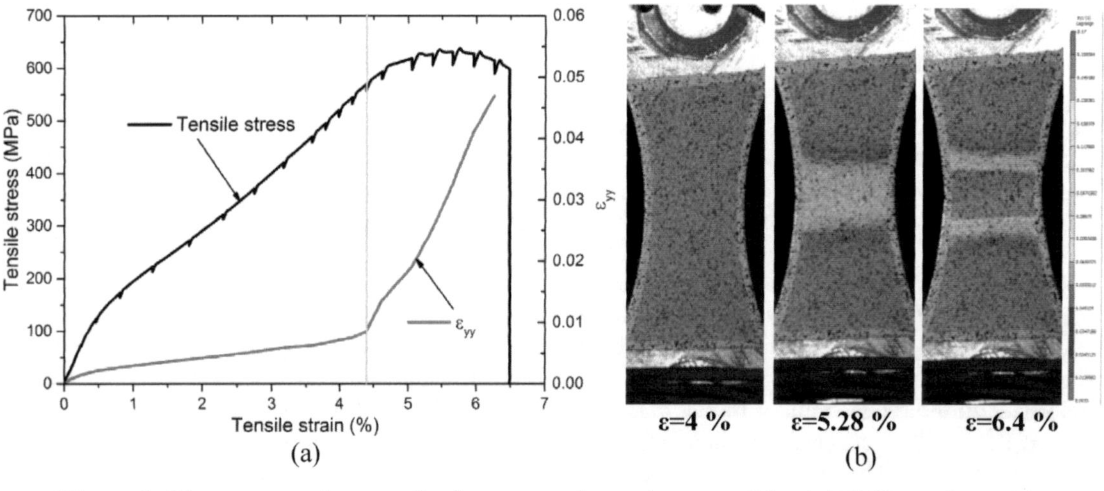

Figure 3. The stress-strain curve for the step mode tension test of the AA7068 specimen (a) and normal strain ε_{yy} measured by the Vic-3D (b).

Figure 3a represents the stress-strain curve of the AA7068 specimen loaded in step mode until fracture: there are 21 points of defined loads and strains where loading was stopped and ultrasound testing was performed. It could be observed that the vertical green line divides the graph on Figure 3 onto the elastic region and the region with plastic deformation where the specimen is deformed irreversibly. The ε_{yy} graph was plotted by averaging of all ε_{yy} values in the calculation area shown on Figure 3b. The ε_{yy} plot is also divided on two stages. The first stage is characterized by the fully elastic straining of the specimen and the images of ε_{yy} field look nearly uniform while at the second stage the plasticity concentrates in the narrowest cross section of the specimen thus the ε_{yy} plot rises rapidly.

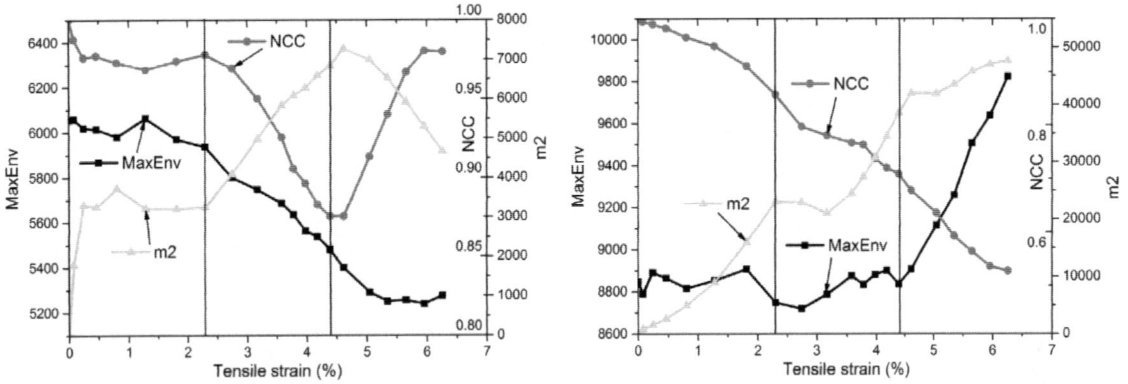

(a) (b)

Figure 4. The plots of maximum of envelope (MaxEnv), normalized correlation coefficient (NCC) and variance μ_2 (m2) by the elongation for initial AA7068 specimen: (a) 60 kHz and (b) 350 kHz.

Figure 4 displays the interdependence between informative parameters and applied strain. For low frequency the max of envelope (MaxEnv) is decreasing during all straining process until fracture with a small region during plasticity where the values remain nearly constant. This change can be associated with the slight variation of the specimen dimensions due to elongation thus the Lamb waves arrive to the sensor at different phase with a wave reflected from boundaries and different interference pattern occurs.

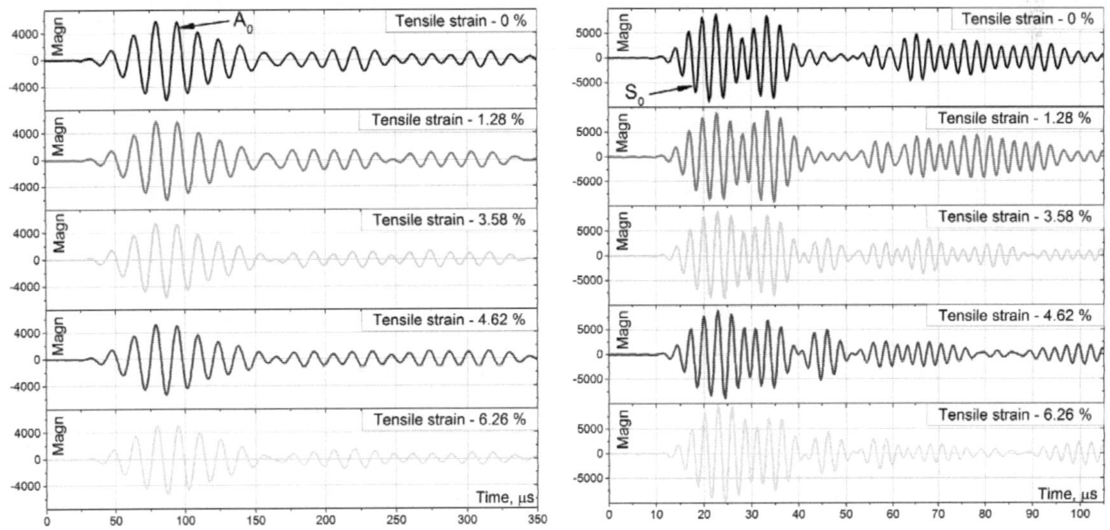

(a) (b)

Figure 5. The images of signal waveforms captured for the specimen without welded joint (a) 60 kHz, (b) 350 kHz.

Two parameters characterizing the waveform changes of the signals (Variance μ_2 and NCC) have nearly the same behavior (considering their reverse nature – the higher the differences of compared signals the higher the μ_2, while NCC will be lower). Such relation is relevant to all of the sets of results presented in the paper so only the NCC parameter will be described and discussed. As for the figure 4a the normalized correlation coefficient after the first stage with constant values decreases but during plastic deforming it goes back thus just before the fracture the NCC is nearly the same as at the beginning of the tensile test. Such controversial result doesn't allow us to make a correct decision on the state of the specimen thus it should be concluded that entirely the low frequency testing provides not reliable results. The informative parameters obtained for the high frequency of testing signal (figure 4b) have a better explainable behavior: the NCC decreases during the entire tensile test while the MaxEnv stays nearly constant until the plastic deformation starts (after 4.4 % elongation).

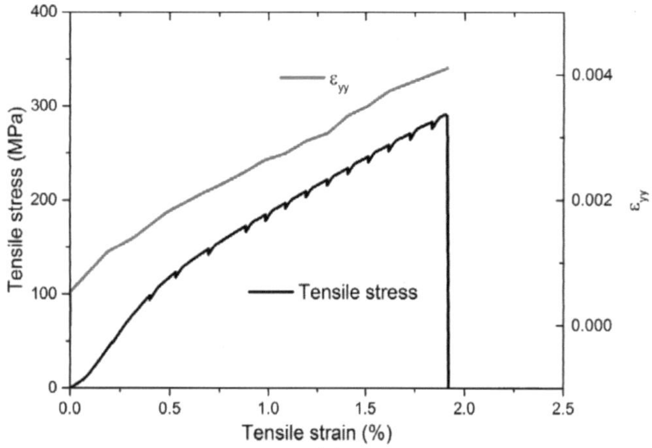

(a)

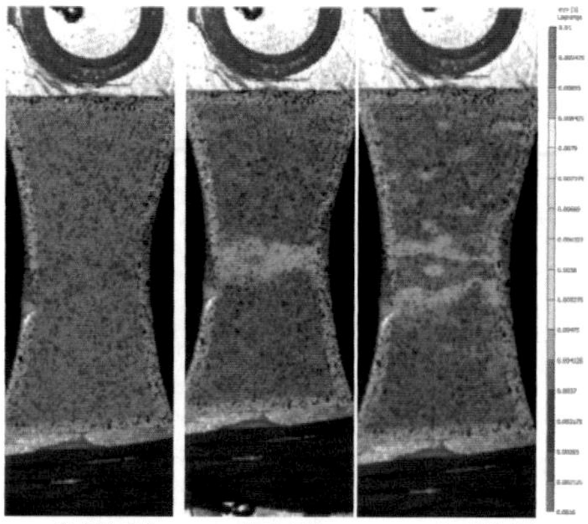

ε=0.32 % ε=1.2 % ε=1.87 %

(b)

Figure 6. The stress-strain curve for the step mode tension test of the welded AA7068 specimen (a) and normal strain ε_{yy} measured by Vic-3D (b).

Figure 5 shows the images of signal waveforms recorded during the tensile test. By the joint analysis of the informative parameters and these plots we can highlight the following regularities: (1) the shape of the first wave packet of low frequency signals is almost unchanged during the test, there are only a minor changes of the signals' "tail"; (2) a close look to the first wave packet figure 5a indicates that the amplitude of received signals decreases that is confirmed by the MaxEnv plot; (3) the first wave packet of the high frequency signal (which corresponds to the S_0 mode) stays nearly unchanged and the MaxEnv increases on the last stage as well; (4) the second wave packet is produced most likely by the side reflections and decrease significantly during the test; (5) the "tail" of the high frequency sensed signals changes drastically throughout the tensile test, but it is quite difficult to make any precise assumption about this due to small size of the specimen and large amount of different reflected waves that interfere with each other.

It can be seen in figure 6 that the welded joint drastically decreases the ultimate strength: σ_u of initial AA7068 alloy is about 627 MPa while for the specimen with welded joint the fracture point is at 290 MPa. Actually this alloy has low weldability by methods of liquid state welding like tungsten arc welding. Present work is aimed on the investigation of the ultrasonic testing technique thus high mechanical properties do not have a significant importance. Instead this weakened welded joint acts like a strain concentrator that can be seen from the stress-strain and ε_{yy} curves. While the specimen without welded joint has two easily identified stages of the ε_{yy} curve due to good plasticity of initial material the ε_{yy} curve for welded specimen is linear showing nearly elastic deformation behavior and brittle fracture. The area of welded joint at fracture exhibits elastic strains due to the lack of plasticity of welded material thus the averaged ε_{yy} curve is nearly linear.

Figure 7 shows the graphs of the informative parameters for the frequency of 60 kHz and 350 kHz. Due to low weldability of AA7068 the specimen fractured at low elongations (within 2-2.5 % of elongation). This leads to the lower stressed state of the specimen just before the fracture compared to the non-welded specimen.

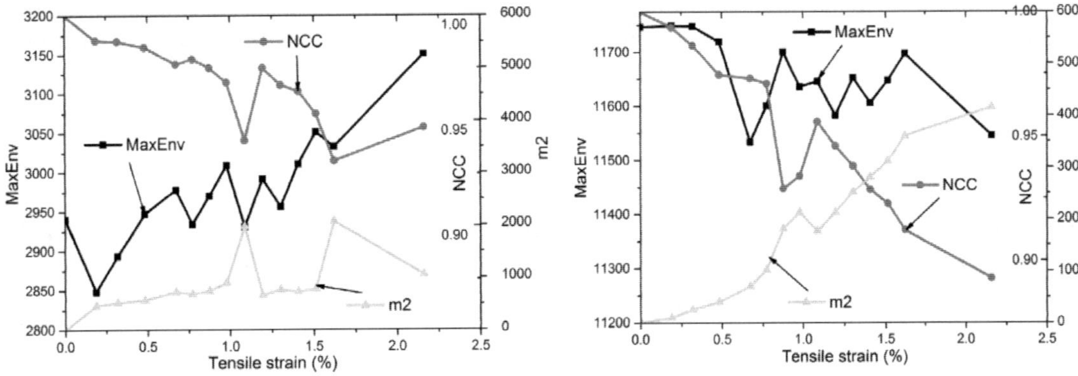

(a) (b)

Figure 7. The plots of maximum of envelope (1), normalized correlation coefficient (2) and variance μ_2 (3) by the elongation for welded AA7068 specimen: (a) 60 kHz and (b) 350 kHz.

The analysis of the data gives the following conclusion: the behavior of the plots of the informative parameters for the welded specimen is nearly the same as for the first stage for the initial specimen without welding joint (this stage is shown by the vertical black line on figure 4). For the high frequency testing the MaxEnv stays nearly the same until fracture and NCC decreases during the entire tensile test. For the low frequency all informative parameters have a sideway trend.

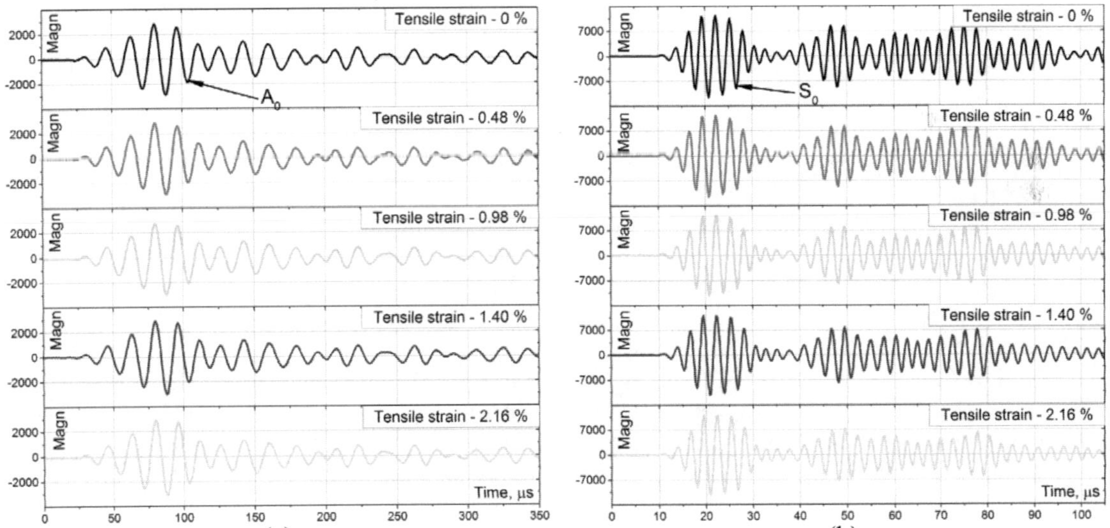

(a) (b)

Figure 8. The images of signal waveforms captured for the specimen with welded joint (a) 60 kHz, (b) 350 kHz and for the specimen with welded joint.

Figure 8 shows the variations of captured signals throughout the tensile test of specimen with welded joint. The shapes of signals compared to those for the specimen without welded joint are nearly the same. In the low frequency plots the A_0 mode is easily distinguished and its shape doesn't changes up to fracture. The "tail" of the signal at 60 kHz also stays constant up to fracture. For the high frequency the first wave packet corresponds to S_0 mode but now the second wave packet is observed separately from the S_0. It confirms the assumption that this wave packet can be associated to the interference of side reflections. Also it can be seen that the signal "tail" has higher amplitude in comparison with the initial specimen signal waveforms.

4. Conclusion

The static tensile tests of AA7068 specimens were performed with acquisition of strain data using Instron load and strain gages, strain distribution fields via VIC-3D system and ultrasonic data by acoustic inspection system. After the digital processing of raw acoustic signal data, the set of informative parameters was calculated including MaxEnv and NCC characterizing the changes in propagated signal amplitude and its shape correspondingly. After the analysis of the results we can conclude the following:

1) Small size of the specimen leads to the very complex and inexplicable results. There are a lot of boundary reflections from sides of the specimen as well as the influence of grips of testing machine which is very difficult to assess.

2) It is seen that low frequency testing (using A_0 mode of Lamb waves) doesn't provide good results of quantitative analysis of signals. High frequency tests (mainly S_0 mode) provide better explainable results allowing it to be used for characterizing the changes of stress-strain state of the specimen during loading.

3) The first wave packet of the signal for all specimens was used to calculate the MaxEnv parameter, while the entire signal was analyzed by NCC and Variance μ_2. But according to the plots (figure 4) the last parameters allow to make better conclusions about the state of the specimen. Thus it is necessary to make a more detailed investigation of variation of the shape of the first wave packet of sensed signal, because now NCC and Variance μ_2 are influenced by the "tail" of the signal which behavior is difficult to explain in contrast to the first travelled wave packet which is less influenced by the side reflections and grips.

Acknowledgments
The work is supported by the fundamental research grant 2013-2020 of Russian Academy of sciences.

References

[1] Hellier C 2001 *Handbook of Nondestructive Evaluation* (New York: McGraw-Hill Professional) 594 p
[2] Diamanti K, Soutis C 2010 *Prog. Aerosp. Sci.* **46** 342–352
[3] Farrar C, Worden K 2007 *Phil. Trans. R. Soc. A.* **365** 303–315
[4] R. Kazys, O. Tumsys, D. Pagodinas 2008 *NDT&E Int.* **41** 457–466
[5] Torres-Arredondo M, Tibaduiza D, Mujica L, Rodellar J and Fritzen C-P 2014 *Struct. Health Monit.* **13** 19-32
[6] Burkov M, Byakov A, Shah R, Lyubutin P and Panin S 2015 *IOP Conf. Series.* **93** 012025
[7] Hahn SL 1996 *Hilbert Transforms in Signal Processing* Norwood, USA, Artech House, 460 p
[8] Giurgiutiu V, Cuc A 2005 *Shock. Vib. Digest.* **37** 783-105
[9] Hyun Woo P, Seung Bum K, Hoon S 2009 *Wave Motion* **46** 451–467
[10] Cho H and Lissenden CJ 2012 *Struct. Health Monit.* **11** 393-404

IOP Conf. Series: Materials Science and Engineering **156** (2016) 012010 doi:10.1088/1757-899X/156/1/012010

Dislocation structure and deformation hardening alloy fcc single crystals at the mesolevel

L A Teplyakova[1], T S Kunitsyna[1,2], N A Koneva[1], E V Kozlov [1], A A Kondratyuk[2], N A Zboikova[1], Yu A Kakushkin[1], A A Iakhin[2]

[1]Tomsk State University of Architecture and Building, 2, Solyanaya Sq., 634003, Tomsk, Russia
[2]National Research Tomsk Polytechnic University, 30, Lenin Ave., 634050, Tomsk, Russia

E-mail: lat168@mail.ru

Abstract. The article presents the evaluation results of impacts of various strengthening mechanisms to flow stress. Such evaluations were made on the basis of the measured parameters of the dislocation substructure formed in monocrystals of [001]-Ni_3Fe alloy deformed by compression within the stage II. It was found that the main impact to deformation resistance in the alloys with net substructure is made by the mechanism of dislocation impediment, which is caused by contact interaction between moving dislocations and forest dislocations.

1. Introduction

The plastic deformation of metal materials takes place in a hierarchically organized wide interval of scales which is closely connected for the with the initial defect structure [1, 2]. As a result the scale-structural deformation level spectrum is formed. During deformation in the most well-known metal materials their multistage deformation strengthening takes place. The single-phase alloys with f.c.c. lattice, particularly monocrystals, are characterized by rather prolonged linear strengthening (stage II) which is most pronounced for the single crystals with symmetric orientations [3]. In such materials the slipping dislocations and moving point defects can be slowed down by various mechanisms and the strengthening has a complex nature. At the low temperatures the deformation is mostly provided by the dislocation movement. The analysis and systematization of all well-known strengthening mechanisms is given in [2]. At present there are about 60 of such mechanisms, 50 of them are the low temperature ones. Only this fact alone allows us to realize that to understand and describe the hierarchically organized multifactor multistage strengthening at the plastic deformation even of the materials with the f.c.c. lattice is a difficult task. It is no wonder that it is still a topical problem.

Identification of the deformation strengthening mechanisms of the metal materials requires detail data about the dislocation structure and its evolution during deformation. So since 50th – 60th of the XX century, the researchers have actively studied the dislocation structure of the deformed materials using the translucent electron microscopy method. Much attention was paid to the f.c.c. monocrystals, first of all to Cu, Al and Ni [4]. Studies on monocrystals let us understand the nature of the dislocation structure formation in "clean" state that is without other factors, for example such as the presence of the grain boundaries. The analysis of the deformation strengthening mechanisms should be logically carried out for the monocrystals of single phase alloy whose dislocation structure was studied in detail and the mechanisms of its evolution in the wide interval of deformation degrees were surely confirmed. Such alloy is the Ni_3Fe alloy [3,5,6].

The given work presents the results of the experimentally determined contributions of various strengthening mechanisms realized on the first mezolevel that is the level of the dislocation substructures according to the classification considered in [2].

2. Materials and methods

In this work the Ni3Fe alloy monocrystals with the short range order were investigated. The loading axis (compression) was parallel to the crystal graphic direction [001]. The monocrystal orientations were determined using the URS – 60 apparatus by the Laue's method. The deviation from the indicated orientations in the initial state did not exceed 1 - 2°. The monocrystals had the form of the parallelepiped with the size of 3x3x6 mm3 and the side faces of the cubic orientation. The investigation of the dislocation structure was carried out on the electron microscope attached with a goniometer at the accelerating voltage of 100 and 125 kV. By the electron microscopy images of the dislocation structures the free segment lengths (λ) between various stoppers along the dislocation lines of the dislocations were measured. Such stoppers are the compound dislocation formed as the result of the dislocation reactions and the thresholds. For measuring the free segment lengths the micrographs of the foils cut parallel the octahedral plane were used. When measuring the distances between the compound dislocations no distinctions were made between the reactions in the slip plane being parallel the foil surface and the reactions with dislocations of other octahedral planes. The thresholds and bends were not distinguished. The measurements were carried out separately for the dense and thin substructure parts.

3. Results and discussion

3.1. Evolution of substructure and its spatial organiozation.

In the [001]-monocrystals of the disordered Ni_3Fe alloy the beginning of plastic deformation (transitional stage) is connected with the formation in the local monocrystal volumes of the flat dislocation clusters in the {111} planes. In the continuation of the stage II replacing it the layer net DSS is formed [6]. In the local volumes corresponding to the shift fragments with the asymmetrical slipping one system of layers with the regular-shaped net section and the increased density of dislocations parallel the primary for the given fragment of the slipping plane is formed (Figure 1). In the fragments with symmetry slipping as a rule two crossing systems of layers are formed. They are parallel each of the octahedron planes (Figure 1, near B) Rarely more than two systems of layers are formed.

3.2. Substructure hardening.

In the net substructure the deformation strength is mostly caused by the contact deceleration of moving dislocations. The following hardening mechanisms make their contribution into the contact stress (τk): 1) dislocation intersection, 2) jog dragging, 3) formation and break of deformation reactions, 4) overcoming the Lomer-Kontrell's, Hirt's dislocation barriers, etc. 5) point defect generation. The first contribution is caused by the contact interaction between the slipping and forest dislocations. Figure 2, 3 shows the histograms of the measured free segments lengths at different deformation degrees (within the stage II). As it follows from these figures the dislocation substructure (DSS) evolution is not the same in the dense and thin sections. In the thin substructure sections the λ distribution is inhomogeneous (Figure 2). The inhomogeneity is kept up to the middle of the stage II. The average length of a free segment between the reactionary junctions, jogs and any kind of these stoppers in the thin sections of the DSS is not practically decreased with the deformation degree increase (Figure.4a). Consequently, the hardening of the investigated alloy cannot be provided by the DSS thin areas. In the dense sections of DSS the λ distribution is changed with deformation (Figure 3). The maximum of the λ distribution shifts towards lower values. Within the stage II in the places with the increased density of dislocations the λ average value is decreased with the deformation increase (Figure 4b).

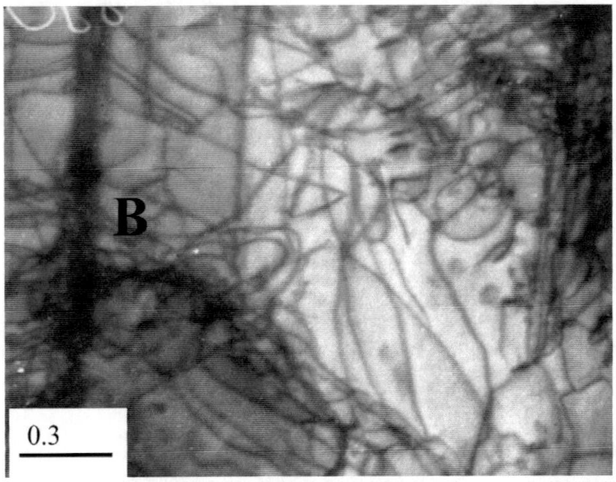

Figure 1. Dislocation structure in monocrystals Ni_3Fe, $\varepsilon = 0.17$ foil plane (111)

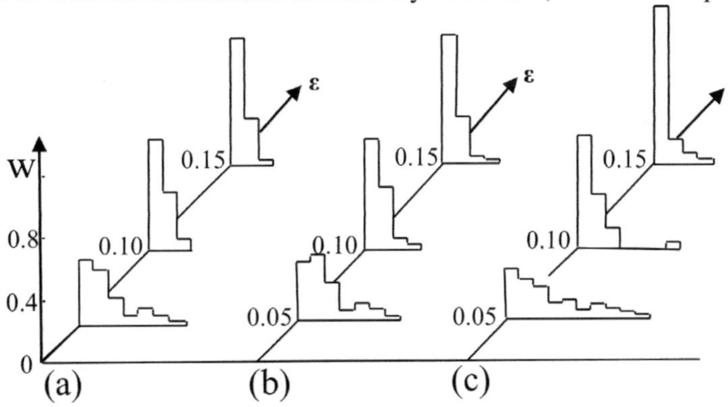

Figure 2. Histograms of the distances along the dislocation line between any stoppers (a), dislocation junctions (b), jogs (c) at various deformation degrees in the DSS dense sections.

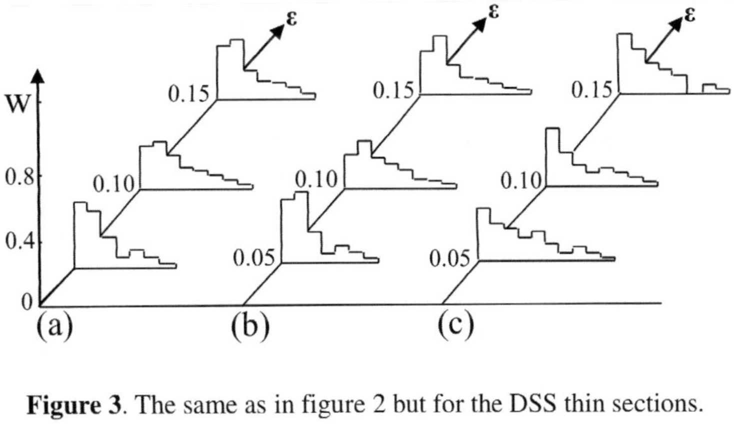

Figure 3. The same as in figure 2 but for the DSS thin sections.

Simultaneously in the crystal the volume part with the high dislocation density is increased. Slipping which is naturally developed in the weakest sections of DSS has to go through the sections with the dense net DSS due to the decrease of their quota. Thus, during deformation in the densest sections of DSS the resistance connected with the reacting forest dislocations and/or the jogs drawing will be increased.

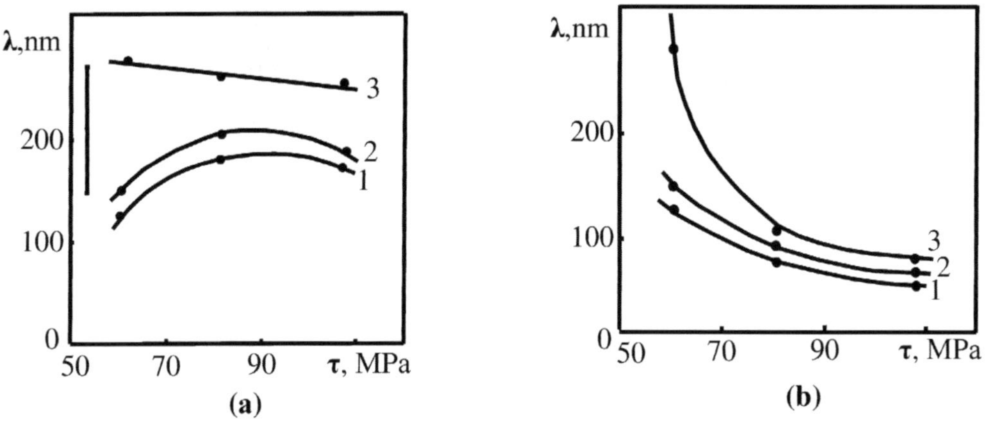

Figure 4. Dependences of the average distance along the dislocation line between any stoppers (cv.1), reactionary junctions (cv.2), jogs (cv.3) on the given stress in dense (a) and thin (b) DSS sections.

3.3. Long range order internal stress fields and elastic dislocation interaction.

The characteristic that is of primary importance in terms of the dislocation movement and multiplication is the internal stress field of the dislocation ensemble. In this paper an attempt was made to determine the magnitude (amplitude) of the long-range stress fields due to contributions from different mechanisms and determine the role of some of these mechanisms. To solve this problem the following DSS parameters were measured: 1) the bending radius of free dislocation segments; 2) the distances between the dislocations slipping both in one and in different planes; 3) the distances between the active slipping planes.

Measurement of the bending radius R let us determine the dislocation ensemble-averaged static fields of the internal stresses (IS) and their fluctuations. These fields are made up of many components: particularly, the fields from the parallel dislocations laying both in the own slipping plane and on the parallel planes; the fields formed by the forest dislocations; the fields of dislocation self-action. The values of the distances between the dislocations obtained in this work allow us to calculate the contributions to the local static field of IS. Using 3) one can determine the dynamic long-range order fields originating when the dislocations pass along the parallel planes.

3.4. Static long-range order stress fields.

The homogeneous stress τ_{st}, acting on a dislocation should bend it in an arc with a radius [7]: $R=Gb/2\tau_{st}$, where G is the shear module (for the alloy Ni_3Fe G=770 MPa), b is Burger's vector (b=0,25 μm). Here

$$\tau_{st} = \frac{Gb}{2R}$$

The bending radius is easily determined provided one knows the segment height h and the distance between the fixing points

$$R = \frac{4h^2 + l^2}{8h}.$$

Figure 5 shows the dependence of the dispersion τ_{st} on the shear stress (because the τ_{st} average value is equal to zero). As it follows from this figure the static long range order fields are significant only at the beginning of the plastic deformation. Their value is decreased with the deformation degree increase.

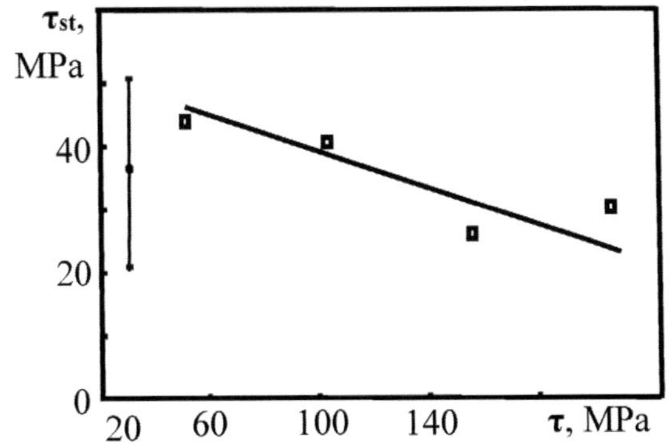

Figure 5. Dependence of the static long range order dispersion on the shear stress.

Thus the static long range order fields do not contribute to the alloy deformation strengthening (hardening). This conclusion agrees with the DSS qualitative pattern: the dislocation tangles present the multipole configurations without the excess dislocation density [2,3,5]. The free dislocation orientations along both sides the tangles are the same. This conclusion made for the high symmetrical orientation [001] where all octahedral planes are equally loaded and the secondary slipping systems are missed cannot be extended to other less symmetrical orientations without special investigation.

In the given paper the attempt was also made to single out the contribution of the elastic dislocation retardation experienced by the nearest neighboring dislocations (τs.pl) and the forest dislocations (τf)

$$\tau_{s.pl} = \frac{Gb}{2\langle L_{s.pl}\rangle}; \quad \tau_f = \frac{Gb}{2\langle L_f\rangle}$$

The values of τ_{cl} and τ_f at the stage II are given in Table 1 where one sees that the elastic short-range interaction of the nearest neighbors may be significant and its role increases with the deformation. However these stresses are compensated by the elastic interaction of the second, third and the following dislocations located along the both sides of this dislocation. At full summation of the contributions of all dislocations these stresses should give zero average value and the dispersion equal to the stress of the static long-range fields.

3.5. Dynamic long-range stresses.

The resistance experienced by dislocations when passing along parallel planes can be estimated as follows:

$$\tau_{dyn} = \frac{k}{2\pi} \cdot \frac{Gb}{\langle X\rangle}$$

where < X> is the average distance between the active slipping planes; k is a factor depending on the type of dislocations that implement deformation (k=2 in case of deformation by the superdislocations, k=1 in case of the deformation by the monodislocations). The dependence of the obtained τ_{dyn} values on the deformation degree is given in Table 1. As one sees the resistance experienced by the dislocation when it moves relatively others moving along the parallel plane in the Ni_3Fe alloy is small (of the order of 30-40 MPa and remains practically unchanged at the stage II (within the spread of values)). At the beginning of the stage II τ_{dyn} may comprise a significant part of the acting stress.

Table 1. To determination of the long-range stresses

Flow stress τ (MPa)	The degree of deformation ε	Static measured average τ_{st} (MPa)	Dynamic average τ_{dyn} (MPa)	In its plane τ_{pl} (MPa)	Forests τ_f (MPa)
60	0.05	3.5	0.5	6	5.2
75	0.10		0.6	6.3	5.4
80	0.11	2.3			
90	0.15		0.7	11	9.7
105	0.17	1.5			
		By bending radius of free dislocation segments	On parallel planes for nearest traces	Elastic short-range interaction for the first neighbours	

Thus, the given evaluations of the τ_{st} and τ_{dyn} values show that in the investigated Ni_3Fe alloy with the short-range atom order the long-range fields do not play a significant role in the deformation hardening. The contact retardation (deceleration) of dislocations plays the decisive role in the hardening of this alloy (the dislocation (dislocation friction). The increase with the deformation of the stopper density along the dislocation line (both the jogs and the reactions) may be the reason of the deformation hardening. In other words, the work hardening of the Ni_3Fe disordered alloy is caused by the increase of deceleration of the dislocation friction type.

3.6. Cut dislocation density.

Measurement results of $<\lambda>$ were used to determine the $\Delta\tau_i$ contributions of the main mechanisms of the dislocation deceleration to the shear stress of the short-range atom order alloy. When determining the contributions the approximations and coefficients given in [8] were used. Table 2 presents the results of $\Delta\tau_i$ determination on the basis of the experimental $<\lambda>$ values, where где $\Delta\tau_{nc.j}$ is the resistance connected with the nonconservative jog dragging, $\Delta\tau_r$ is the resistance connected with overcoming the reacting "forest" dislocations, $\Delta\tau_d$ is the resistance connected with overcoming the long-range stress fields. Here one can see the $\Sigma\Delta\tau_i$ total values conditioned by acting of these mechanisms as well as the values of the acting shear stresses. As follows from Table 2 with the deformation degree increase the contributions caused by overcoming the resistance of reacting "forest" dislocations and jogs on the dislocations are also increased (in the DSS dense sections). At the stage II the total stress $\Sigma\Delta\tau_i$ of all mechanism contributions appears to be comparable with the stress acting in the slipping plane.

Table 2. Contributions of the main deceleration mechanisms to the shear stress.

mechanism	$\Delta\tau_r$, (MPa)	$\Delta\tau_{nc.j}$, (MPa)	$\Delta\tau_d$, (MPa)	$\Sigma\Delta\tau_i$, (MPa)	τ_{acting}, (MPa)
$\varepsilon = 0.05$	26	37	35	98	60
$\varepsilon = 0.10$	39	39	23	102	80
$\varepsilon = 0.15$	42	49	15	106	105

4. Conclusion

Based on the obtained results one can imagine the following scheme of the investigated alloy deformation. At the beginning of the plastic deformation when the stresses are not large only the longest free dislocation segments will take part in deformation and the dislocation slipping will occur in the less dense areas. With the increase of the stoppers density on the unit of dislocation density the longest free dislocation segments do not provide the deformation course and the slipping dislocations start cutting the areas with greater dislocation density. As a result the greater stresses should be applied for implementation of further shear. At the same time more part of the length distribution λ is involved in the deformation. Simultaneously with the deformation degree increase the "cut" dislocation density approximates the <ρ> average value. The evaluations show that in the described scheme the dislocation slipping runs either by separation from the stoppers or by the action of free dislocation segments as the Frank-Read' sources.

References

[1] Panin V E, Grinyaev Yu V, Elsukova T F and Ivanchin A G 1982 *J. Proc. of H. Education. Phys.* **6** 5-27
[2] Kozlov E V 2002 J. Materials Questions 29 1 50 – 112
[3] Teplyakova L A, Kunitsyna T S, Koneva N A and Kozlov E V 2004 *J. Bulletin of the Russian Academy of Sciences: Physics* 68 10 1456-1461
[4] Nabarro F R N, Basinski Z S and Holt D B 1967 The Plasticity of single crystals (Moscow, Metallurgy)
[5] Koneva N A, Teplyakova L A, Kunitsyna T S and Kozlov E V 2006 Structure and properties of advanced materials (Tomck, Scientific and technical literature) p 323
[6] Teplyakova L A, Kunitsyna T S, Starenchenko V A and Poltaranin M A 2014 *J. Proc. of H. Education. Phys.* ***57*** 2 61-70
[7] Seeger A 1980 Dislocations and mechanical properties of crystals (Moscow, Publishing foreign literature)
[8] Czernikow I and Marcinkowski M I 1971 J. *Met. Trans.* **2** 3217-3222

IOP Conf. Series: Materials Science and Engineering **156** (2016) 012011 doi:10.1088/1757-899X/156/1/012011

Experimental study of residual stresses relaxation in ring details during multicyclic loading

A V Korolev[1], **A A Korolev**[1], **A F Balaev**[1], **S A Savran**[1] **and A S Yakovishin**[1]
[1]Yuri Gagarin State Technical University, Saratov, Russia

E-mail: bandrey.sc@gmail.com

Abstract. The paper presents the results of experimental studies of vibro-mechanical mechanism of residual stresses relaxation in ringed parts. There was described the mechanism of vibro-mechanical relaxation that includes machining of ringed parts among three rotating rolls under pressure. This leads to multicyclic loading due to which there occurs relaxation of residual stresses. To study the process of vibro-mechanical relaxation a complete factorial experiment was carried out. As a result of experiments we obtained empirical dependence of residual stresses on processing factors.

1. Introduction

Ringed parts have a wide application in the industry. The highest demands are made to geometric accuracy of rings in rolling bearings. Manufacturing of bearing rings is associated with high material and time expenditures which are caused by a high accuracy of form and ring fragility. Even small residual stresses occurring in the workpieces of bearing rings during machining and heat treatment result in significant deformation of rings and loss of geometrical accuracy during the manufacturing process. During the operation of rolling bearings under the cyclic loads there occurs residual stress relaxation, accompanied by elastic-plastic deformation and a corresponding change in the contact conditions of rolling elements, in particular the contact area and the amount of preload. Changes in operating conditions lead to a decrease of bearing life. Therefore, to reduce the internal residual stresses before the final machining, all rings are subjected to process step of size stabilization.

2. Relevance

To reduce the residual stresses and stabilize geometrical parameters of metal products, methods based on different physical action find their applications in the production. The most common ones include: method of natural aging, thermal, chemicothermal, electron-beam and mechanical methods of stabilization.

Natural aging [1] has low productivity that is measured in years; temperature and cold tempering [2] in addition to poor performance have high cost of the equipment. The last one is also accompanied by a significant distortion of geometric shape of the product during the tempering, which does not provide a complete stress relief.

The laser and electron-beam methods [3] provide sufficiently high stability in the distribution of residual stresses due to the high controllability of the process; however, along with the high performance they have high cost of the equipment. Therefore, the reasonability of these methods is justified only during the stabilization of precision products with complex shape.

The method of application of nanostructured coatings and chemicothermal methods [4] have high cost and low performance compared to mechanical methods of stress stabilization. They are used to provide the required physical and mechanical properties of the surface layer of the workpiece.

The methods of mechanical stabilization [5] do not require expensive equipment, while ensuring high stability of distributed residual stresses with a relatively high productivity. Among the mechanical methods, vibro-mechanical one has the best performance. Due to multicyclic loading of a product, the performance of this method can be tens of seconds [6], which is much higher in comparison with other methods of stress relief.

3. Description of stabilizing method

The method of vibro-mechanical stabilization consists of repeated cyclic loading of products [6, 7]. In the device for residual stresses relaxation of ringed parts the basing of machined rings is carried along the cylindrical surface of the rolls with equal diameters and crossed rotation axes at an angle α. Crossed axes of the rolls allow longitudinal movement of the rings at a speed V. At mutually opposite turning of the support rolls at an equal angle α along the vertical plane the trajectory deviates from a straight line. Kinematics of the centerless running in cylindrical products during vibro-mechanical stabilization is similar to kinematics of the centerless superfinishing described in detail in [8, 9]. From the work [9] it is known that the curvature of the trajectory is greater, if the angle of the rolls rotation α is greater.

After installation of the ring between the rolls, radial load does not change throughout the whole treatment cycle consisting of a set of cycles for the running-in of parts.

Multiple running-in of parts leads to an accumulation of internal energy in the areas of residual stresses. When this energy reaches critical values, in these areas there begin shears of material micrograins, accompanied by relaxation of residual stresses. This low speed energy storage for plastic flow of the grains does not have time to cause a change in the shape and size of the processed ring. The radial load P, produced by the third roll, is applied to the ring at a point equidistant from the places of contact between the ring and basing rolls. One of the basing rolls imparts rotary motion n to the ring, resulting in a uniform elastic deformation of the material along the entire circumference of the ring by the value of process induced distortion Δ.

4. Experiment description

To assess the influence of processing factors on the quality of relaxation of residual stresses and stabilize the geometrical parameters, the method of complete factorial experiment was used. The following processing factors were selected as independent ones: the deformation value of stabilized part - Δ, conditioned by the radial load applied by the work roll to the processed ring; the rotation frequency of the roll - n; the angle of mutual turning of the rolls in the vertical plane - α, conditioning the value of parts traverse and the time of processing cycle. Based on the equipment capabilities, for the selected processing factors there was defined an area boundary in the form of the upper and lower levels of variation that are presented in Table 1.

Table 1. Studied factors with real values.

Levels	Process factors in units		
	Δ, mm	n, min^{-1}	α, grad
Lower	0.02	20	1
Main	0.0325	28	1.25
Upper	0.045	36	1.5
Variability interval	0.025	16	0.5

There were selected 24 outer bearing rings SH20.01 GOST 3635-78 as experimental samples subjected to multicyclic loading. Bearing rings were made of steel SH-15 and went through all the process steps prior to stabilization.

The measurement of residual stresses was conducted on ring deformation after the cut and the subsequent calculation using the formula (1):

$$\sigma_O = \frac{2 \cdot \delta \cdot E \cdot J}{\pi \cdot r^2 \cdot W} \qquad (1)$$

where E=2,1·10^5 MPa is the material modulus of elasticity; J=76 mm^4 is the inertia moment of the sample section; W=36 mm^3 is the moment of bending resistance of the ring section; r=14,5 mm is the radius of a neutral layer of the ring; δ, mm is the ring deformation in cross-section that is perpendicular to the cutting line of the ring along the generating line.

To evaluate the deviation of the optimization parameter from the average value, the error mean square was calculated according to data obtained from the parallel observations of planning matrix at each point of the experiment plan. Verification of the homogeneity of variance hypothesis was conducted using the Cochran's Q test for significance level equal to 5%. The error mean square was determined assuming that there is a uniform experiments variance.

5. Experiment results

Confidence limits of the regression coefficients were determined in order to select statistically significant coefficients. The comparison of the confidence limits of the regression coefficient with the values of the coefficients revealed that the statistically significant coefficients are b0, b1, b2.

Thus, for the found regression coefficients the empirical model took the following form (2):

$$\sigma_O = 5{,}43 - 290 \cdot \Delta + 10{,}51 \cdot \alpha \qquad (2)$$

After finding the variance of calculated values for optimization parameter at each point of the experiment plan, the model adequacy was verified using the F-test.

Based on the obtained regression model (2), there were plotted graphs showing the dependency of the ellipticity on the intersection angle of the rolls (Figure 1, a) and the value of process induced distortion of the ring (Figure 1, b).

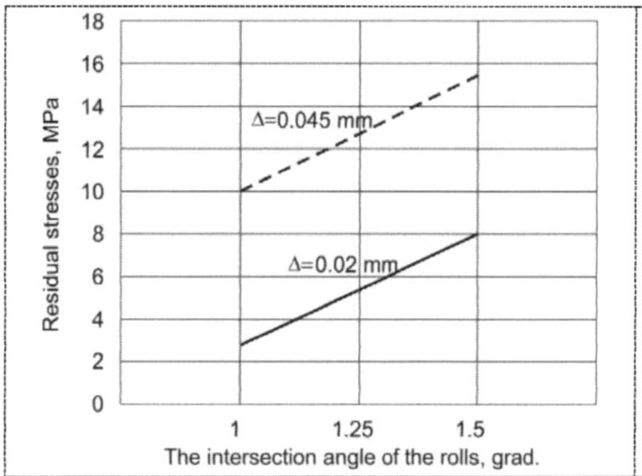

(a)

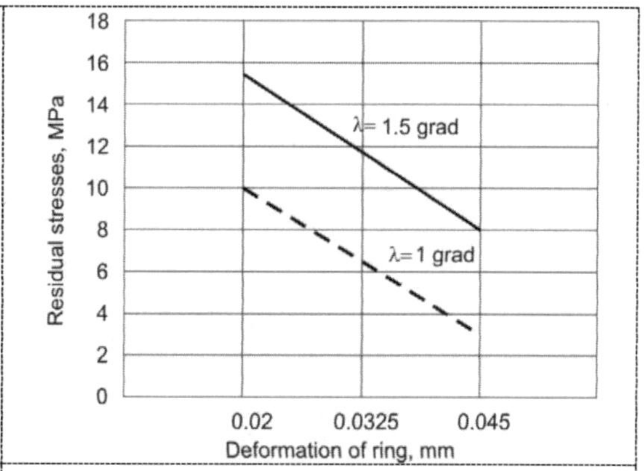

(b)

Figure 1. Dependency of residual stresses on: a) the intersection angle of the rolls; b) the value of process induced distortion.

Figure 1a shows that increasing the intersection angle of the rolls leads to an increase of residual stresses. It may be due to the fact that an increase in the angle of the rolls rotation leads to an increase in the radius of curvature of the path. Thus, the action of originally tuned value of the process induced distortion will be in the lower part of the path. For this reason, the maximum effect of the deformation influence on the residual stresses will take less time relative to the duration of the entire cycle of vibro-mechanical processing.

Analyzing the behaviour of the residual stresses depending on the process induced distortion of the ring, figure 1b shows that with the increase of the initial value of the process induced distortion, residual stresses are reduced. Thus, this confirms the findings based on figure 1 that the value of relieved residual stresses depends on the overall work process induced distortion at vibro-mechanical processing. It is defined by both the maximum value of the process induced distortion and the duration of its influence on the product.

Figure 2 shows the dependence of the residual stresses in the experimental samples on the value of their deformation in the area of rolls axis crossing during vibro-mechanical processing.

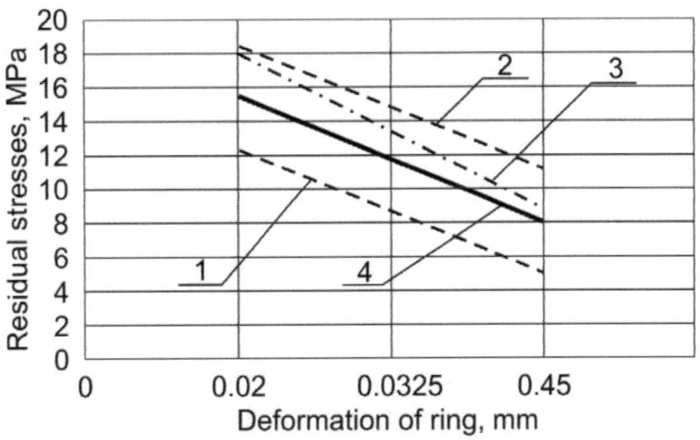

Figure 2. Theoretical and experimental dependencies of residual stresses on the value of ring deformation: 1, 2 represent the lower and upper confidence limits of the experimental dependency, respectively; 3 – theoretical dependency; 4 – experimental dependency.

Considering the regression model (2) it can be concluded that the rate of process induced distortion in the selected ranges of processing factors variation that are limited by upper and lower levels of variation, practically has no effect on the residual stress relief.

Firm line 4 on figure 2 shows the regression dependence (2) obtained during the experiment. Dotted lines 1 and 2 represent the lower and upper confidence limits of the experimental dependency, respectively. Chain line 3 shows the theoretical dependence, built on the basis of the mathematical model of vibro-mechanical stabilization presented in [10].

As it can be seen on Figure 3, the theoretical dependence is within the confidence limits of the experimental values of residual stresses. This suggests that the theoretical dependence is adequate to experimental data.

6. Findings and Conclusion

The experimental studies have established a link between processing factors of vibro-stabilizing treatment and the amount of residual stresses. On the basis of the experiment graphs (Figure 1) and the empirical dependency (2) the following conclusions can be drawn:

- the increase of intersection angle of the backup rolls leads to the increase of residual stresses;
- the increase of initial value of process induced distortion leads to the decrease of residual stresses;

- the value of relieved residual stresses depends on the overall work process induced distortion at vibro-mechanical processing. It is defined by both the maximum value of the process induced distortion and the duration of its influence on the product;
- rotation frequency of the rings has practically no effect on the change in residual stresses;
- experimentally obtained dependency of residual stresses on the process induced distortion confirms the theoretical dependency.

As a result of experiments there was established dependency of residual stresses of rings in bearings on processing factors. This proves the validity of the theoretical model of vibro-mechanical stabilization.

Acknowledgments
This paper has been developed under the state order of the Ministry of Education and Science of the Russian Federation (No. 9.896.2014/K).

References

[1] Zhang J and Dong P 2000 Residual stresses in welded moment frames and implications for structural performance *Journal of Structural Engineering* **126** **3** 306–315
[2] Berezhnyts'ka M P 2001 *Materials Science* **37** **6** 933–939
[3] Nosov N V Gusev A A and Nyrtsov I V 2013 Optimization of laser strenghtening parameters *Vestnik SamGTU* **2** 124–130
[4] Korolev A V Korolev A A and Zhuravlyov M M 2015 The Mechanism of Hardening and Internal Stress Stabilization in the Process of Laser Treatment *Journal of Russian Laser Research* **36** **4** 356–363
[5] Kolbasnikov N G Mishin V V Naumov A A and Zabrodin A V 2014 Research into structure and rheological and relaxation properties of nanocrystalline beryllium at temperatures of hot rolling and research into stress relaxation kinetics in different sorts of beryllium *Nanotechnologies in Russia* **9** **7** 430–440
[6] Korolev A V and Korolev A A 2011 Residual stresses relaxation method Patent 2478031 RU **2**
[7] Korolev A V Chistyakov A M Krivego V A and Moiseev V G 1999 Vibroaging method of parts Patent 2140842 RU **8**
[8] Zakharov O V Brzhozovskii B M and Balaev A F 2006 Generalized Geometric-Analytical Model for Calculating the Shaping Trajectory in Centerless Superfinishing *Izv. Vyssh. Uchebn. Zaved., Mashinostr.* **10** 11–15
[9] Brzhozovskii B M Zakharov O V and Balaev A F 2006 Kinematic and Mechanical Aspects of Centerless Superfinishing *Stanki Instrumenti* **11** 2–5
[10] Balaev A F Zakharov O V Korolev A V Korolev A A and Kochetkov A V 2015 Modeling the mechanism of stress relaxation of ring parts at high-cycle loading *Proceedings of 2015 International Conference on Mechanical Engineering, Automation and Control Systems, MEACS 2015* 1–4

Investigation impact of stressed state conditions and thermomechanical parameters on the texture and structure evolution in 1565ph aluminium alloy

V V Yashin[1], E V Aryshensky[2], R F Kawalla[3], V N Serebryany[4] and S V Rushchits[5]
[1]ZAO Alcoa SMZ, 29, Alma-Atinskaya str., Samara, 443051, Russia
[2]Samara State Aerospace University, 45, Moskovskoye Shosse, Samara, 443086, Russia
[3]Technische Universität Bergakademie Freiberg, Bernhard-v.-Cotta-Str. 4, 09599 Freiberg, Germany
[4]A.A.Baikov Institute of Metallurgy and Material Science RAS 49, Leninskiy prospect Moscow, 119991, Russia
[5]South Ural State University, 76, Lenin prospect, Chelyabinsk, Russia

E-mail: vasiliy.yashin@alcoa.com

Abstract. The paper is devoted to study of the impact stress condition and thermomechanical treatment parameters on the structure and texture evolution of new 1565 ph aluminum alloy. For that purposes, we use test on Gleeble equipment, FM calculation, optical microscopy and x ray diffraction texture analysis. The dependency between the deformation texture components development and strain rate value was established. Differences in the texture evolution at uniaxial compression stress and plain strain mode were revealed.

1. Introduction

Despite of the new types of materials emergence, aluminum alloys still maintain their competiveness and remain much in demand, especially, in aerospace industry. Aluminum alloys are so attractable because of combination low weight with high formability and acceptable strength [1]. That is why new alloys development and existing aluminum alloys improvement is continuously ongoing. Zirconium or scandium addition to aluminum alloys have become one of the key practices of its creation over the recent 10 – 15 years [1 - 3]. Its representative is 1565ph alloy (Al-Mg-Zn-Zr), recently developed by the Russian scientists [2]. This alloy is designed for low temperature applications, i.e. for aerospace field, cryogenic liquids transportation and the extreme north environment; it also demonstrates high strength and good weldability.

However, properties get in laboratory condition cannot be automatically achieved during real industrial process. It is well known that in order to obtain necessary properties level, one should control the texture and structure evolution and other important parameters like temperature and mechanical behavior during all manufacturing process [4 - 8]. Investigation of mechanical properties evolution was performed for this alloy in [9,10]. The texture evolution during hot rolling in the continuous mill was studied in [10]. However, we still need to establish impact of main technology parameters as well as stressed state condition on the texture evolution during thermomechanical treatment. It allows us to create reliable models and get possibility to achieve the required texture and structure in any combination of technological conditions.

In the present paper, the texture and structure evolution for two different type of stress condition and wide range of temperature and strain rate values is studied. Two different with different stress condition

2. Experimental procedure

All deformations are carried out on plane-strain and axial compression modes. Temperature and strain rate parameters for each mode are presented in the Table 1, in all cases true strain did not exceed 1. The specimens had the following dimensions: for the plane strain: h = 10 mm, w = 26 mm and l = 28 mm, and for axial compression: h = 15 mm, d = 10 mm. The samples were collected after the hot rolling in industrial reversing mill and had fully recrystallized structure [10].

Table 1. Temperature strain rate parameters for two different stress condition.

	Plane strain mode								
Temperature, °C	350	400		440			490		
Strain rate ($\dot{\varepsilon}$), s^{-1}	10	10	1	10	50	0,1	1	10	90
	Axial mode								
Temperature, °C	350	400			450			490	
Strain rate ($\dot{\varepsilon}$), s^{-1}	50	50	0,1	1	10	50		50	

After the deformation, the samples were quenched in water for texture and structure fixation. Then for the further X-ray structural analysis specimen was cut out from the middle section. Some specimens were cut crosswise for microstructure analysis in polarized light in Axiovert-40MAT microscope.

The texture, represented by four incomplete pole figures (PF) {111}, {200}, {220} and {311}, was analyzed by „reflection“ method in C_oK_α-radiation, using x-ray diffractometer DRON-7. Pole figures view plane was parallel to the rolling direction. The following angle ranges were applied: inclination α = 0–70 ° and rotation β = 0–360 ° with 5 ° step. The defocusing-induced decrease in the intensity at the periphery of a PF was corrected using coefficients calculated for PF recording conditions [11]. The orientation distribution function (ODF) and the volume fractions of the main orientations were calculated using the method of approximation by a large number (2000) of Gaussian normal distributions. For convenience of orientations distribution, function application three-dimensional function may be approximately represented by a sum of standard functions (texture components, perfect orientations dispersion); the combination of these texture components was selected based on analyses of sections of orientations distribution function, computed based on experimental pole figures. Newly created orientations file with three Euler angles and preliminary functions parameters, was processed by Texxor [12].

We also analyzed the stress components distribution and the accumulated strain field into workpiece for both deformation modes. For its calculation we use the DEFORM Software. The computation was performed for all possible process regimes presented in Table 1.

3. Experimental results

Results of microstructural analysis for different combinations of strain rates and temperature entities are shown in Figure 1.

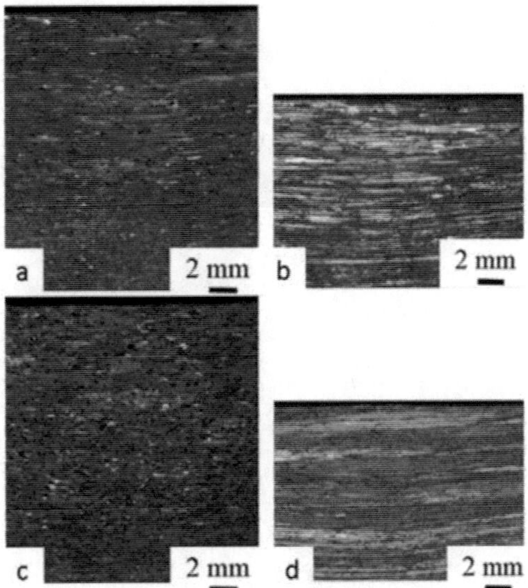

Figure 1. Comparison of microstructure analysis results: top part of cross section (a) axial compression (T = 450 C^0, $\dot{\varepsilon}$ = 10 s^{-1}); (b) plain strain mode (T = 440 ^{0}C, $\dot{\varepsilon}$ = 10 s^{-1}); (c) axial compression (T = 440 ^{0}C, $\dot{\varepsilon}$ = 50 s^{-1}); (d) plane strain mode (T = 440 ^{0}C, $\dot{\varepsilon}$ = 50 s^{-1}).

It should be noted that we show results mainly for regimes with high strain rates, as they present the interest for development thermomechanical processes with large deformation speed. Significant difference in structure (Figure1) is observed for two types of stressed state. Elongated deformed grains are clearly seen in the entire specimen sections in the case of plane strain. In the case of axial compression, the structure has mixed nature with recrystallized grain close to the near-surface zone and deformed macrostructure in the specimen mid-section. Such difference is explained by the presence of large sticking friction zone in cylindrical specimen. That zone is clearly seen in Figure 2, which shows accumulated strain distribution into workpiece for axial compression mode.

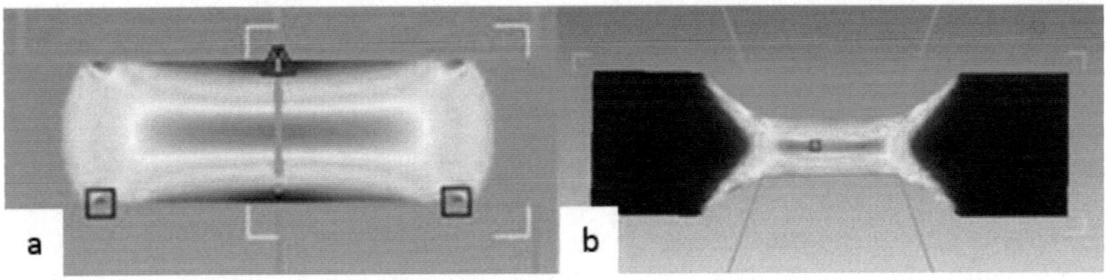

Figure 2. Accumulated strain distribution for different stress conditions, which was calculated in DEFORM: (a) axial compression (T = 450 ^{0}C, $\dot{\varepsilon}$ =10 s^{-1}); (b) plane strain condition (T = 440 C^0, $\dot{\varepsilon}$ =10 s^{-1})

As seen from Figure 2 a, in near-surface zone workpiece actually doesn't get any deformation. As a result, this zone almost fully corresponds to the area with recrystallized grain in real cylindrical workpiece. Actually, these grains don't get any deformation during axial compression and save from hot rolling process. In the opposite way, in the case of plane strain condition the accumulative strain is more uniformly distributed across the workpiece section. The different in accumulative strain distribution, explains the distinction in the observed microstructure. It should be noted, that the strain rate increasing in this investigation conditions has almost no effect on microstructure evolution in both deformation modes.

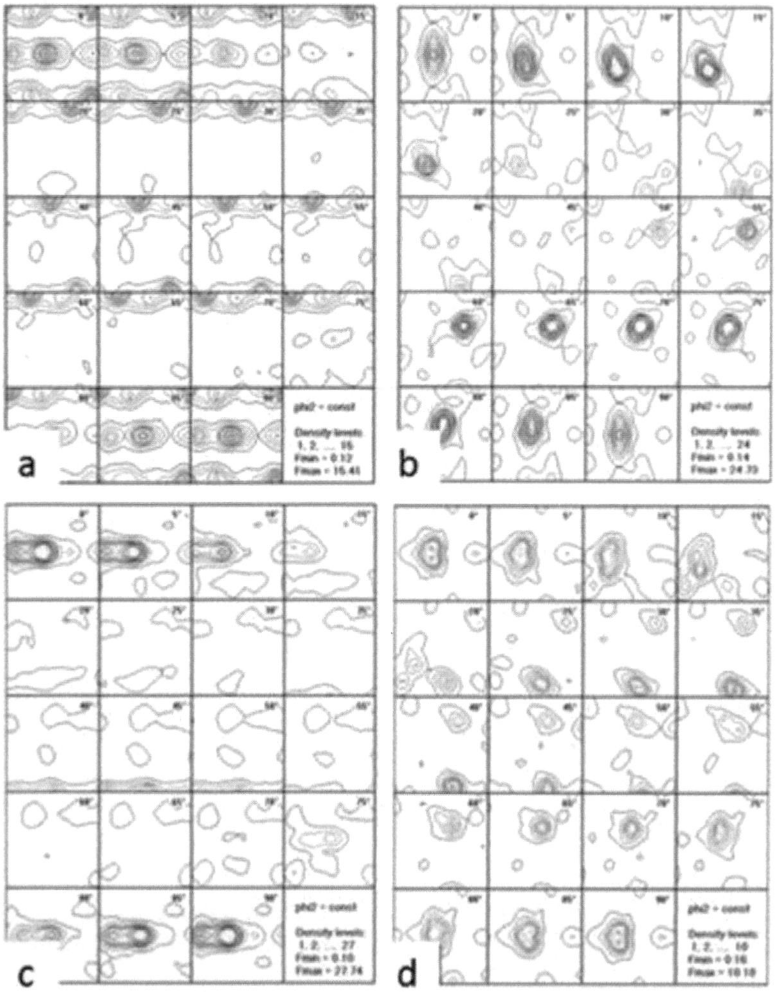

Figure 3. Comparison of the X-ray texture analyzes results:
(a) Axial compression (T = 450 ^{0}C, $\dot{\varepsilon}$ = 1 s^{-1}); (b) Plane strain condition (T = 450 ^{0}C, $\dot{\varepsilon}$ = 1 s^{-1});
(c) Axial compression (T = 450 ^{0}C, $\dot{\varepsilon}$ = 10 s^{-1}); (d) Plane strain condition (T = 440 ^{0}C, $\dot{\varepsilon}$ = 10 s^{-1}).

Comparison between texture analysis results for axial compression and plane strain condition are illustrated in Figure 3 a, b. In the first case we have mixed texture, including both cube and deformation components. In the second case, mostly deformation texture components are observed. In plane strain condition in some areas of Euler space the deformation texture is strongly pronounced, however, in general, it is volume is not very large.

Comparison of the texture analysis results (Figure 3 c, d) for higher strain rates demonstrates, that clearly pronounced deformation texture are observed in axial compression case. In case of the plane strain condition the deformation texture is not so pronounced, however it be could observe it in more Euler space areas then after axial compression.

The differences in the texture components can be easily explained by distinction in stressed state conditions (Figure 4), which have significant effect on their evolution.

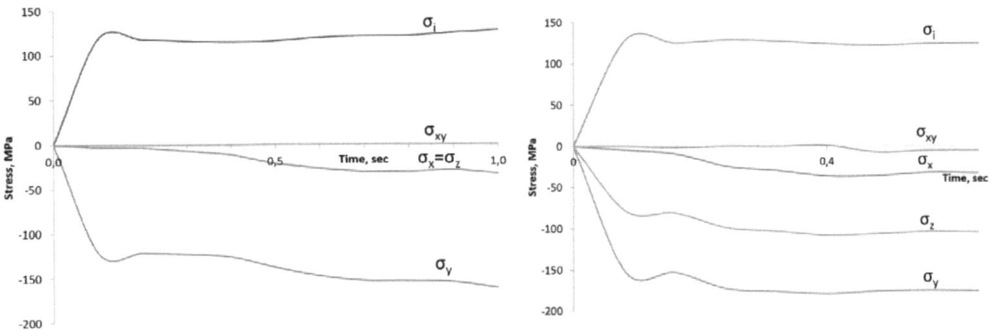

Figure 4. Stress distribution in the samples middle section: (a) axial compression (b) plain strain condition.

In conclusion, the strain rate influence on deformation components growth is investigated (Figure 5).

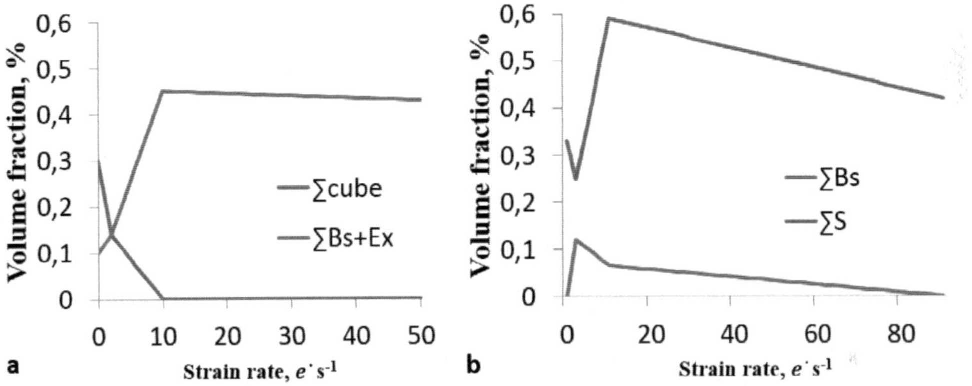

Figure 5. The texture component dependence on strain rates: (a) for axial compression; (b) for plane strain condition.

The similar picture is observed in both cases. The strain rate impact on texture evolution drastically increase, in the range of 0 to 10 s^{-1}. Then its, reaches peak and start decreasing: slowly in case of axial compression and more fast in plane strain condition. We can also conclude from Figure 3 a, b, analyze that logarithmic strain about 1 is no enough for full transformation of the cube texture components into deformation texture in case of low strain rate entities. These are significant results, which support the necessity of the high-speed hardening [13 – 14] consideration in the texture evolution models for such type of alloys.

4. Summary

1) The alloy 1565ph demonstrates the absence of bending to recrystallization. However, in axial compression mode not only deformed, but also recrystallized structure is observed. Simulation in DEFORM software show, that the main reason for such type of structure is the presents of sticking friction zone. Grain in this zone actually do not change during deformation process.

2) Distinction between main texture components, which are observed in different treatment modes, could be explain by various stress condition. Strain rate with combination in stress also has significant effect on the texture evolution, as result this influence must be accounted in mathematical model of texture evolution for such group of alloys in two different test mode.

Reference

[1] Davydov V G, Elagin V I, Zakharov V V and Rostova T D 1996 Alloying aluminium alloys with scandium and zirconium additives *J. Metal Science and Heat Treatment* **38 VII-VIII** 347-352

[2] Oryschenko A S, Osokin E P, Barakhtina N N, Drits A M and Sosedkov S M 2012 Aluminum – magnesium alloy 1565 for cryogenic application *J. Tsvetnye Metally* **11** 84-90

[3] Røyset J and Ryum N 2005 Scandium in aluminium alloys *International Materials Reviews* **50 I** 19-44

[4] Gorelova S, Schaeben H and Kawalla R. 2014 Quantifying texture evolution during hot rolling of magnesium twin roll cast strip *Materials Science and Engineering A* **602** 134-142

[5] Hirsch J 2005 Texture evolution and earing in aluminium can sheet *Materials Science Forum* **495-497 II** 1565-157

[6] Aryshenskii E V, Aryshenskii V Y, Grechnikova A F and Beglov E D 2014 Evolution of Texture and Microstructure in the Production of Sheets and Ribbons from Aluminum Alloy 5182 in Modern Rolling Facilities *Metal Science and Heat Treatment* **56 VII-VIII** 347-352

[7] Nam A, Prüfert U, Eiermann M and Kawalla R 2016 Modelling the temperature evolution during hot reversing strip rolling of magnesium alloys *Materials Science Forum* **854** 140-145

[8] Kawalla R, Graf M and Tokmakov K 2010 Simulation system of multistage hot rolling process of flat Products *J. Metalurgija* **49-III** 175-179

[9] Rushchits S, Aryshenskii E, Kawalla R and Serebryany V 2016 Investigation of texture structure and mechanical properties evolution during hot deformation of 1565 aluminum alloy *Materials Science Forum* **854** 73-78

[10] Rushchits S V, Aryshensky E V, Sosedkov S M and Akhmed'Yanov A.M. 2016 Modeling the hot deformation behavior of 1565ch aluminum alloy *Key Engineering Materials* **684** 35-41

[11] Serebryany V N, Kurtasov S F and Litvinovich M A 2007, Analysis of ODF errors upon the reversion of polar figures with the use of the statistical method of ridge estimations, *Zavod. Lab., Diagn. Mater.* **73 IV** 29– 35

[12] Kurtasov S F 2007, Quantitative analysis of the rolling textures of materials with a cubic symmetry of crystal lattice, *Zavod. Lab., Diagn. Mater.* **73 VII** 29–35

[13] Asaro R J 1983 Micromechanics of Crystals and Polycrystals *Advances in Applied Mechanics* **23** 1-115.

[14] Beaudoin A J, Mathur K K, Dawson P R, Kocks U F 1995, *J. Plast.* **11** 501-521

Construction of layered structures on valve metal alloys by microplasma oxidation

T A Baranova, A K Chubenko, A I Mamaev, V A Mamaeva and Ya B Kovalskaya
National Research Tomsk State University, 36 Lenina ave., Tomsk, 634050, Russia

E-mail: konstantinova.ta9@gmail.com

Abstract. Process of layered structure materials creation based on aluminum alloys is presented. Microplasma texturing method, microplasma oxidation method and chemical metallization method were used to create these structures. Non-conductive nonmetallic inorganic coatings were produced by microplasma oxidation method. Obtained structures showed high durability under thermal stress loads due to substrate metal – non-conductive nonmetallic inorganic coating phase boundary texturing.

1. Introduction.

Construction of laminated functional materials that are able to work under mechanical and thermal loads is an actual issue nowadays.

In this connection relatively new class of coatings – nanostructured non-metallic inorganic coatings – represents specific interest [1, 2]. These coatings can be obtained under conditions of high energy flows localization at the metal – electrolyte solution interface usually followed by microarc charges.

Based on this non-metallic inorganic type of coatings there is a capacity to create laminated composite materials with different constitutions and applications.

Applying these materials as construction materials appears to be very perspective for electronics, microelectronics, micromachines and also for space equipment, military equipment and general engineering.

Melting point of nanostructured non-metallic inorganic layers formed under microplasma influence is 2680 °C for zirconium oxides and 1780 °C for titanium oxides. Besides, such coatings can be classified due to their structure as microporous that allows to consider these coatings as practically ideal sublayers.

Imposition of successive layers like metals, for example, on local places of the surface provides functional properties of obtained coatings.

Real operating conditions of proposed materials applications require ensuring of maximum adhesion to the substrate metal surface as well as durability of the formed structures under mechanical and thermal loads.

Paper [3] based on mathematical modeling shows that texturing of substrate metal surface provides increase of nanostructured non-metallic inorganic coatings and successive connected layers' adhesion to substrates.

Mechanical loads result in laminated materials deformation and cracks formation. The way of cracks distribution depends on texture type: cracks can be orthogonal or circular. Circular cracks formation on the textured coating surface leads to elimination of loads and prevents trunk cracks formation and exfoliation of laminated non-metallic inorganic coating material. As a result, durability of the coating material itself under deforming conditions significantly increases.

Valve metal surface texturing can be achieved using microplasma discharge effects [1, 2]. Microplasma texturing contains significant amount of regulating factors including electrolyte composition, duration of exposure and methods of process installation.

Let us consider formation of non-metallic inorganic coating with textured metal-coating phase boundary using formation of such coating on aluminum.

2. Materials and methods

Samples produced in form of plates of D16 aluminum alloy were initially grinded to 10th class of roughness in order to eliminate oxide layer and to get unique surface. Then organic contaminants were removed from the surface by rinsing with 1 M Na_3PO_4 solution at a temperature not lower then 60 °C. Treated samples were washed with distilled water, then with ethanol and with distilled water again.

Microarc coatings deposition was carried out for all aluminum samples in fluorine–phosphate solution with the following composition:

- NaF – 5 g/l,
- $Na_2B_4O_7 \cdot 10H_2O$ – 7 g/l,
- HBO_3 – 10 g/l,
- Na_2HPO_4 – 18 g/l.

Treated aluminum alloy sample served as anode; stainless steel tub with water cooling casing served as cathode.

Coatings obtained in electrolyte of described composition predominantly consisted of aluminum oxide and aluminum phosphate:

$$Al - 3e \rightarrow Al^{3+},$$
$$Al^{3+} + PO_4^{3-} \rightarrow AlPO_4,$$
$$2\,Al^{3+} + 3O^{2-} \rightarrow Al_2O_3.$$

Coatings were chemically etched with 3 % NaOH solution for 5 minutes at a room temperature, then thorough and intensive rinsing with distilled water was carried out. Process was conducted repeatedly in order to determine influence of deposition and etching cycles amounts on metal texture.

Roughness and sample's profilograms after texturing were scrutinized by contactless 3D Micro Measure Station produced by STIL. Maximum roughness (Ra = 1.015 μm) was achieved after 3 cycles of coating/etching (Table 1).

Table 1. Aluminum samples surface roughness

Coating/etching cycle quantity	Ra, μm
1	0.787
2	0.962
3	1.015
4	0.815
5	0.797

Obtained surface texture data were used in order to form laminated structures with predetermined textured phase boundary on aluminum substrates.

Next stage of laminated material forming was introducing of metal copper on non-metallic inorganic coating oxide layer surface. Metallization was provided by chemical method.

Chemical deposition of the metal results in sewing the oxide layer along its longitudinal and transverse pores with formation of monolithic composite metal-oxide coating. This fact also provides smoothing of oxide layer imperfections and thus friction coefficient decreasing. Surface conductivity increases, that makes it possible to deposit subsequent metal layers using electroplating techniques.

Moreover, the given approach allows to equalize thermal stresses under thermal loads and to increase the whole durability of laminated material in case of local thermal loads. Melting of copper layers with the following filling of micropores and microcracks under significant thermal loads is permitted and leads to the «healing» effect in relation to re-appeared imperfections.

Sensitization and activation operations were successively conducted within the framework of preparing the surface for copper deposition [4].

Chemical deposition of copper was carried out in tartrate copper solution. Formalin served as reducing agent. Copper was reduced during the copperplating process [4]:

$$2H_2CO + Cu^{2+} + 4OH^- \rightarrow Cu + H_2 + 2HCOO^- + 2H_2O.$$

Thickness of chemically deposited copper layer was 1 - 3 μm. Obtained samples were investigated by semi-quantitative elemental composition assessment using EDAX Genesis microanalyzer. Results of investigation showed that percentage of copper on the surface layer amounted from 98 to 98.8 wt.%.

3. Results and discussion

Micrograph of layer material surface which has copper outer layer has been obtained by raster scanning microscopy method (Figure 1).

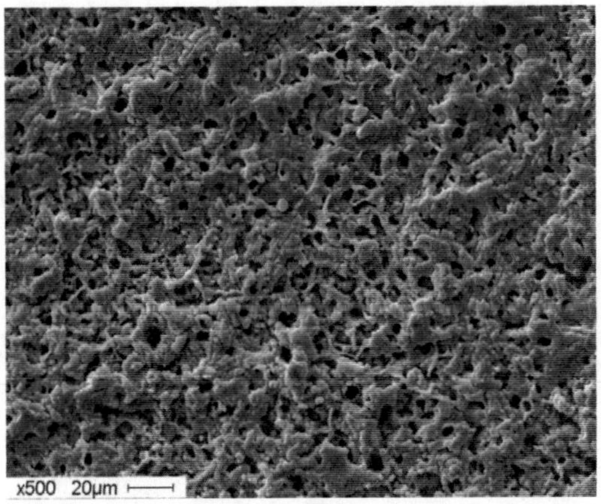

Figure 1. Surface micrograph of three-layered laminated structure on aluminum alloy

Coating destruction process was analyzed on obtained samples of three-layered laminated structures with predetermined metal-coating phase boundary texture. Texturing influence on destruction character was revealed under thermal stress loads. Thermal stress tests were carried out within temperature values from - 60 °C to + 360 °C. Exposure time at extreme temperature values was 2 hours, 10 test cycles were carried out.

Scanning electron microscopy results presented in Figure 2 confirmed that thermal loads exceeding the material breaking point led to successive load removal by microcracks formation in porous coating. Circular microcracks relaxed and eased down on themselves during their growth. Emergence of circular microcracks eliminated the load within the coating and prevented trunk cracks resulting in coatings exfoliation. Further increase of the load caused macrocracks and then exfoliation of the upper part of the coating [3].

IOP Conf. Series: Materials Science and Engineering **156** (2016) 012013 doi:10.1088/1757-899X/156/1/012013

Figure 2. Surface micrograph of three-layered laminated structure on aluminum alloy substrate after critical thermal loading

4. Conclusions

Thus, microplasma oxidation method allowed to obtain nanostructured non-metallic heat-resistant coatings with predetermined phase boundary texture on valve metal alloys. Metal - coating phase boundary texturing provided durability of the coating, also coating exfoliation and deformation under external impacts was not generated.

Acknowledgement

This work was financially supported by the Ministry of Education and Science of Russian Federation, Assignment # 16.461.2014 / K for performing research work in the framework of project of the state assignment in the field of scientific activity.

References

[1] Lukiyanchuk I V, Rudnev V S, Tyrina L M and Chernykh I V 2014 *Applied surface science* **315** 481–489

[2] Nawrat G, Wierzbinska M, Wieczorek T, Nieuzyla A, Krzakala A, Maciej A, Simka W and Raga R 2015 *Solid state phenomena. International Scientific Conference Corrosion* **227** 151–154

[3] Konstantinova T A, Mamaev A I, Chubenko A K, Mamaeva V A and Beletskaya E Yu 2015 *Procedia chemistry* **15** 174–179

[4] Melashchenko N F 1987 *Galvanic coating of dielectrics: Directory* (Minsk: Belarus') 176

Cladding of the carbon fiber on the steel base using electron beam in the air atmosphere

A Losinskaya, E Lozhkina, A Bardin, N Stepanova
Department of Material Science in Engineering, Novosibirsk State Technical University, 20 Prospekt K. Marksa, Novosibirsk, 630073, Russia

E-mail: anna.losinskaya@mail.ru

Abstract. The formation of the high-carbon layers on the low-carbon steel (0.18 % C) using the method of electron-beam partial melting of the carbon fibers is considered. A 1.4 MeV electron beam extracted into air was used. The features of the cladded layers formation using different binders for a reliable fixation of the cladding material are studied. It is revealed that the best results are obtained using the phenol-formaldehyde glue as the binder. A 3 mm thickness layers with 2.2 % C are shown to be formed.

Keywords: Steel, Carbon fiber, Electron beam cladding, Surface hardening.

1. Introduction

In the up-to-date industry various kinds of high-impact effect on materials such as laser, plasma, electron-beam treatment and high-frequency current treatment gained widespread currency. These methods provided the basis for such technological processes as welding, cutting and various types of the surface hardening (cladding, sputtering, surface alloying, surface quenching) [1-4].

At the present day the surface hardening by high-impact stimulation techniques is one of the most effective mechanisms of ensuring the required performance properties to the material (strength, wear resistance, corrosion resistance, etc.). Due to the surface hardening, increasing the reliability of machine parts and mechanisms and cost reduction of metal products in general is achieved. In addition, high-impact techniques allow replacing thermal and chemical-thermal treatment operations. This work is devoted to surface hardening of steel parts by non-vacuum electron-beam partial melting of carbon fibers.

The advantages of electron-beam treatment are as the follows: high efficiency (over 75%), the ability of the electron-beam scanning by the magnetic field, the lack of impact of the optical properties and surface roughness on the degree of energy absorption, high power electron-beam mount, treatment parameter control flexibility, treatment of large products (during non-vacuum electron beam treatment).The method of electron-beam cladding of carbonaceous materials is a relatively simple process that does not require time-consuming preparation of the surface of the work piece.

In previous works the peculiarities of high-carbon layers formation on steel by deposition of powder materials on flat bars have been studied [5]. However, it should be noted that the consolidation of the powders on the details of complex configuration causes certain difficulties. In this regard the problem was set: exploration the possibility of formation of high-hardened layers during the treatment of carbon fibers and fabrics using the binders.

2. Materials and methods

In the capacity of work pieces for cladding, steel plates with the thickness of 4.5 mm were used. The chemical composition of the used steel work pieces is presented in the table 1.

Table 1. The chemical composition of the used steel work pieces.

Elements content, %					
C	Mn	Si	Cr	S	P
0.18	0.5	0.23	0.25	0.04	0.035

As a material for forming the cladded high-carbon layers, a carbon fiber GG 210-P was used. The surface density of the material was 0.017 g/cm^2, the mass fraction of carbon - 99.9%. Due to the fact that carbon does not melt, but is subjected to sublimation at the temperature of 3300 °C, in order to facilitate its dissolution and prevent burnout carbonyl an iron powder OSCH6-2 (TU 6-09-3000-78) was used. The average particle size of the powder was 64 µm. Table2 shows the chemical composition of the iron powder. Also, to protect the molten pool from the atmospheric agents MgF_2 flux was used. The density of the flux was 0.3 g/cm^3, melting temperature - 1263 °C.

Table 2. The chemical composition of the iron powder.

Elements content, %				
C	Ni	Si	Mg	Cu
0.02	0.02	0.0005	0.0001	0.0001

Iron and flux powders were mixed with a binder. Three types of binders were used: polyvinyl acetate emulsion in water, an alkaline aqueous sodium silicate solution and phenol-formaldehyde glue.

Before applying the cladding material, the work piece surface was cleaned from the oxide and degreased.

The work considers the cladding of high-carbon layers with the density of the cladding materials 0.13 g/cm^2 and 0.26g/cm^2. In the first case, the carbon fiber fabrics was applied in a single layer, in the second case - in two layers, and a mixture of a binder with an iron powder and a flux was applied between the fiber fabrics layers and on top of them (Figure 1.). In all cases, the composition of the deposited material was identical: 13% (wt.) C, 37% (wt.) Fe and 50% (wt.) MgF_2, 100% (wt.) binder.

The work pieces with the applied cladding materials were dried in a furnace at 40 °C until complete binder drying out.

To form high-carbon coatings an industrial electron accelerator ELV6-type, designed and produced by the staff of the Budker Institute of Nuclear Physics SB RAS (Novosibirsk) was applied.

The surface of the work piece during treatment was at a distance of 90 mm from the outlet. Electronbeam energy was 1.4 MeV, beam current - 8 and 10 mA. The treatment took place in the track mode, at 10 and 25 mm/s.

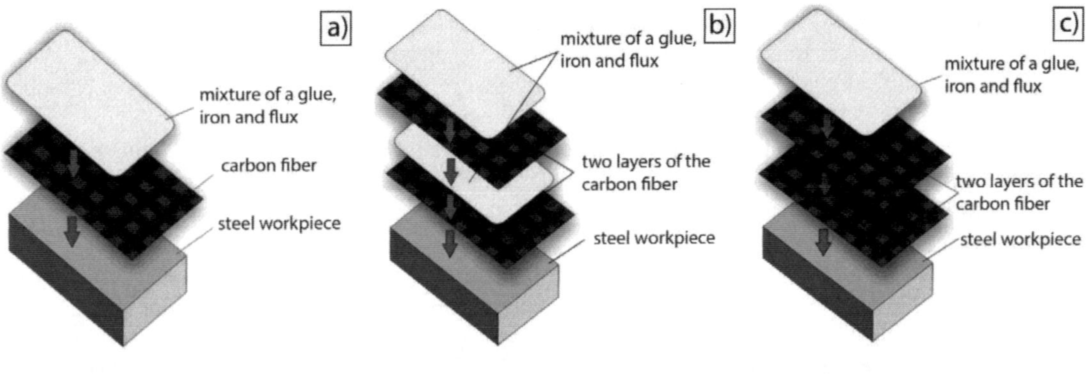

Figure 1. The model of the cladding materials application

To determine the chemical composition of the resulting coatings an optical emission spectrometer ARL 3460 was used. For carrying out metallographic studies an optical microscope Axio Observer Alm(Carl Zeiss) was used. In the capacity of the subject of research the metallographic sections were prepared using standard techniques, including grinding and polishing operations. To reveal the coatings micro structure, the test method of chemical etching using 5% nitric acid solution in ethanol was used.

3. Results and discussion

Primarily during the experiment, the work pieces with the aqueous alkaline solution of sodium silicate as a binder were treated. The experiment showed that the use of this binder is inefficient, due to the formation of poor quality coatings (Figure 2)

Figure 2. Appearance of the specimen with an aqueous alkaline solution of sodium silicates a binder

The appearance of the samples after electron-beam treatment with a water emulsion of polyvinyl acetate and phenol-formaldehyde glue has no significant differences (Figure 3a, b). It should be noted that when removing the cladded material residues on the sample surface with an emulsion of polyvinyl acetate, corrosion products were observed (Figure 3, c).

Also during the experiment, it was found that applying the mixture of the binder with the powders of iron and flux through the layer of carbon fiber fabrics (Figure 1b) is inefficient, since in this case the wetting components are not sufficient for encapsulation of the top fiber fabrics layer, whereby it is not activated in coating formation (Figure 3, d).

It is worth noting that when heated by high-impact source, the temperature distribution in the material (from the surface layer in depth) varies smoothly. Thus, the highest temperature has a surface layer, and temperature decreases if and when going from the surface into the material. In connection with this, the structure formed after the surface layer after the high-impact treatment is also nonuniform. Several zones can be distinguished in it: the 1st zone is a cladded layer; the 2nd zone is the transition zone; the 3rd - the heat affected zone; the 4th - the base metal.

Using the cladded material with a density of 0.13 g/cm^{2}and a processing speed of 25 mm/s provides a coating with a thickness of 1.5 mm and a carbon concentration of up to 1.95%. It was revealed during the metallographic investigations that the cladded layer is a hypereutectoid steel, consisting of pearlite and secondary cementite (Figure 4a). And in the local areas the presence of zones with quenching structures (retained austenite and martensite) was recorded (Figure 4b). The transition zone is composed of perlite, its thickness is 150 μm. Heat affected zone includes pearlite and wide manstatten ferrite.

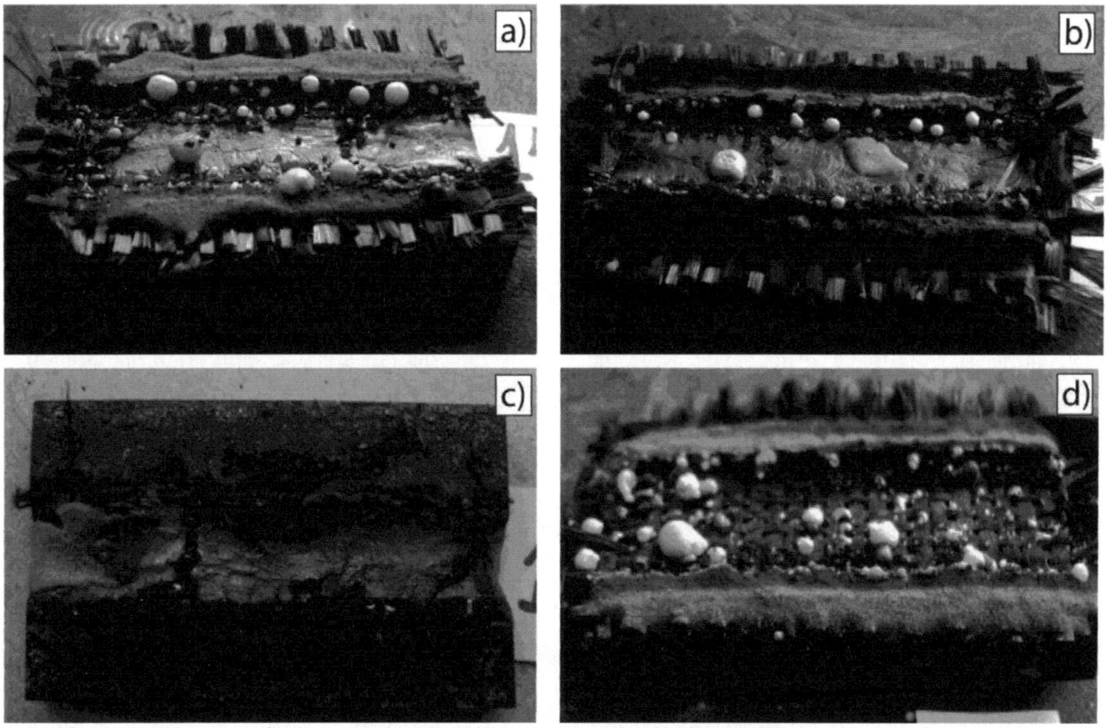

Figure 3. Appearance of the specimen with a) a polyvinyl acetate emulsion in water; b) phenolformaldehyde glue; c) a polyvinyl acetate emulsion in water after removing the cladded material residues; d) phenol-formaldehyde glue with the applied powders of iron and flux through the layer of carbon fiber fabrics

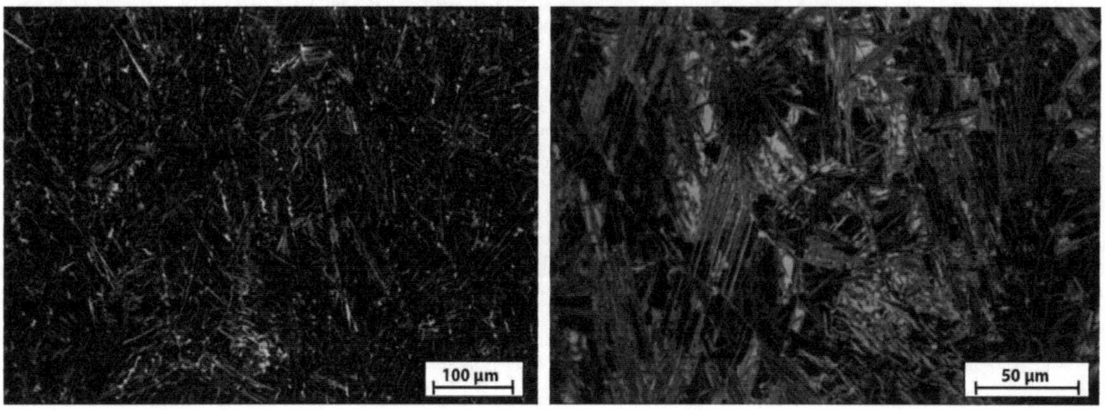

Figure 4. The structure of the cladded layer, obtained at the cladding material density of 0.13 g/cm^2 and a treatment rate of 25 mm/s

To ensure good fusion penetration and mixing of the materials with high density of the cladded material up to 0.26 g/cm^{2}is a reasonable to reduce the treatment speed to 10mm/s. As a consequence coatings with a 3 mm thickness, in which the carbon concentration was 2.2%were obtained. In this case the structure of the cladded layer along with secondary cementite and pearlite has a ledeburite (Figure 5).

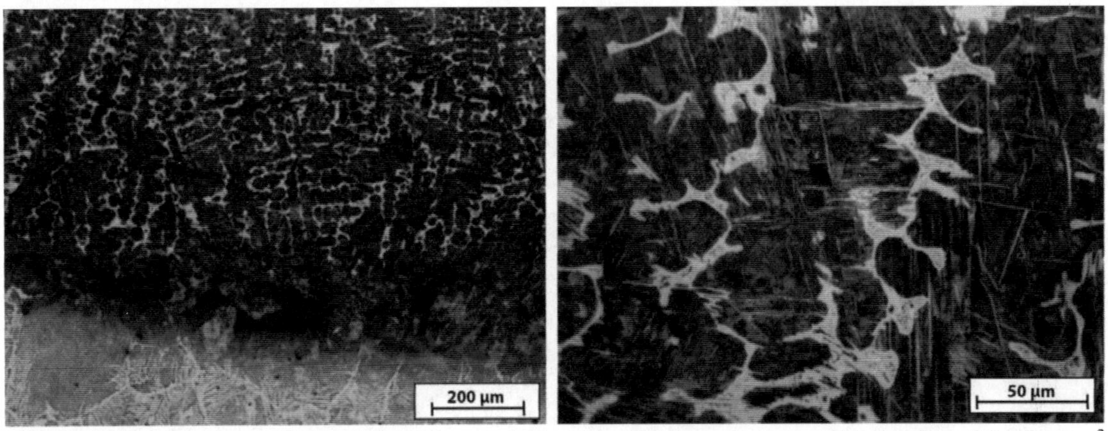

Figure 5. The structure of the cladded layer obtained in the deposited material density of 0.26 g/cm^2 and a treatment rate of 10 mm/s

4. Conclusions

A high capacity and efficiency of electron-beam mounts allows obtaining hardened layers of large thickness on large-sized items made of carbon and low-alloyed steels. This method allows to create high-carbon layers with thickness up to 2.7 mm. The main structural components of the deposited layers are: ledeburite, perlite and widemanstatten type secondary cementite.

The best quality of high-carbon layers on curved surfaces by electron-beam partial melting of carbon fiber fabrics were obtained using phenol-formaldehyde glue as a binder.

Acknowledgements
The study was financially supported by RFBR in the framework of the research project number 1633-00141mol_a

References

[1] Stepanova T Yu 2009 *Tekhnologii poverhnostnogo uprochneniya detalej mashin: ucheb. Posobie* (Ivanovo: Ivan.gos.him. – tekhnolog. un-t) p 64
[2] Yang L J 2001 Plasma surface hardening of ASSAB 760 steel specimens with Taguchi optimization of the processing parameters *Journal of Materials Processing Technology* **113** 521-526
[3] Shin H J, Yoo Y T, Ahn D G *et al* 2007 Laser surface hardening of S45C medium carbon steel using ND: YAG la-ser with a continuous wave *Journal of materials processing technology* **187–188** 467–470
[4] Rudnev V 2009 Single-coil dual-frequency induction hardening of gears *Heat treating progress* **10** 9-11
[5] Bataev I A, Golkovskii M G, LosinskayaAA, *et al* 2014 Non-vacuum electron-beam carburizing and surface hardening of mild steel *Applied Surface Science*. **322** 6-14

IOP Conf. Series: Materials Science and Engineering **156** (2016) 012015 doi:10.1088/1757-899X/156/1/012015

Effect of the silicon content in steel on the hot-dip zinc coating microstructure formation

O S Bondareva [1] and A A Melnikov [2]
[1]Department of Metal Technology and Material Science, Samara National Research University, Samara, Russia
[2]Department of Metal Technology and Material Science, Samara National Research University, Samara, Russia

E-mail: osbond@yandex.ru

Abstract. The aim of this study was to clarify the effect of the silicon content in steel on the structure of hot-dip galvanized zinc coating. It was found that in steel samples containing the silicon in an amount of about 0.1% and 0.5% the increased thickness coating was formed. This fact was associated with structural features of the ζ-phase. Energy dispersive microanalysis had shown the maximum concentration of silicon in the coating was observed in ζ-phase on St3 steel (Si = 0.1%) and in the fine mixture of δ and ζ-phase on 09G2S steel (Si>0.5%). This phenomenon was analyzed using a Zn-Fe-Si system diagram and its polytermic sections. It was found that there were eutectic reactions of decomposition the liquid to mixture (ζ+η+ FeSi) phases at the content of silicon in steel about 0.1% and more than 0.5%. Particles of FeSi-phase were involved in the dissolution of Γ and δ phases, which led to a direct contact of the melt and the steel substrate. This process was accompanied by the intensive ζ-phase formation and the rapid growth of coating thickness.

1. Introduction

The zinc coating structure consists of several intermetallic layers forming by interdiffusion of zinc and iron. The formation of the coating is directly related to the main production parameters: temperature and chemical composition of the melt, the holding time in the melt, the chemical composition of galvanized steel, especially of silicon content in steel [1].

Silicon is a cheap and very effective reinforcing alloying element in steel. Many sheets of construction steel contain small quantities of silicon because it is used in them as a deoxidizing component in production. The sudden forcing of the reaction between the steel with silicon content about 0,06-0,10% and molten zinc was described by Sandelin in the last century [2, 3]. This reaction leads to the abnormal growth of zinc coating thickness and is called in the literature "Sandelin effect". Furthermore, low-alloy steels containing silicon as a cheap alloying element in an amount up to 1% are widely used in reliable constructions. It is known that the presence of silicon in the steel in amounts more than 0.5% can impact negatively on the quality of the zinc coating and leads to its discontinuity, increasing the thickness and impairing the adhesion to the base metal. [4-7]. Despite the great number of studies, the mechanism of the silicon effect on the interdiffusion processes between Fe and Zn in the coating formation is still not completely clear.

Thus, the main aim of this work was research the influence of silicon in the steel on the phase transformations taking place during the zinc coating formation.

2. Materials and methods

Samples for the study were steel plates with different contents of silicon. The chemical composition of the samples was determined on the optical emission analyzer Foundry-Master XPR and was presented in Table 1.

Table 1. The chemical composition of the samples

Grade of steel	C	Si	Mn	P	S	Cr	Ni
St3	0.176	0.028	0.327	0.0037	0.0140	0.089	0.0560
St3	0.172	0.085	0.339	0.0070	0.0073	0.040	0.0226
St3	0.175	0.227	0.406	0.0070	0.0026	0.038	0.0276
09G2S	0.122	0.767	1.530	0.0094	0.0094	0.033	0.0569

Before galvanizing samples were subjected to degreasing, pickling, fluxing and drying. Galvanizing was carried out in an experimental bath in the temperature range from 438°C to 468°C with steps of 5°C. The holding time in zinc melt was 4 minutes. Lowering speed and lifting speed from the bath were the same for all parts of samples. Melt in the bath was alloyed with 0.002-0.005% aluminum and 0.028-0.031% nickel. The measurement of the coating thickness was carried out by magnetic and metallographic methods using the Axiovert 40 MAT microscope. The study of the coating microstructure was carried out by TESCAN Vega SB scanning electron microscope. The zinc coating phase composition was determined in monochromatic CuKα radiation on the automated ARL X'TRA X-ray diffractometer. The identification of the coating microstructural parts were conducted by determining their local elemental composition using INCAx-act energy dispersive microanalysis detector.

3. Results

Studies have shown that the thickness of the coating depends on the silicon content in the steel. Coatings with maximum thickness are formed on the steels with a silicon content of about 0.1% (Sandelin's steel) and 0.5% (high-silicon steels). This effect depends on temperature (Figure 1).

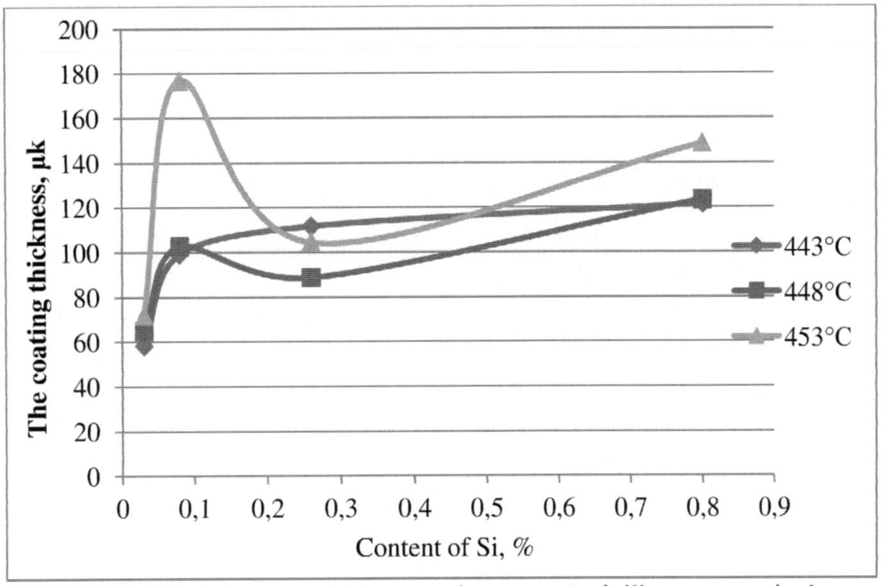

Figure 1. The dependence coating thickness (micrometers) of silicon content in the steel (%)

On St3 with silicon content 0.028% and 0.227% at galvanizing temperature in the range of 438-468^0 C, coating with a constant thickness less 100 μk are formed. On the high-silicon steel with a silicon content about 0.767% coating thickness sharply increases with increasing temperature and is reached 200 μk at 468^0 C (Figure2).

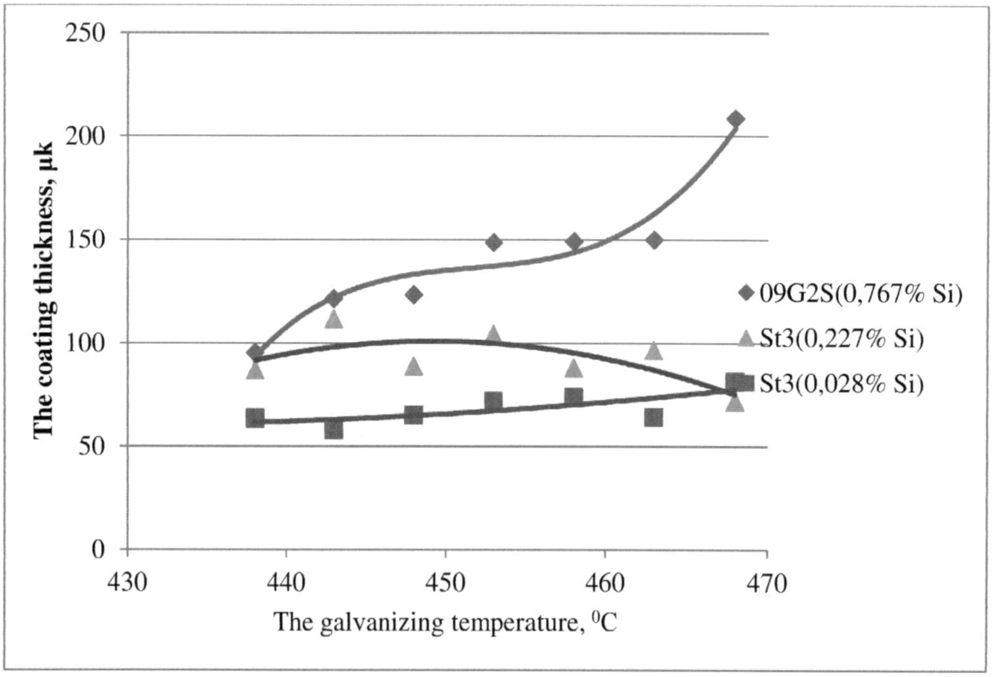

Figure 2. The dependence coating thickness (μk) of the galvanizing temperature (^{0}C)

On "Sandelin's steel" St3 (Si=0.085%) coating with a polythickness from 40 to 180 microns are formed. The difference between the maximum and the minimum coating thickness increases with the growth of galvanizing temperature (Figure3).

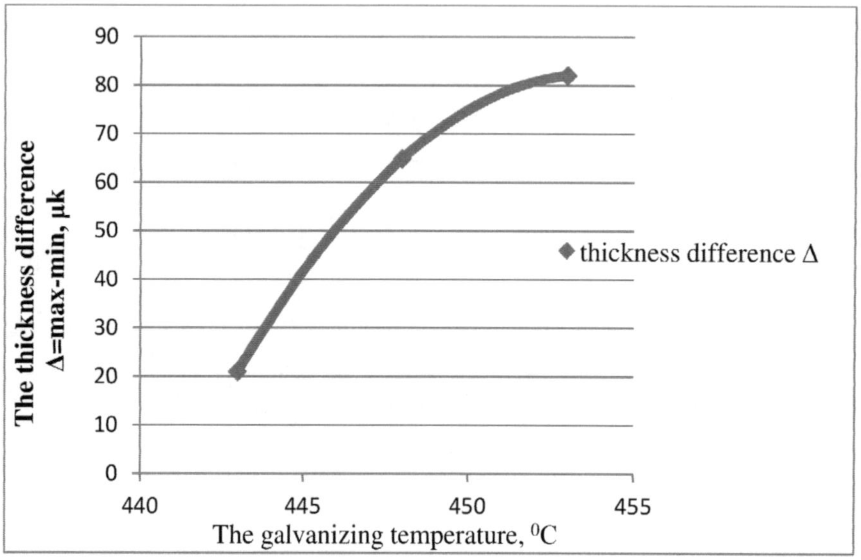

Figure 3. The influence of galvanizing temperature on the polythickness of coatings on steel St3 (Si = 0.085%)

Increasing the coating thickness is undesirable because it leads to increasing zinc consumption and deterioration of the surface quality.
Studies have shown that the structure of the main phase (Γ+Γ1, δ, ζ, η) of the zinc coating depends on the silicon content in the steel (Figure4).

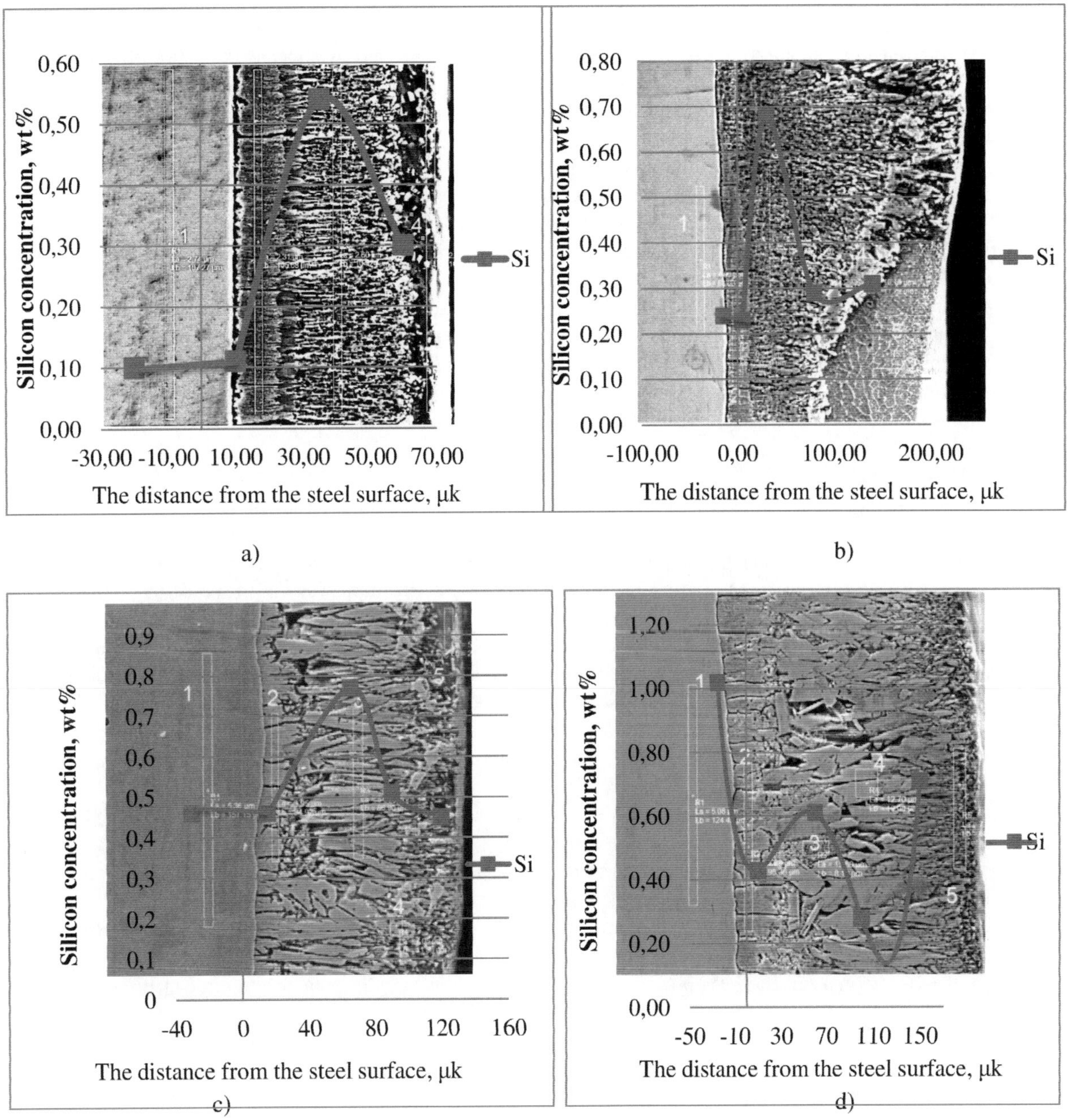

Figure 4. The microstructure of the coatings obtained by T=450^0 C, τ=4 min:
a - on steel St3 (Si=0.028%); b - on steel St3 (Si=0.085%);
c - on steel St3 (Si=0.227%); d – on steel 09G2S (Si=0.767%).

In coating on steel with a minimum silicon content (0.028%), all phases according to the diagram Fe-Zn were found by X-ray analysis. The δ-, ζ- and η phase ratio is approximately the same over the entire temperature range. Phases Γ and Γ1 are observed in the form of a thin black stripe. Phase δ is columnar, compact, homogeneous. Phase ζ has the fine structure and the minimum thickness (Figure4a).

On "Sandelin's steel" with a silicon content of 0.085% the coating has a maximum thickness and a significant variation in thickness. The sites of coating with maximum thicknesses represent the large ζ-phase dendrites, the size of which is 2-3 times greater than the thickness of this phase at adjacent sites. With increasing galvanizing temperature the ζ-phase dendrites grow through the whole thickness of the coating and are come on its surface, giving it the gray color. In places of the large dendrites, ζ-phase location δ-phase is more thin or absent. There is no pure zinc (η-phase) in those places. (Figure 4b).

The coating on St3 steel with a silicon content of 0.227% Γ and Γ1-phases are not visible. The δ-phase has a columnar structure of constant thickness, the ζ-phase has porous branched dendrites. The main feature of the coating on that steel is the presence of areas with compact structure in the upper part of the ζ-phase dendrites, possibly constituting the additional phase layer. These areas may be an additional phase layer (Figure 4c).

The structure of coating on 09G2S steel with a silicon content of 0.767% Γ and Γ1-phases are not visible. The δ-phase has a columnar structure without clear boundaries. It gradually changes to a fine mixture of phases supposedly eutectic nature. Very wide ζ-phase dominates in the coating. It is represented by large crystals of cubic and rectangular shapes. Close to the coating surface the ζ-phase crystallites become denser and more elongated in the direction of crystallization. (Figure4d).

4. Discussion

For an understanding of the silicon effect, analysis of its distribution across the coating thickness was carried out. (Figure4). Studies have shown that the silicon concentrations in the different phases of the coating are very different. The amount of silicon in the steel surface layer (α-phase) is almost three times more than its average content in the steel in all tested samples. Some researchers explain this effect the accumulation of (FeO2*SiO2) double oxides in the surface layer of steel [8]. The coating on St3 silicon concentration reaches its maximum values in ζ-phase. According to the theory of diffusion layers formation, the maximum concentration of the element indicates the formation of a chemical compound (phase). It allows us to assume that silicon integrates into ζ-phase, forming a three-component Zn-Fe-Si compound. As opposed to St3 in the coating on steel 09G2S maximum values of the silicon concentration are observed in the fine mixture of phases δ and ζ, and in the ζ-phase layer near the surface.

Analysis of the Zn-Fe-Si phase diagram has shown that the silicon presence in the steel leads to the appearance of a new phase - FeSi in the zinc coating (Figure5) [9]. The mechanism of formation of FeSi particles is described by nonvariant reactions that are demonstrated in the polytermic sections of the Zn-Fe-Si system (Figure6). When the content of silicon of about 0.1% (Sandelin's steel) eutectic reaction of the liquid phase disintegration to mixture $\zeta + \eta$ + FeSi takes place in the system. FeSi phase is formed in a wide range of compositions less than 9.1 wt.% Fe. When the silicon content of about 1% (high-silicon steel) the FeSi phase amount becomes significantly greater (Figure 6) [9].

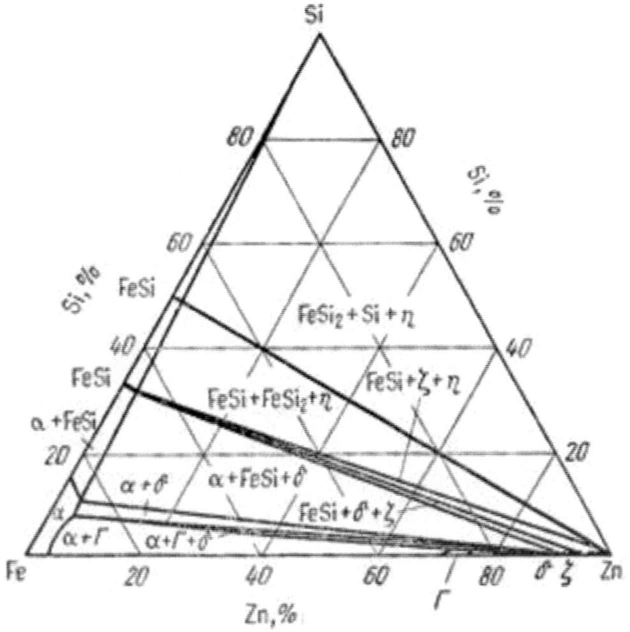

Figure 5. The phase structure of Zn-Fe-Si system alloys at ambient temperature

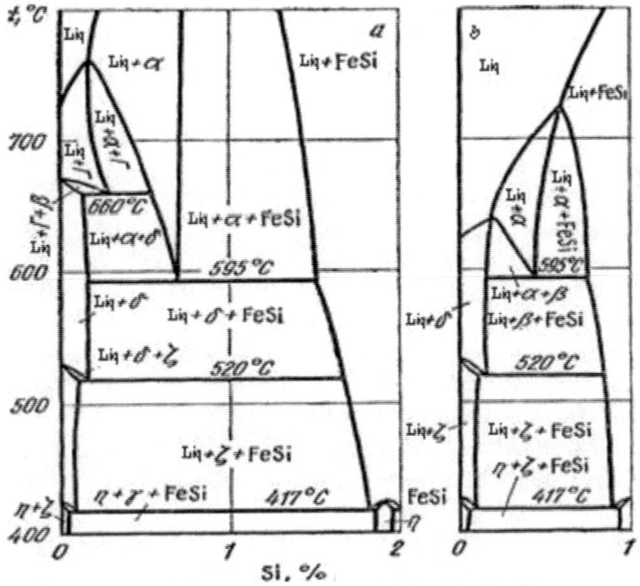

Figure 6. the polytermic sections of the Zn-Fe-Si system:
a – 95% Zn, b – 98% Zn [9]

Foreign studies have also confirmed the existence of an equilibrium state between the liquid, the ζ-phase, and FeSi-phase, which prevents the equilibrium between the liquid phase and the δ-phase [10, 11]. This phenomenon destroys the integrity of the coating layers; liquid zinc has the direct contact with the steel base, which accelerates the interdiffusion of iron and zinc. On reaching the surface coating, the iron stream is converted into ζ - phase crystallites in zinc.

5. Conclusion

Metallographic studies have concluded that the maximum thickness of the zinc coating is observed in steel samples with silicon content about 0.1% (Sandelin's steel) and 0.5% (high-silicon steel). This fact was associated with structural features of the ζ-phase. Energy dispersive microanalysis had shown the maximum concentration of silicon in the coating was observed in ζ-phase on St3 steel (Si = 0.1%) and in the fine mixture of δ and ζ-phase on 09G2S steel (Si> 0.5%). Analysis of Zn-Fe-Si system diagram and its polytermic sections has shown that there are eutectic reactions of decomposition the liquid to mixture (ζ+η+ FeSi) phases at the content of silicon in steel about 0.1% and more than 0.5%. Particles of FeSi-phase were involved in the dissolution of Γ and δ phases, which led to a direct contact of the melt and the steel substrate. This process was accompanied by the intensive ζ-phase formation and the rapid coating thickness growth.

Acknowledgement

This work was supported by the Ministry of education and science of the Russian Federation in the framework of the implementation of the Program of increasing the competitiveness of SSAU among the world's leading scientific and educational centers for 2013-2020.

References

[1] Foct J, Reumont G, and Perrot P 1993 The Morphology of Zinc Coatings *Phys. Met. of Zn Coated Steel* 1-9
[2] Sandelin R W 1940 Galvanizing characteristic of different types of steel *Wire and wire product.* **15** 655-660
[3] Kozdras M S and Niessen P 1989 Silicon-induced destabilization of galvanized coatings in the Sandelin peak region *Metallography* **22** 253-267
[4] Bondareva O S, Melnikov A A and Amosov A P 2014 Influence of Hot-Dip Galvanizing Temperature on Formation of Zinc Coating on a Steel with a High Silicon Content *Adv. Environ. Biol.* **8** 943–948
[5] Che C, Lu J, Kong G and Xu Q 2009 Role of silicon in steels on galvanized coatings *Acta Metallurgica Sinica (English Letters)* **22** 138-145
[6] Foct J, Perrot P and Reumont G 1993 Interpretation of the role of silicon on the galvanizing reaction based on kinetics, morphology and thermodynamics *Scripta Metallurgica et Materiala* **28** 1195-1200
[7] Guttman H and Niessen P 1972 Reactivity of silicon steels hot-dip galvanizing, *Canadian Metallurgical Quarterly* **11** 609-614
[8] Inoue J, Miwa S and Koseki T 2014 Effect of Si content in steel on formation of Fe-Zn intermetallic compound layer at pure Zn melt/steel interface *Tetsu-To-Hagane/Journal of the Iron and Steel Institute of Japan* **100** 390-396
[9] Koester W 1969 Ternary system IRON-SILICON- ZINC *Metallurgia* **80** 219-229
[10] Tang N Y 2008 Control of Silicon Reactivity in General Galvanizing *Journal of Phase Equilibria and Diffusion* **29** 337-344
[11] Uwakweh O, Jordan A and Maziasz P 2000 Thermal transformations in mechanically alloyed Fe-Zn-Si materials *Metallurgical and Materials Transactions A: Physical Metallurgy and Materials Science* **31** 2747-2754

Enhancing thermal barrier coatings performance through reinforcement of ceramic topcoat

V I Bogdanovich, M G Giorbelidze
Samara National Research University, 34, Moskovskoe shosse, Samara, 443086, Russian Federation

E-mail: bogdanovich@ssau.ru

Abstract. This paper studies structure of thermal barrier coatings applied to hot gas path components in gas turbine engines and produced in a number of ways, and its impact on performance. Methods of structural reinforcement for ceramic topcoat in thermal barrier coatings are considered.

1. Introduction

At present, increasing component life for hot gas path in gas turbine engines (GTE) is inherently dependent on development of highly efficient system of thermal barrier coatings (TBC). Recent years have witnessed a growing interest therein, as advanced GTE appeared with gas temperature approaching to or exceeding the operating maximum of existing heat-resistant alloys, more advanced equipment was designed, and spray materials for industrial applications were produced. TBC are being successfully used for protection of first stage GT nozzle and rotor blades, liners, and combustion chambers [3,9]. Available capacities for enhancing TBC performance, however, still have a long way to go to be fully mobilized. One of such capacities lies in controlling TBC structure during coating formation and following treatment of the finished item.

Among multi-layer TBC, the most efficient system proved to be the one consisting of an external ceramic coating (CC) layer, composition ZrO_2-Y_2O_3, and intermediate one or two-layer damping metallic ceramic coating, composition Me–Cr–Al–Y with different proportions of components. In course of research, this coating system was deposited by electron beam and plasma thermal spraying techniques (ceramic and metallic ceramic coatings), as well as by plasma-enhanced vacuum deposition technology with high-energy ions (metallic ceramic coatings).

2. Impact of coating technique on its structure

Studies revealed significant impact of coating technology used in TBC formation on component life and also demonstrated a contradiction in requirements applicable to its structure. On one hand, increase in density and cohesive strength of coating materials augments the overall system resistance to high-temperature gas corrosion, on the other hand, reduces its thermal endurance and endurance limit. Findings showed that the external oxide layer produced by different techniques has distinct degree of "transparency" for ions of oxygens and does not prevent exchange processes on the border between CC and MCC. This is conditioned by the fact that microstructure of the coating produced by electron-beam technique is columnar (Figure 1 and Figure 2). Each single column contains several crystallites (Figure 1) (column boundaries are marked with a solid line, crystallite boundaries, with a dotted line) that have a relatively strong adhesive connection with metallic ceramic coating along the bases and a weak cohesive connection with other columns.

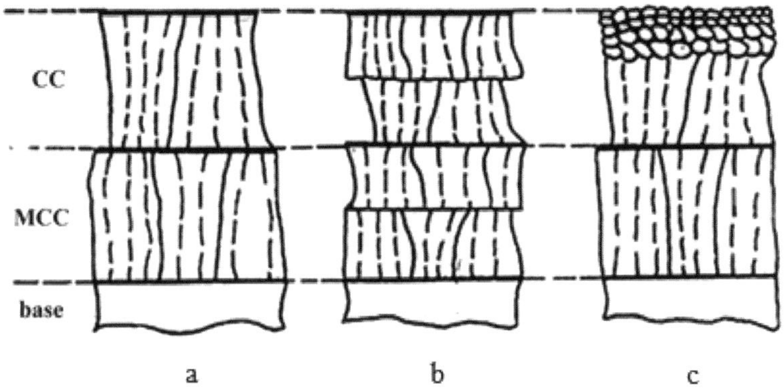

Figure 1. Diagram of a thermal barrier coating produced by atomic layer deposition technique (a) and its alteration in case of elimination of continuous intercrystalline boundaries (b) and heat pulse modulation of the external surface (c)

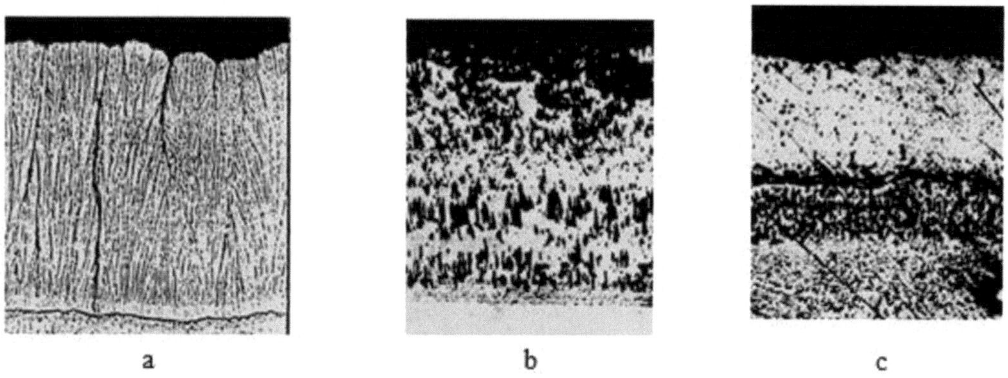

Figure 2. Structure of a thermal barrier coating produced by atomic layer deposition technique (*a*) and its alteration in case of elimination of continuous intercrystalline boundaries (*b*) and heat pulse modulation of the external surface (*c*), x500

Boundaries between columns and crystallites that originate on the external surface of ceramic coating and end on the border with metal base, serve as channels for entry of oxygen ions. Formation of such boundaries is determined by technological legacy when coatings are being grown from atomic layers [4]. At the same time, all techniques aimed at elimination of continuous intercrystalline boundaries within ceramic coating (Figure 1b) increase the system resistance against high temperature gas corrosion, and in case of thick coating layers significantly reduce the endurance limit of the system [1].

In coatings produced by plasma spraying, intergranular boundaries have an essentially different structure (Figure 3), therefore, speed of oxygen ions penetration to the oxide - metal border is reduced. Besides, due to their microstructure, plasma sprayed coatings [1,2] provide much better thermal protection. Pores in coatings produced by plasma spraying, are relatively large, disk-shaped and located parallel to the base surface [6,8]. Pores in coatings produced by atomic layer deposition, are small, elongated and located perpendicular to the base surface. Experiments have shown that plasma sprayed coatings have less thermal conductivity than TBC produced by atomic layer deposition.

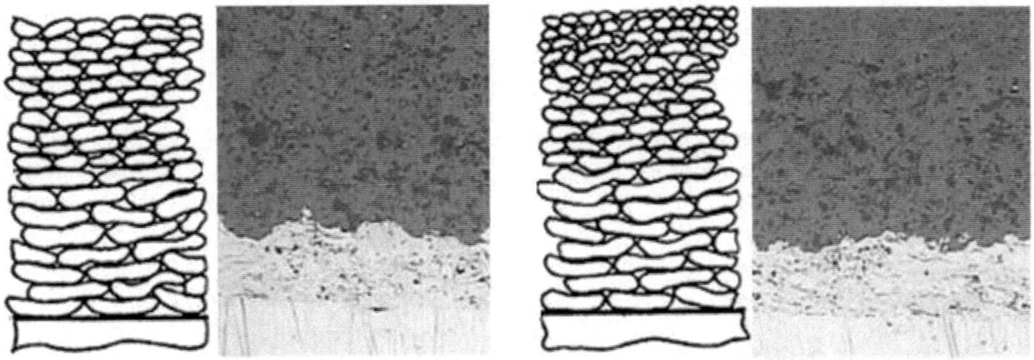

Figure 3. Diagram and structure of a thermal barrier coating produced by thermal and plasma-enhanced powder deposition (*a)* and its alteration in case of fine-grained powders introduction at the final stage of the external layer formation (*b*) scale ×500

It was also shown that increase in density of coatings and in strength of intergranular boundaries overall through the system results in reduction of thermal endurance and endurance limit due to weakening of damping properties in thermal barrier coating layers. Therefore, a certain consensus shall be maintained within the system structure in order to enhance its resistance against volumetric high temperature gas corrosion and to increase its resistance to multi-cycle intermittent thermal and mechanical loads.

With due regard to promising potential of the described TBC system in aviation applications for new generations engines experiencing considerably higher temperatures in combustion chamber, as well as to the exposed mechanisms that reduce the coatings service life, a series of further studies on plausible directions for improvement have been performed. The underlying idea for this research and development was essential feasibility of considerably enhancing coatings resistance against high temperature gas corrosion by creating a denser fine-grained topcoat with no continuous boundaries (Figure 1c, Figure 3b). At the same time, structure of the other layers within ceramic or metallic ceramic coating shall not undergo any significant changes, otherwise the coating would lose its damping properties.

Such a structure of thermal barrier coatings might be created on top of a formed base coat by potent thermal impulses (Figure 1c), and on top of a forming base coat - by using of finer-grained powders for plasma thermal (Figure 3b), or by using ions with variable kinetic energies in order to reduce the negative technological legacy while growing coatings from atomic layers (Figure 1b) [1,4,5].

Research was started along the three plausible lines: in topcoat structure refinement through impulse plasma deposition, works have advanced up to the pilot production stage; as for laser surface enhancement, studies are still going on in laboratories; and research on fine-grained powders introduction has reached its final stage. In what follows we consider some of the findings.

3. Topcoat structure reinforcement through impulse plasma deposition

The first challenge of creating a dense fine-grained layer within the formed thermal barrier coating on the item was to determine whether such technology is technically feasible. This is explained by the fact that the temperature of fused external surface on the ceramic topcoat shall exceed its melting temperature $T_{M,1}$= 2910 K. At the same time, temperature on the border between CC and MCC shall not exceed the approximate value of $T_{C,2} = 0.8 \cdot T_{M,2}$= 1500 K ($T_{M,2}$ being the temperature when MCC material starts to melt). This restriction for $T_{C,2}$ is conditioned by the fact that at higher temperatures, processes of gas release and sublimation begin on the MCC surface, which results in a cohesive failure of ceramic coating continuity. As the literature review revealed no data on analysis of similar situations, we developed a specific mathematical model and analyzed, on the basis of a numerical simulation, the potentiality of matching thermal impulse parameters (t_1 – time interval for which an impulse acts, q – density of energy flux) that would make the situation above possible. The mathematical model was conceived in terms of a unidimensional, flat, nonlinear problem of thermal conductivity within a three-layer composite body, while its external surface is being exposed to a thermal impulse, and it is being cooled via heat radiation and evaporation of material. Heating, melting, crystallization, and cooling stages were sequentially calculated.

Numerical simulation showed that heating time t_M of the oxide surface up to melting temperature $T_{M,1}$ is more or less proportional to q^{-2} (Figure 4). At time point $t = t_M$, thermal impact depth of an impulse is concentrated in a narrow layer adjacent to the surface. Starting from the time point t_M, a molten mass is formed on the surface, and its melting front drifts into the depth of the material. At the same time, the temperature of the oxide surface is rising rather quickly, but as it approaches the oxide boiling temperature T_B = 4573 K, its rising rate slows down and the temperature is stabilized around that value. Temperature distribution per system thickness during the impulse effect is characterized by a steadily decreasing function (Figure 4).

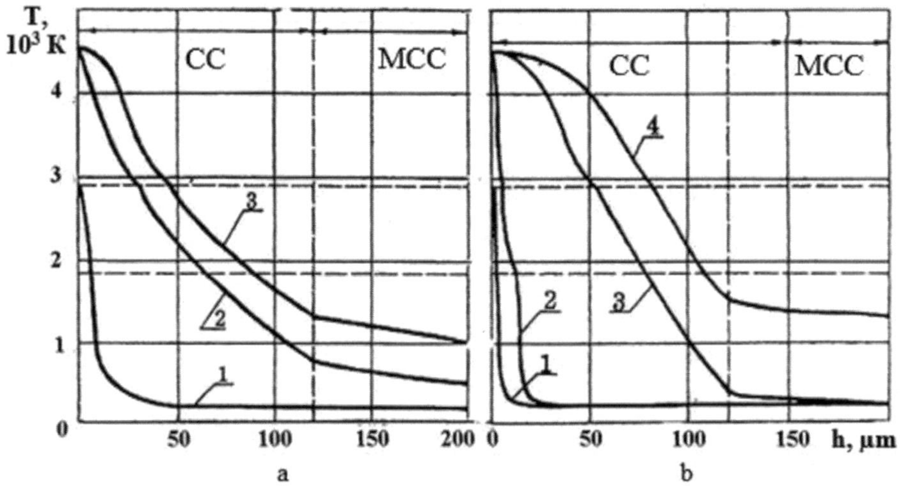

Figure 4. Temperature distribution per system thickness ($ZrO_2 + Y_2O_3$) -(NiCoCrAlY) – (base) with pulse effect q=10^8 W/m^2 (a); q=10^9 W/m^2 (*b*) and ti>$9.2 \cdot 10^{-3}$ at time: (a): *1* – $4.2 \cdot 10^{-3}$ s, *2* – $7.8 \cdot 10^{-3}$ s, *3* – $9.4 \cdot 10^{-3}$ s and (*b*) *1* – $7.8 \cdot 10^{-3}$s, *2* – $5 \cdot 10^{-3}$ s, *3* – $7.8 \cdot 10^{-3}$ s, *4* – $9.4 \cdot 10^{-3}$ s

Numerical analysis of the exposed situation allows to choose q and t parameters for a pre-set system thickness, so that maximum or set CC melt-through value would be reached, while temperature on the border between CC and MCC rests below the pre-set critical value. At the same time, as the grain size in the crystallization zone decreases along with the solidification rate increase and molten mass zone reduction, it was suggested to treat the thermal barrier coating by several sequential impulses. The first impulse shall provide for penetration of (20 – 50) % of the CC depth, and the next impulses, for a more reduced penetration, which shall create a finer-grained structure adjacent to the external surface of the TCC without altering the rest of its structure.

This treatment technique was implemented in an impulse plasma coaxial accelerator for gas plasma. Turbine blades, blades simulators and witness samples with thermal barrier coatings were placed into the vacuum chamber, assisted by special tools, and treated with an impulse flow of gas plasma with energy flux density of ($2 \cdot 10^8 \ldots 2 \cdot 10^9$) W/m^2. Blades were treated sequentially from both sides with two impulses, with a pause between them lasting 12 minutes.

Studies of TBC specimens produced by electron-beam technique confirmed the hypothesis about alterations in ceramic topcoat structure resulting from impulse plasmadynamic treatment. Adequacy of the mathematical model was assessed by comparing theoretical calculations of penetration depth and thickness of structurally altered coating zone in cross-sections.

Studies showed that surface roughness is reduced after treatment, from R_a=1.25 μm down to R_a=(0.8...0.63) μm; open channels along the column boundaries within the structure disappear, and partially reduced zirconium oxide produces zirconium on the surface . Color of the coating surface changes from matte white into lustrous dark metallic. Depending on the treatment mode, cross-section demonstrates substantial alterations in the layer adjacent to the surface; alteration zone width varies from 20 to 60 μm. At the penetration depth of the oxide layer, columns and crystals boundaries, characteristic for the surface, disappear. The very structure of the layer has the appearance shown in Figure 1c, and demonstrates fine-grained equiaxial pattern, with no trees nor pronounced directions of solidification. Maximum grain size within the structure varied in the range of (1...3) μm. Presence of such a structure implies predominantly volumetric solidification of the molten mass zone.

Specimens and turbine blades were tested for thermal endurance by thermal shock (heating in 3 minutes up to T = 1404 K, hold time of 10 minutes, cooling for 40 seconds down to T = 473 K), cyclical thermal loading consisting of $2 \cdot 10^7$ cycles (T_{Max} = 1233 K, T_{min} = 473 K, cycle duration of 20 minutes), with subsequent study of structure alteration and emerging defects. Studies showed that number of cycles required for defects to emerge was at least 1.8 times higher, as compared to similar coatings or coatings of different type, but without surface reinforcement. However, testing of reinforced blades were discontinued by thermal shock after 1300 cycles, before micro and macro defects had time to emerge. Heat resistance in specimens with reinforced coating increased 1.5 – 2.2 times in relation to coatings without reinforcement. Long-term high temperature strength in air (T = 1248 K, σ = 270 MPa) exceeded long-term high temperature strength of the base material, and endurance limit at normal temperatures complied with the allowance range for base material endurance limit.

All the above demonstrates the potential of continuous works in this direction and future opportunities for implementing this technology in gas turbine engines manufacturing and upgrading. At the same time, industrial operation and maintenance of the impulse plasma coaxial accelerator were shown to be complicated. Therefore, research on possible use of laser technique for ceramic coatings reinforcement via dynamic focus laser beam configurations has been started.

4. Topcoat structure reinforcement through fine-grained powders introduction at the final stage of the external layer formation

Nowadays, general-purpose plasma sprayed coatings are widely used in gas turbine engines applications, including thermal barrier coatings [1,3-5]. Existing equipment, specificity of powder materials production and storage, specificity of its heating and transportation within the plasma jet, as well as specificity of its interaction with the base material surface, determined the applicable particle size range for these powders of 40...200 μm. In general, for most technical problems addressed by this technique, emphatic issue of finer-grained fractions use seems irrelevant. Therefore, the challenge formulated above is rather specific, but very important from the practical point of view. Should this technology be implemented in production, it can generate substantial economic profits, due to its simplicity, low energy consumption and high operational efficiency.

Studies of plasma coatings properties in relation to the particle size of the powder used (40...200) μm show [3] that particle size reduction improves coating filling, increases its density, reduces porosity, and the structure generally becomes more homogeneous [7]. However, during storage, powder materials with particle size under 10 – 20 μm form conglomerates consisting of various numbers of small particles, the reasons for which being humidity, extended surface and molecular cohesive forces. It results in unstable flow rate of powder material supplied to the heating zone. The conglomerates spend a very short time period in the plasma jet nucleus, so that the particles they contain acquire rather different velocities and temperatures, and, therefore, the coatings they grow possess a wide range of operational characteristics. Besides, due to their low inertia, small particles are carried away by the gas flow passing along the surface, so that they do not end up on the surface, or if they do, their incidence angle is rather big. It reduces both adhesive and cohesive strength of the coating, increases its porosity, and results in low materials utilization rate.

Therefore, the challenge of ceramic topcoat reinforcement called for creation of a technique for fine powders preparation prior to the deposition, for development of specific ultrasound dosing devices and plasma guns, as well as for optimization of spraying modes, firstly in relation to flow rates of plasma forming gas and spraying distance.

Studies used fractions of (20 – 40) µm and (10 – 20) µm of ZrO_2+Y_2O_3 powder. Those fractions were applied both independently and in a mechanical mixture with a standard fraction of (40 – 60) µm. Best results were obtained in deposition of a topcoat (40 – 60) µm thick from the fraction (10 – 20) µm. Use of this fraction allowed to reduce maximum value of geometric surface deflection from 30 µm to 10 µm, reduced open porosity by almost 3 times and reduced total porosity as determined on metallographic specimens down to 4 %. The coatings have passed a series of standard tests in accordance with the above-described procedure, and thermal endurance, long-term strength and endurance limit thereof were not reduced, as compared to coatings produced by previously developed techniques, while they presented an increase in heat resistance of almost 50 %. At the same time, when spraying this fraction, materials utilization rate resulted almost twice as low in comparison to spraying of (40...60) µm fraction.

5. Results and Conclusions

Thus, we used the example of thermal barrier coatings for hot gas path components in gas turbine engines in order to show that materials with dense fine-grained crystal structure might prove to be a significant resource in increasing performance, but only in case of its justified and rational use. Rational scope of application for TBC is restricted to the external part of the ceramic layer.

Creation of a fine-grained crystal structure in the external part of the ceramic layer requires development of specific techniques: as such, impulse plasma deposition and laser surface reinforcement were used for the formed layer, and plasma spraying was used for deposition of coatings from fine-grained powders. Cited results demonstrate the promising potential of further research and development in this area.

References

[1] Barvinok V A 2005 *Plasma in Technology: Reliability and Opportunities* (Moscow: Nauka i technologii) p 456

[2] Barvinok V A, Bogdanovich V I 1999 *Physical Principles and Mathematical Modeling of Vacuum-Assisted Ion Plasma Spraying* (Moscow: Mashinostroyenie) p 309

[3] Bratuhin A G, Yazov G K and Karasev B E 1997 *Modern technology in the production of gas turbine engines* (Moscow: Mashinostroyenie) p 408

[4] Dokukina I A 2012 Selection of the design and application of plasma technology protective coatings parts gas turbine engines *Vestnik of the Samara State Aerospace University* **3 (34)** 167-172

[5] Abusdel A M, Ilinkova T A and Lunev A N 2005 Application of thermal barrier coatings in contemporary gas turbines. I. Thermal barrier layer *Russian Aeronautics* **48** **1** 60-64

[6] Dokukina I A 2009 Theoretical investigation of the mesostructure-ordered cluster structures formation in plasma coatings *Engineering and Automation Problems* **4** 106-112

[7] Dokukina I A 2012 Increase of components operational characteristics due to the application of plasma thermal spray coating with cluster structure *Vestnik of the Samara State Aerospace University* **5** 49-54

[8] Bogdanovich V I, Giorbelidze M G 2016 Model of powder material plastic transformation during plasma coating application *Key Engineering Materials* **685** 685-689

[9] Bobrov G V, Ilin A A and Spektor V S 2014 *Theory and technology of inorganic coatings formation* (Moscow: Alfa-M) p 925

Investigation of Ni-Cr-Si-Fe-B coatings produced by the electron beam cladding technique

T A Zimogliadova[1], E A Drobyaz[1], M G Golkovskii[2], V A Bataev[1], V.G. Durakov[3] and N Yu Cherkasova[1]

[1] Material Science in Mechanical Engineering Department, Novosibirsk State Technical University, 20 Prospekt K.Marksa, Novosibirsk, 630073, Russia
[2]Budker Institute of Nuclear Physics of Siberian Branch Russian Academy of Sciences,11 Prospekt Akademika Lavrentieva, Novosibirsk, 630090, Russia
[3]Institute of Strength Physics and Materials Science of the Siberian Branch of the Russian Academy of Sciences, 2/4 Prospekt Akademicheskii, Tomsk, 634021, Russia

E-mail: zimogliadovatatiana@gmail.com

Abstract. This paper presents the results of structural investigations and results of tribological and microhardness tests of the coating obtained by electron beam cladding of a Ni-Cr-Si-Fe-B self-fluxing alloy on low-carbon steel. After electron beam treatment high-quality dense layer with a thickness of 1.2-1.8 mm was obtained. The structure of the coating consisted of dendrite crystals based on γ-Ni-solid solution and eutectic with complex composition. Microhardness of the coating achieves 370 HV. Wear-resistance of the coating obtained by electron-beam cladding technique was 1.6-fold higher than that of low-carbon carburized steel.

Keywords. wear-resistant coatings, electron-beam cladding, microstructure, microhardness.

1. Introduction

One of the most advanced directions of the modern material science is protective coatings processing which provides significant increase of lifetime of machine components working at high temperatures and abrasive wear. Currently, surface hardening techniques based on highly-concentrated energy sources are developed extensively. Among these techniques, electron-beam treatment should be particularly emphasized. Such advantage as high efficiency (up to 75 %), high thickness of hardened layers, high adhesion between a substrate and a coating singularizes this technique from others technologies of surface hardening based on highly-concentrated heating sources [1-2, 6-7].

Ni-Cr-Fe-Si-B self-fluxing alloys are widely used for formation of coatings and for building-up of worn-out parts [3,4]. These alloys are characterized by high stability to abrasive and impact wear. Superior wear-resistance of cladded layers can be explained by the presence of hard strengthening phases such as carbides, borides, silicides [5]. High Ni and Cr content have a positive effect on heat-resistance of these alloys. Si and B reduce melting point (up to 1050 °C) and act as fluxing components. These chemical elements appear to be the active chemical reducing agents for majority of metal oxides. B and Si generate a thin glassy layer, which protects the coating surface from oxidation [3,5].

The aim of this work consisted in structural investigation and evaluation of tribotechnical characteristics of the Ni-Cr-Fe-Si-B self-fluxing coating obtained by the electron-beam cladding technique on a low-carbon substrate.

2. Materials and methods

To form the coating, electron-beam cladding in vacuum was used. Cladding was realized by the electron-beam device ELU-5, developed in Institute of Physics ofStrength and Material Science of the Siberian Branch of the Russian Academy of Sciences (Tomsk) and equipped with a system of powder delivery. A powder material was directly supplied to the electron beam operating area. Thickness of the layer cladded for one path achieved 0.4…0.7 mm. Coating formation was carried out in four paths according to the following technological parameters: current (I) – 40 mA, accelerating voltage (U) – 27 kW, speed of workpiece movement relatively to the outlet (v) – 3 mm/sec.

Low-carbon steel plates with dimensions of 100×25×10 mm were used as the base material. Commercial Ni-based self-fluxing powder with particle size in the range 40 – 100 μm was used as the surfacing material. Chemical compositions of the substrate and surfacing materials are presented at the Table 1.

Table 1. Chemical composition of base and coating materials, wt. %

Material	Ni	Cr	B	Si	Fe	C	Mn	S	P
Low-carbon steel (base material)	-	-	-	0.28	Bal.	0.12	0.50	0.04	0.03
Ni-based self-fluxing powder	Bal.	15	2	3	3	-	-	-	-

Structural researches of obtained coating were carried out using Carl Zeiss Axio Observer A1m optical microscope (OM). To reveal the microstructure of coating, chemical agent consisted of ferrous chloride ($FeCl_3$), nitric (HNO_3) and hydrochloric (HCl) acids was used. For more detailed investigations the Carl Zeiss EVO 50 XVP scanning electron microscope (SEM) was utilized. Hardness of the obtained materials was estimated by the Wolpert Group 402MVD Vickers tester. The load on the tetrahedral diamond indenter was 0.98 N. Wear-resistance of the cladded coating was evaluated by friction test against embedded abrasive grains in accordance with standard GOST 17367-71. Silicon carbide with 80-100 μm particle size was used as an abrasive material. Low-carbon (0.2 % C) carburized and quenched steel (with subsequent low tempering) was used as a reference sample. Amount of wear was evaluated by mass loss measurement using Pioneer DA214C balance.

3. Results and discussion

The above described cladding conditions lead to formation of high-quality surface layer. The thickness of the cladded layer is 1.2-1.8 mm. At the Figure 1a a general view of obtained coating is presented. The key aspect of the coating, obtained by electron-beam cladding technique, is formation of gradient structure, which is characterized by gradual change of structure from the surface to initial material. Structure of the cladded coating can be divided into three significant zones: melted layer, transition zone and heat-affected zone, obtained because of the heat-dissipation into unheated layers of base material.

At the Figure 1b, c micrographs of surface and conversion zone of formed coating are presented, respectively. The formation of a dendritic structure is typical for the coating obtained by cladding of the self-fluxing powder. At the boundaries of dendritic crystals, eutectic structure is observed. According to the literature data, eutectic in this alloys may consist of borides, carbides, complex carboborides, silicides with γ-Ni [5]. More detailed studies confirmed the presence of precipitations of hardening phases with size below 30 μm (Figure 1 d). At the near-border area Ni-base columnar grains appeared [3-5], elongated in heat-dissipation direction. It should be noted, that on boundary "coating-substrate" appears entire band, presented by γ-solid solution of Ni, enriched with Fe. An increase of the concentration of iron is associated with diffusion processes, occurred during the coating formation. The thickness of this area varied from 7 to 12 μm (Figure 1 c).

The results of the hardness tests are shown in Figure 2 a. Microhardness of the coating reaches 370 HV. At the direction from surface to base material the hardness of the coating decreases. This could happen due to decrease in volume fracture of reinforced particles and due to dilution of coating material with base material at the bottom part of the coating.

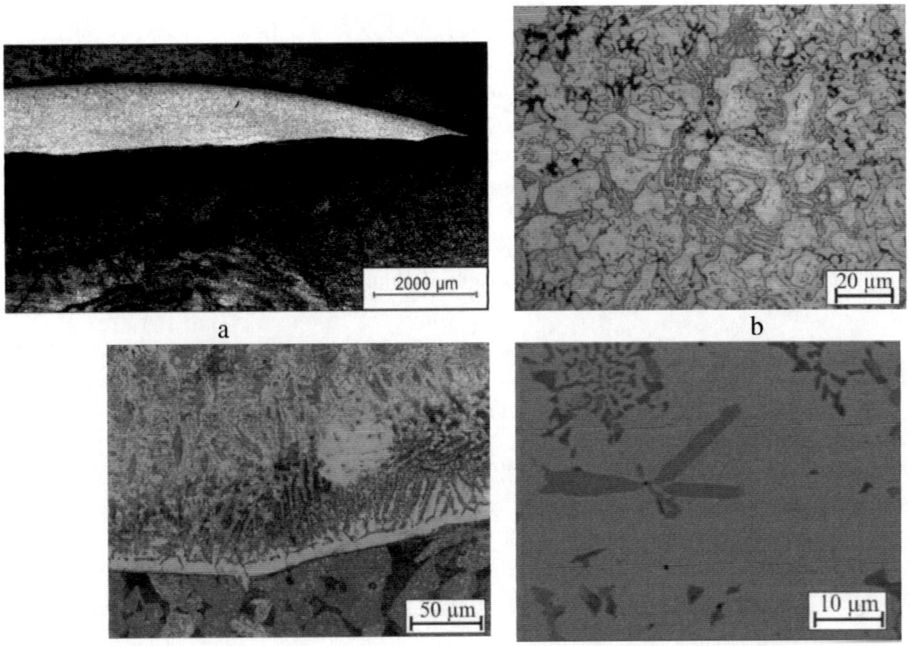

Figure 1. Structure of the coating, obtained by electron-beam cladding technique (a) - general view of obtained coating, (b) - surface structure of the coating, (c) - transition zone ("coating-heat affected zone") structure, (d) - SEM-image of hardening phases.

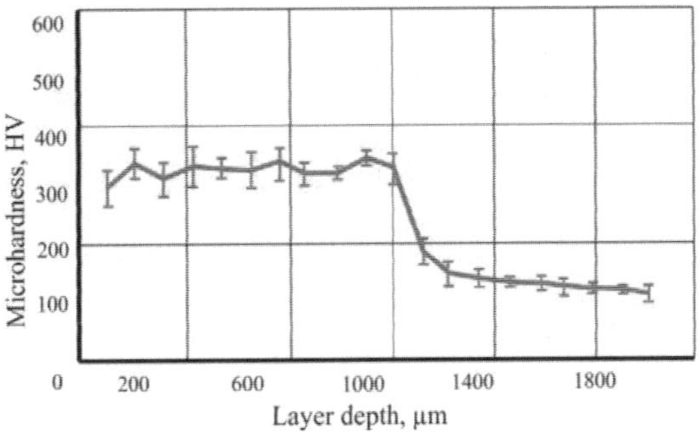

a

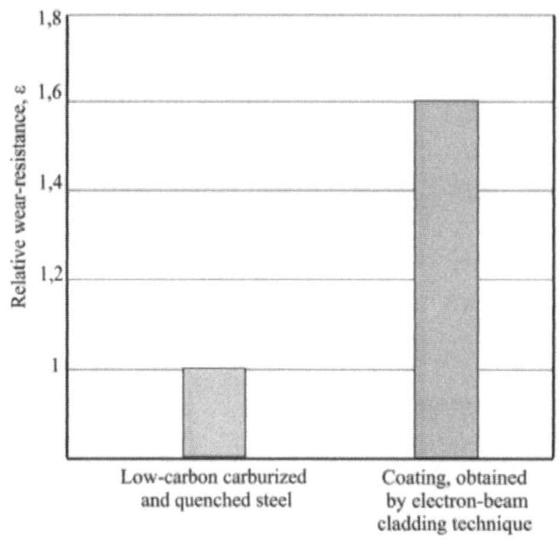

b

Figure 2. (a) The distribution of the microhardness in the cladded sample, (b) mass loss and relative wear resistance of the cladded layer

Tribological tests were carried out to evaluate the performance of the cladded layers. Ideally, the conditions of the test should be close to the working conditions of material. In this case the results of the test can provide valuable information of materials behaviour in the real situation. Since the coatings of this type frequently work in abrasive environment, the wear resistance of the materials was evaluated by friction against fixed abrasive particles. The results of the tests are presented at Figure 2 b. The relative wear resistance of the reference material was equal to 1. The wear resistance of the coating obtained by electron-beam cladding technique is 1.6 times higher in comparison with the reference sample. These data indicate efficiency of electron-beam cladding technique for fabrication of wear resistant materials.

4. Conclusion

Consequently, electron beam cladding is a perspective technique for obtaining dense highly-alloyed wear-resistant coatings. High treatment velocity allows to form surface layers possessing high-level mechanical properties.

Acknowledgements
This work was supported by the Ministry of Education and Science of the Russian Federation. The unique ID of the project is RFMEFI60414X0135.

Reference

[1] Mul D O, Drobyaz E A, Chakin I K, Samoylenko V V, Lozhkin V S and Dostovalov R A 2013 *Obrabotka metallov (tekhnologiya, oborudovanie, instrumenty) (in Russian)* **60** 115-120
[2] Dampilon B V, Durakov V G, Eryomina L V, Losinskaya A A and Mochalina N S 2011 *Obrabotka metallov (tekhnologiya, oborudovanie, instrumenty) (in Russian)* **52** 80-84
[3] Hemmati I, Ocelik V, Csach K and Hosson J Th M 2014 *Metall. Mater. Trans. A* **45** 878-892
[4] Kornienko E, Smirnov A and Kuzmin V 2015 *Applied Mechanics and Materials* **698** 405-410
[5] Han H, Baba S, Kitagawa H, Suilik S A, Hasezaki K, Kato T, Arakawa K and Noda Y 2005 *Vacuum* **78** 27-32
[6] Lenivtseva O G, Bataev I A, Belousova N S, Golovin E D and Zimoglyadova T A 2013 *Obrabotka metallov (tekhnologiya, oborudovanie, instrumenty) (in Russian)* **61** 49-57
[7] Drobyaz E, Zimoglyadova T and Gromov V 2015 *Applied Mechanics and Materials* **698** 419-423

Complex-shaped hardened parts fatigue limit prediction according to the witness sample study results

Yu N Surgutanova, N N Mikushev, N A Surgutanov, P E Kiselev, P A Shlyapnikov and A A Meshcheryakova

National Research University of Samara named after S P Korolyov (University of Samara), Samara city, Russia

E-mail: dogg199191@mail.ru

Abstract. The aim of this study is to investigate the possibility of assessment of the effect of preparatory surface plastic deformation by hydraulic shot blasting on the fatigue strength of cylindrical parts of different diameters (10-40 mm) of D16T alloy with circular notches of semicircular section, based on measurements of residual stress (initial deformations) of a witness sample. The residual stresses of smooth parts were used to calculate the residual stresses of parts with stress raisers. These were used to predict the increment of these parts fatigue limit caused by hardening hydraulic shot blasting. It was found that the highest compressive residual stresses in the smooth parts obtained through calculations differ from the observed values not more than by 7%, and in notched parts by 8%. Using the criterion of mean integral residual stresses, we calculate the increments of the fatigue limit of parts due to superficial hardening. The discrepancy between the experimental and calculated increment values of the fatigue limit of hardened parts with raisers does not exceed 17%.

Key words: Residual stresses, criterion of mean integral residual stresses, fatigue limit, surface hardening.

1. Introduction

Improvement of reliability and durability of machine parts is one of the most important problems, a full and comprehensive solution of which accompanies the production and operation of all products of modern technology. It is known that the main type of destruction of machine parts is fatigue failure that leads to severe consequences, as this type of failure develops suddenly.

One of the most important characteristics of machine parts by their fatigue strength is their fatigue limit, the predicted value of which largely depends on the reliability of the calculations performed on the stages of design and development of design documentation. This is why the selection of design, materials, geometric parameters of parts with stress concentrations, and their production techniques pose very sophisticated challenges.

Witness samples are commonly used to monitor the processing procedure of part hardening by various methods of superficial plastic deformation (SPD). These samples of a certain shape and size undergo the whole technological cycle of processing together with a standard part. Generally, processed parts are costly and labor-intensive products, so this non-destructive method of quality control hardening has an undeniable advantage over other methods, primarily in terms of cost efficiency and especially in mass production.

The use of witness samples is based on the assumption that the part and the witness sample, after having been processed together, receive the same initial strain at SPD. We studied the results of fatigue tests and the results of experimental determination of residual stresses of cylindrical samples of D16T aluminum alloy, given in [1]. Figure 1 shows a sample with the external diameter D, the hole diameter *d*, and the diameter of the weakest section of the D_1 sample.

Before notching, smooth specimens were subjected hydraulic shot blasting (HSB) for 8 minutes with 2 mm shots in oil at a pressure of 0.19 MPa (D16T alloy). As witness specimens in a study of the above-described batches of specimens, we used cylindrical sleeves with the dimensions of 51.5x45 mm/mm, which are commonly used for determination of residual stresses through the depth of the hardened superficial layer by the method of rings and strips [2].

2. Residual stresses' calculating

The computational part of the study was carried out by the method of finite element modeling, using PATRAN/NASTRAN software package. Finite-element models in an axially symmetric form represented a quarter section of the sample, with imposition of appropriate boundary conditions. In the simulation, a flat triangular 2D-Solid element with six nodes was used.

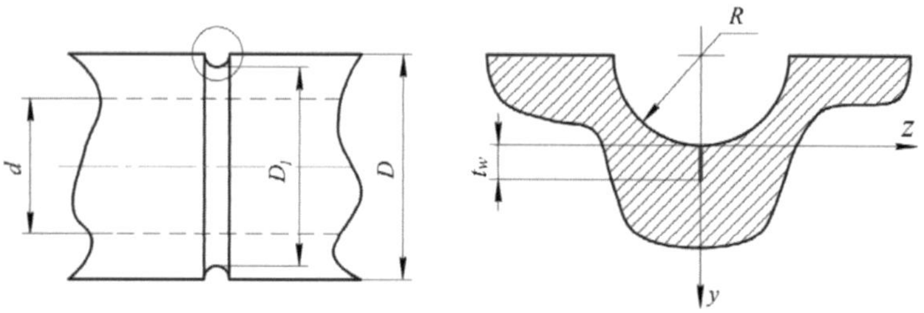

Figure 1. Cylindrical specimen with a non-propagating fatigue crack

When determining the initial strain, we used the observed distribution of the axial σ_z residual stresses in thickness *a* of the hardened superficial layer of a witness sample (sleeve) as the source data. It was established that, under the assumptions made, three iterations are sufficient to achieve agreement of the distribution of the values of the axial residual stress, calculated on the basis of the initial strain and the experimental values in the witness specimen.

The next stage of the calculations was performed using finite-element models of the smooth test specimens, on the basis of the initial deformation of the witness specimen. It is known [1, 4] that axial residual stresses are principal in assessment of the incremental fatigue limit. For that very reason the comparison of calculated and experimental distributions of residual stresses through the depth of the hardened superficial layer for the test specimens was performed for the axial component. The experimental results [1] and the calculation of distribution of residual stresses σ_z through the depth of the superficial layer for the smooth test specimens of D16T alloy are shown in Figure 2.

The maximum difference between the calculated and experimental values of the largest axial compressive residual stress was 10% at *D/d* = 10/0 mm/mm, in other cases, the difference did not exceed 2%.

Prior to fatigue tests, circular notches of semicircular section with the radius *R* = 0.3 mm were made on both non-hardened and hardened smooth samples. Axial σ_z residual stresses in thickness *a* of the thinnest (weakest) section of the notched samples, were determined by analytical and numerical methods, using calculated distributions of residual stresses of smooth specimens.

It should be noted that the residual stresses calculated with the two methods were in good agreement.

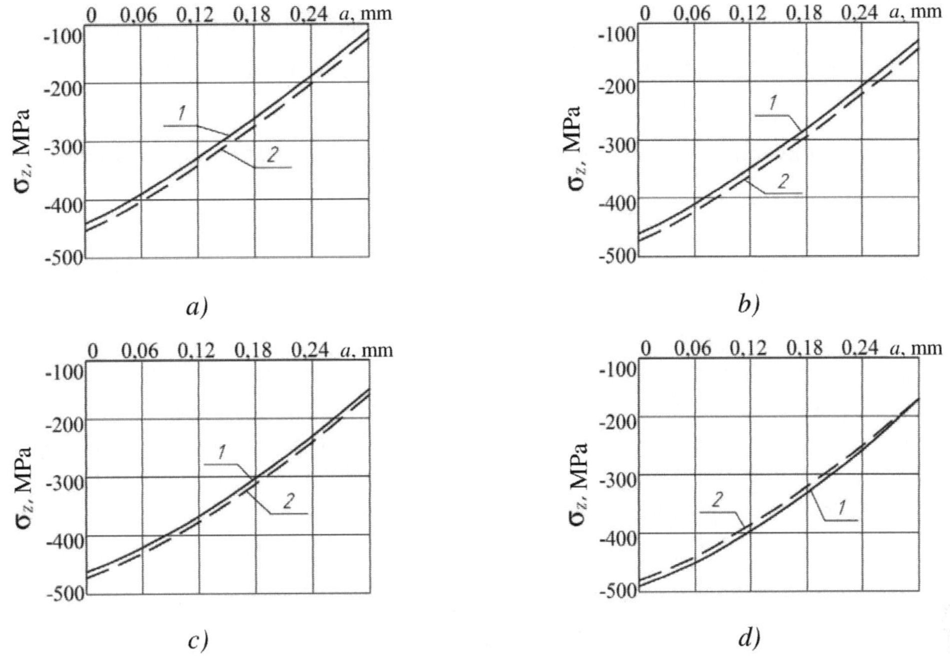

Figure 2. The distribution of axial σ_z residual stresses after HSB determined by experimental (1) and calculation (2) methods in smooth samples of D16T alloy with the following diameters: $a - D = 10$ mm, $b - D = 15$ mm, $c - D = 25$ mm, $e - D = 40$ mm

3. Residual stresses' rating

Prediction on the effect of superficial hardening on the increment of the fatigue limit of notched parts at bending in the case of a symmetric processing cycle $(\Delta\sigma_{-1})_{calc}$ was made based on the criterion of mean integral residual stresses $\bar{\sigma}_{res}$ [12]

$$(\Delta\sigma_{-1})_{calc} = \bar{\psi}_\sigma |\bar{\sigma}_{res}|, \tag{1}$$

where $\bar{\psi}_\sigma$ is the factor of effect of surface hardening on the fatigue limit at bending according to the criterion $\bar{\sigma}_{res}$;

$$\bar{\sigma}_{res} = \frac{2}{\pi}\int_0^1 \frac{\sigma_z(\xi)}{\sqrt{1-\xi^2}} d\xi; \tag{2}$$

$\sigma_z(\xi)$ are axial residual stresses in the weakest section of the sample (part) with a stress raiser through the thickness of the superficial layer a; $\xi = a/t_w$ is the distance from the surface of the weakest section of the sample (part) to the layer in question, expressed as a fraction of t_w; t_w is the critical depth of a non-propagating fatigue crack occurring in a hardened sample (part) with a stress raiser at operation at the fatigue limit (Figure 1).

The critical depth t_w of a non-propagating fatigue crack was determined from the dependence, established on the basis of numerous experiments described in [1-5]

$$t_w = 0.0216D_1\left[1 - 0.04\left(\frac{d}{D_1}\right)^2 - 0.54\left(\frac{d}{D_1}\right)^3\right], \tag{3}$$

where D_1 is the diameter of a dangerous section of the sample (part) with a stress raiser.

The values of the criterion of mean integral residual stresses $\bar{\sigma}_{res}$ were calculated using the formula (2), and the thickness of the superficial layer of the weakest section of notched samples t_w using the calculated residual stress distribution. The factor $\overline{\psi}_\sigma$ of the effect of superficial hardening on the fatigue limit of the criterion $\overline{\sigma}_{res}$ was calculated using the dependence established in [19]

$$\overline{\psi}_\sigma = 0.612 - 0.081\alpha_\sigma, \tag{4}$$

where α_σ is the theoretical stress concentration factor, which was determined using diagrams from the reference book [4].

Further, according to the formula (1), the values of the increment of the fatigue limit $(\Delta\sigma_{-1})_{calc}$ at HSB hardened notched samples were calculated and compared with the experimental values $(\Delta\sigma_{-1})_{exp}$ given in [5].

4. Conclusions

The difference between the calculated and experimental values of the increment of the fatigue limit does not exceed 17%. Therefore, using the results of the determination of residual stresses in witness samples, it is possible to predict the fatigue strength of superficially hardened parts in conditions of stress concentration with a precision acceptable for high-cycle fatigue.

References

[1] Pavlov V F 2012 *Prediction of the fatigue strength of surface hardened components for residual stresses* (Samara: Publishing house of the Samara Scientific Center of the Russian Academy of Sciences)
[2] Birger I A 1963 *Residual stresses* (Moscow: Mashgiz)
[3] Sazanov V P, Chirkov A V, Samoilov V A and Larionova Yu S 2011 Modeling redistribution of residual stresses in hardened cylindrical samples after preparatory superficial plastic deformation *Bulletin of Samara State Aerospace University* **3(27) 3** 171-174
[4] Peterson R E 1977 Stress concentration factors (Moscow: Mir)
[5] Kirpichev V A, Bukatyi A S, Filatov A P and Chirkov A V 2011 Prediction of the fatigue limit of surface hardened parts with varying degrees of stress concentration *Bulletin of Ufa State Aviation Technical University* **15 4(44)** 81-85.

Aluminum surface layer strengthening using intense pulsed beam radiation of substrate film system

A A Klopotov[1,2], Yu F Ivanov[3,4], V A Vlasov[2,3], A A Kondratyuk, A D Teresov[4], V V Shugurov[4] and E A Petrikova[4]

[1] National Research Tomsk State University, Tomsk, Russia
[2]Tomsk State University of Architecture and Building, Tomsk, Russia
[3]Tomsk Polytechnic University, Tomsk, Russia
[4] Institute of High Current Electronics, Tomsk, Russia

E-mail: klopotovaa@tsuab.ru

Abstract. The paper presents formation of the substrate film system (Zr-Ti-Cu/Al) by electric arc spraying of cathode having the appropriate composition. It is shown that the intense beam radiation of the substrate film system is accompanied by formation of the multi-phase state, the microhardness of which exceeds the one of pure A7 aluminum by ≈4.5 times.

1. Introduction

Aluminum and its alloys are widely used in industry due to quite high specific strength, satisfactory corrosion resistance and good mechanical processability. However, aluminum is characterized by low stiffness and durability [1]. To improve the above properties, the surface modification methods are used, among which are hard, extra-hard and wear-resistant coatings, surface treatment by concentrated energy streams, as well as surface saturation with metal and gas atoms [1-3]. All these surface modification methods are applicable for treatment of products made of aluminum and its alloys. It shall be noted that each method has its benefits and drawbacks.

Spraying of coating allows obtaining layers with set stoichiometric composition on the surface, which enables to forecast functional properties. The peculiarity of surfaces obtained with the help of this method is their structure-phase state of the plane interface between the original product surface and coating layers [4]. Presence of such interface raises the issue of the obtained layers adhesion to the substrate.

When obtaining hard layers on the products of aluminum and its alloys the adhesion issue is especially topical because of oxides that are always present on aluminum surface. The second problem caused by the interface plane is a sharp border between the hard layer and the substrate.

Use of concentrated energy streams allows modifying the treated surface by its high-speed melting and further high-speed cooling. Using such approach enables to decrease roughness of material surface and to increase its physico-mathematical, chemical, electrophysical and tribological characteristics by forming multiphase submicro- and nanocrystalline structure [3, 5, 6]. It is worth mentioning that the effect that occurs during treatment with concentrated energy streams is ultimately manifested in complex alloyed materials treatment, especially with regard to modification of mechanical, tribological and fatigue material properties [7, 8]. The above effects are inconsiderably manifested during treatment of aluminum alloys, and even less during treatment of pure aluminum.

The purpose of the paper is to find out and analyze regularities of structure formation and properties of the surface layer of pure aluminum subjected to modification and alloying by melting the film substrate system.

2. Materials and experiment method

Pure A7 aluminum was used as material for modification. Combined aluminum treatment was done in the following way. At the first stage, Zr-5 at % Ti-5 at % Cu film 1 μm thick was sprayed on the aluminum surface. The film was obtained with the help of updated "Kvinta" unit for ion and plasma spraying. At the second stage, the formed film (Zr-Ti-Cu alloy) (A7) substrate system was irradiated with intense pulsed electron beam using "SOLO" unit. The structure and element composition of the modified surface were investigated using scanning electron microscopy methods; phase composition was investigated using X-ray diffraction analysis; surface layer properties were defined based on the extent of microhardness.

3. Investigation results and discussion

Thermodynamic analysis of possible phase formation that occurs in equilibrium conditions in the Zr-Ti-Cu system (film sprayed on aluminum) and in Zr-Ti-Cu-Al system (film (Zr-Ti-Cu alloy) (A7) substrate irradiated with intense pulsed electron beam) is showed on Figure1 – Figure 3. The Figures illustrate binary composition phases; possible formation of triple phases is given below.

Al–Cu–Zr System [9]. In triple Al-Cu-Zr system within isothermic triangulated surface 10 intermetallic compounds with narrow homogeneity ranges were found (Figure 1a, phases are marked) and the following triple compounds were singled out: τ1(14Al-71Cu-15Zr), τ2(13Al-14Cu-73Zr), τ3(21Al-28Cu-51Zr), τ4(ZrCu2Al), $ZrCu_2Al$, τ5(ZrCuXAl2–X), Zr25Cu25Al50, ZrCuAl, τ6(ZrCu4Al3), τ7(Zr2CuAl5), τ8(Zr6Cu16Al7).

Al–Cu–Ti System [10]. In Al-Cu-Ti triple system intermetallic compounds were found in Al–Ti and Cu-Ti system. Within isothermic triangulated surface the extended homogeneity ranges with the following composition were found (Figure1b): τ1 (TiCu2Al), τ2 (TiCuAl),τ3 (Ti2CuAl5).

Cu–Ti–Zr System [11]. In Cu–Ti–Zr triple system within isothermic triangulated surface only one intermetallic compound τ1 (ZrTiCu2) was found. It had a certainly defined location with narrow homogeneity range (Figure 1c). At the same time, this system contains an extended range of combination with the C11b structure along the Zr2Cu–Ti2Cu section, in which isomorphous substitution of Ti atoms with Zr atoms occurs.

Thus, in equilibrium conditions quite a big number of double- and multi-element compounds can form in Cu-Zr-Ti ternary system and Zr-Ti-Cu-Al quaternary system.

Investigation of element composition and structure of film (Zr-Ti-Cu) (Al) substrate system formed by electric arc spraying of Zr-Ti-Cu cathode and generated plasma deposition on the pure A7 aluminum samples was done using scanning electron microscopy methods. The analysis results are given on Figure 2.

The analysis of the results given on Figure 2 shows that Zr, Ti, Cu (the sprayed cathode) and Al (substrate) are the key elements of the investigated system. The obtained coating contains a big number of microdroplets (Figure 2 b), which is peculiar to electric arc method of coatings and films application. The microdroplets element composition is similar to the one of the coating itself. Intense pulsed electron beam radiation of the film substrate system in the mode of substrate melting treatment without changing element composition of the coating causes changes in its element concentration. Manifold increase (by ≈3.5 times) of aluminum concentration was seen in the analyzed surface layer.

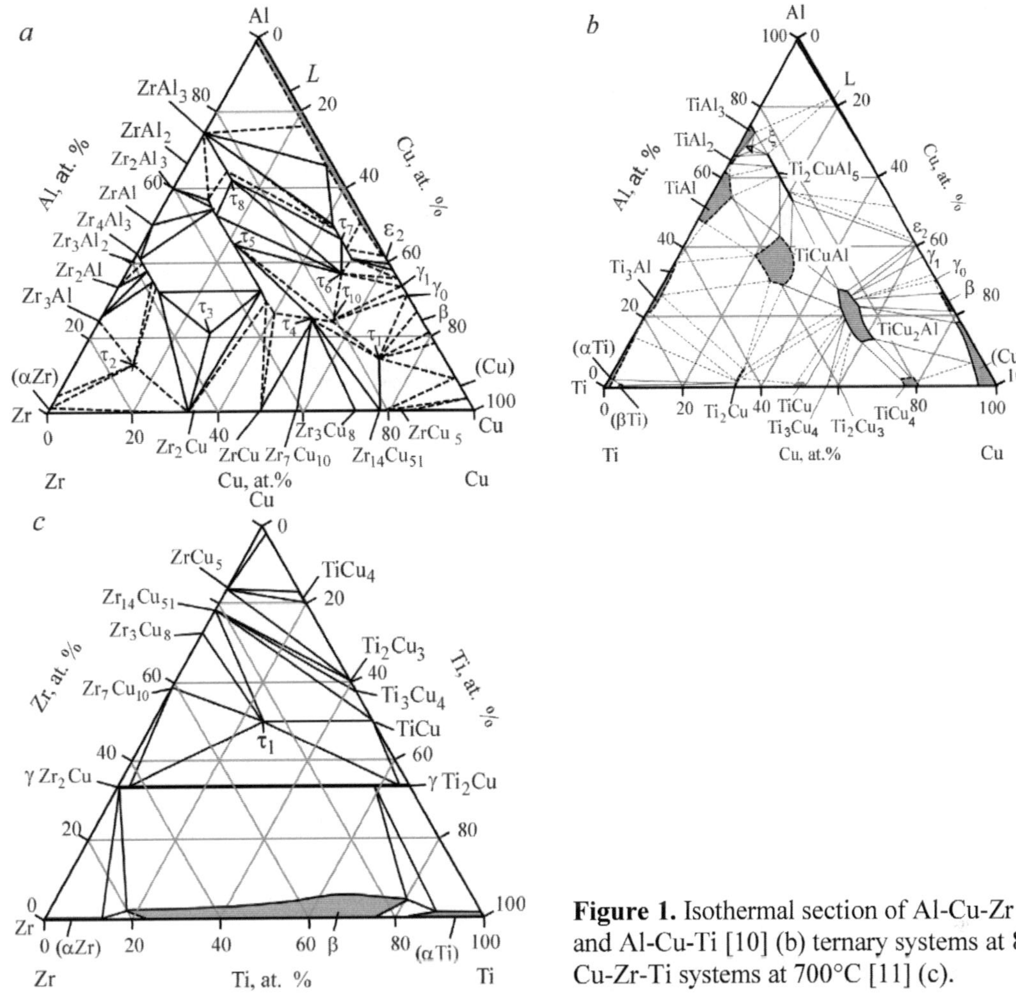

Figure 1. Isothermal section of Al-Cu-Zr [9] (a) and Al-Cu-Ti [10] (b) ternary systems at 800°C; Cu-Zr-Ti systems at 700°C [11] (c).

This fact is the evidence of coating melting into the substrate and aluminum seepage through microcraters and microcracks.

The phase composition of the film substrate system was inve stigated using X-ray diffraction analysis. It was found that coating spraying was accompanied by multi-phase state. The main phases are Al (face-centered cubic lattice, a = 0.40722 nm) (90 mass %), Zr (β-modification, body-centered cubic crystalline lattice, a = 0.35967 nm) (9.6 mass %) and Zr_2Al_3 (0.4 mass %). Zr_2Al_3 formation is the evidence of the substrate alloying with Zr. After intense pulsed electron beam radiation of the film substrate system the phase composition of the surface layer is represented by Al (a = 0.40521 nm; 87.5 mass %), Zr (α-modification, a = 0.32517 nm, c = 0.51814 nm; 6.8 mass %) $ZrAl_3$ (5.7 mass %). Manifold increase of $ZrAl_3$ abundance ratio in the surface layer of the film substrate system confirms the above speculation on the coating and substrate melting in the process of intense pulsed electron beam radiation. Remarkable decrease of aluminum crystalline lattice parameter in the electron beam irradiated material shall be noted. This might be the evidence of the alloying elements' (Zr, Ti, Cu) withdrawal from aluminum crystalline lattice under electron beam thermal action.

The film substrate system was mechanically tested before and after electron beam radiation. The experiment showed that microhardness of the untreated film substrate system doubles compared to the one of pure A7 aluminum sample. And the microhardness of the system after radiation increases by ≈4.5 times.

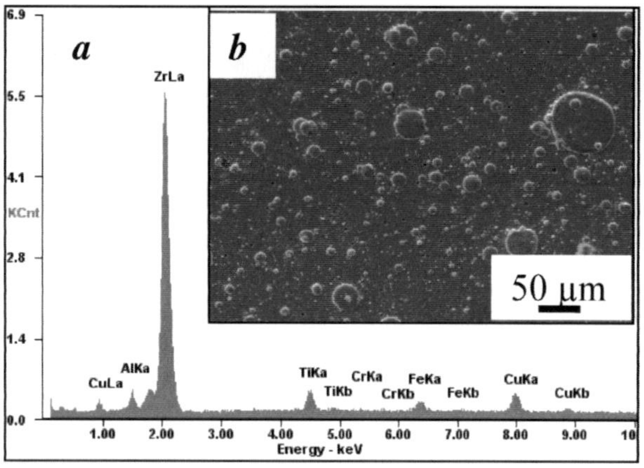

Element	*Wt%*	*At%*
AlK	03.84	11.00
TiK	04.77	07.70
CrK	00.31	00.46
FeK	02.58	03.57
CuK	06.35	07.72
ZrK	82.15	69.57

Figure 2. Energy spectra (a) defined in the process of structure investigation of film (Zr-Ti-Cu) (Al) substrate system (b) using scanning electron microscopy methods. The table presents the results of spectra quantitative analysis (a).

4. Conclusions

As a result of the conducted research the possibility of film (Zr-Ti-Cu) (Al) substrate system formation by electric arc spraying of a certain cathode composition was established. It was found out that intense electron beam radiation of the film substrate system is accompanied by multi-phase state, the microhardness of which is ≈4.5 times higher compared to the one of pure A7 aluminum.

Acknowledgement

The research was supported by the Russian Foundation for Basic Research grant (the project No. 16-58-00075 Bel_a) and carried out during the project implementation (No.8.1.42.2015) within the Tomsk State University Academic D.I. Mendeleev Fund Program in 2015-2016.

References

[1] Belov N A 2010 *Phase composition of industrial and perspective aluminum alloys* (Moscow: MISIS Publishing House) p 551 (in Russian)
[2] Belov N A 2009 *Phase composition of aluminum alloys* (Moscow: MISIS Publishing House) p 392 (in Russian)
[3] Laskovnev A P, Ivanov Y F, Petrikova E A et al 2013 *Modification of eutectic silumin structure and properties by electron-ion-plasma treatment* (Minsk: Belarus. Navuka) p 287 (in Russian)
[4] Hoking M, Vasantasri V and Sidki P 2000 *Metallic and Ceramic Coatings* (Moscow: Mir) p 516
[5] Poate J, Foti G and Jacobson D 1987 *Surface modification and alloying by laser, ion, and electron beams* (Moscow: Mashinostroenie) p 424
[6] Gribkov V A, Grigoryev F I, Kalin B A et al 2001 *Perspective beam radiation technology for metal treatment* (Moscow: Kruglyi Stol) p 528 (in Russian)
[7] Rotshtein V, Ivanov Y and Markov A 2006 *Surface treatment of materials with low-energy, high-current electron beams Materials surface processing by directed energy techniques*, Chapter 6 (Ed. by Y. Pauleau: Elsevier) pp 205-240
[8] Koval N N, Ivanov Y F 2008 Nanostructuring of surfaces of metalloceramic and ceramic materials by electron beams *Russian Physics Journal* **5** 60-70 (in Russian)
[9] Soares D, Castro F 1997 Study of Phase Equilibria in the Al-Cu-Zr System at the Al-Rich Part *J. Chim. Phys.* **94** 958-963
[10] Markiv V Y, Burnashova V V and Ryabov V R 1973 Ti-Fe-Al, Ti-Ni-Al and Ti-Cu-Al Systems, *Metallofizika* **46** 103-110
[11] Teslyuk M Y 1969 *Intermetallic Compounds with Structure of Laves Phases* (Moscow: Nauka) pp 1-138 (in Russian)

IOP Conf. Series: Materials Science and Engineering **156** (2016) 012020 doi:10.1088/1757-899X/156/1/012020

Structural features of Ni-Cr-Si-B materials obtained by different technologies

E E Kornienko[1], A A Nikulina[1], N S Belousova[1], D V Lazurenko[1], A S Ivashutenko[2] and V I Kuz'min[3,4]

[1] Faculty of Mechanical Engineering and Technologies, Novosibirsk State Technical University, 20 Prospekt K. Marksa, Novosibirsk, 630073, Russia

[2] Department of Industrial Electric Power Supply, Institute of Power Engineering, Tomsk Polytechnic University, 30 Lenin Avenue, Tomsk, 634050, Russia

[3] Khristianovich Institute of Theoretical and Applied Mechanics SB RAS, Institutskaya str. 4/1, Novosibirsk, 630090, Russia

[4] Siberian State University of Water Transport, Schetinkina str. 33, Novosibirsk, 630099, Russia

E-mail: kornienko_ee@mail.ru

Abstract. This study considers the structural features of Ni-Cr-Si-B (Ni – base; 15.1 % Cr; 2 % Si; 2 % B; 0.4 % C) materials obtained by different methods. The self-fluxing coatings were deposited by plasma spraying on the tubes from low carbon steel. Bulk cylinder specimens of 20 mm diameter and 15 mm height were obtained by spark plasma sintering (SPS). The structure and phase composition of these materials were investigated by optical microscopy (OM), scanning electron microscopy (SEM), transmission electron microscopy (TEM) and X-ray diffractometry (XRD). The major phases of coatings and sintered materials are γ-Ni, Ni_3B, CrB and Cr_7C_3. We demonstrate that the particle unmelted in the process of plasma spraying or SPS consist of γ-Ni-Ni_3B eutectic and also CrB and Cr_7C_3 inclusions. The prolonged exposure of powder to high temperatures as well as slow cooling rates by SPS provide for the growth of the structural components as compared to those of plasma coatings materials. High cooling rates at the plasma spraying by melted particles contribute to the formation of supersaturated solid solution of Cr, Si and Fe in γ-Ni. The structure of the melted particles in sintering material has gradient composition: the core constituted of Ni grains of 10 μm with γ-Ni-Ni_3B eutectic on the edges. The results of the experiment demonstrate that the sintering material has a smaller microhardness in comparison with plasma coatings (650 and 850 MPa, respectively), but at the same time the material has higher density (porosity less than 1 %) than plasma coatings (porosity about 2…3 %).

1. Introduction

The self-fluxing alloys of Ni-Cr-Si-B system possess high wear resistance [1-7] and corrosion resistance [2-5, 8-10] at room and higher temperatures and also high heat resistance [1, 3]. Due to these properties, machine details made of these materials are widely used in aerospace industry and aircraft construction [11], atomic [3, 12, 13], oil and chemical industry, and metallurgy [14].

There are many techniques of coatings deposition stating from the self-fluxing alloys, for example CVD and DVD [1], different kind of welding (laser cladding [15, 16], vacuum-arc deposition [17])

IOP Conf. Series: Materials Science and Engineering **156** (2016) 012020 doi:10.1088/1757-899X/156/1/012020

and thermal spraying (plasma spraying [6, 10, 18], flame spraying [1, 6, 7], high velocity oxy-fuel spraying [8, 9]). The technology of plasma spraying is widely used [6, 10, 18-23]. It allows to form coatings of any materials including composite coatings on surface of any complexity as well as restore the worn out machine parts surfaces. Unfortunately the disadvantages of plasma coatings are high porosity (up to 20 %) [7] and low adhesion [4, 5, 7, 18]. It may limit their use in corrosive mediums [4, 5, 8] and decrease their wear-resistance. The solution to of this problem is subsequent melting of self-fluxing coatings [2, 5, 7, 18].

SPS is perspective technique of high-pressure sintering. This method allows taking dense fine-grained material in short time [24]. In future this technique will be used for sintering of self-fluxing powders onto detail surface. The authors of study [25] demonstrate that dense coatings with high adhesion and high operation properties may form in that case. Nowadays significant amount of studies devoted to SPS were carried out [26-31], but there is no research into Ni-Cr-Si-B alloys sintering. Thus the questions about the structure and properties of these materials obtained by SPS require investigation.

In this study we consider the structural features of self-fluxing material of Ni-Cr-Si-B system obtained by plasma spraying and SPS.

2. Materials and methods

Coatings and bulk materials were formed from self-fluxing powder of Ni-Cr-Si-B system. Chemical composition of the powder: Ni – base; 15.1 % Cr; 2 % Si; 2 % B; 0.4 % C. The particle of self-fluxing powder had spherical form of 40…100 μm.

Air-plasma spraying was carried out at the Institute of Theoretical and Applied Mechanics SB RAS with the help of plasmatron "PNK-50" with an annual injection of powder unit. The powder was sprayed on tubes from low carbon steel 20 (0.2 wt. % C). The inner diameter of tubes was 60 mm, the wall thickness was 3 mm. The regimes of plasma spraying were: current arc was 140 A, and voltage was 265 V. The blend of air and propane-butane was used as a carrier, focus and protective gas. The air was used as plasma gas. The distance of plasma spraying was 170 mm. Grit blasting of the substrate surface was performed before the spraying. Then the 75Ni15Al interlayer with thickness 10…15 μm was deposited on the surface tubes to increase adhesion.

SPS was carried out at Tomsk Polytechnic University on the SPS10-4 apparatus. The regimes of sintering were: heating rate was 100 °C/min., heating temperature was 1030 °C, sintering pressure was 65 MPa and the high-pressure treatment period at the sintering temperature was 5 min. We obtained the cylindrical specimens with diameter 20 mm and height 15 mm.

The microstructure of coatings and bulk material was identified by the optical microscope Carl Zeiss Axio Observer A1m and scanning electron microscope Carl Zeiss EVO50 XVP. The sample for analysis and microhardness measurement were cut by the cross-section and prepared by standard methods: mechanic grinding and polishing with colloidal Al_2O_3. To reveal eutectic the surface of the samples we etched it with the solution of 10 ml HCl, 0.1 ml HNO_3, and 10 g $FeCl_3$. Thin foils were cut from the middle of the coating for structural analysis by transmission electron microscope Tecnai G2 FEI. The XRD analysis was done using ARL X'TRA with Cu K_α radiation. Microhardness was estimated by Wolpert Group 402MVD with pressure of 10 g.

3. Results and discussion

The structure of plasma coatings is presented in figure 1. We can observe that the material of coating contains slightly deformed and significantly deformed particles with the initial geterophase structure (A in Figures 1b, c). The major of powder particles melt in the process of plasma spraying. The supersaturated solid solution γ-Ni fixes on substrate or already consolidation material of coatings during crystallization process (B in Figure 1b, c) [32].

Figure 2 shows the material structure formed by SPS. The structural research allowed establishing that the most of particles do not change their form. It is clearly seen that the particles have contacts along the whole particle surface, providing for the material low porosity. The majority of particles save their geterophase structure (C in Figure 2b). Small particles which are located in the joints of coarse particles are melted (D in Figure 2b).

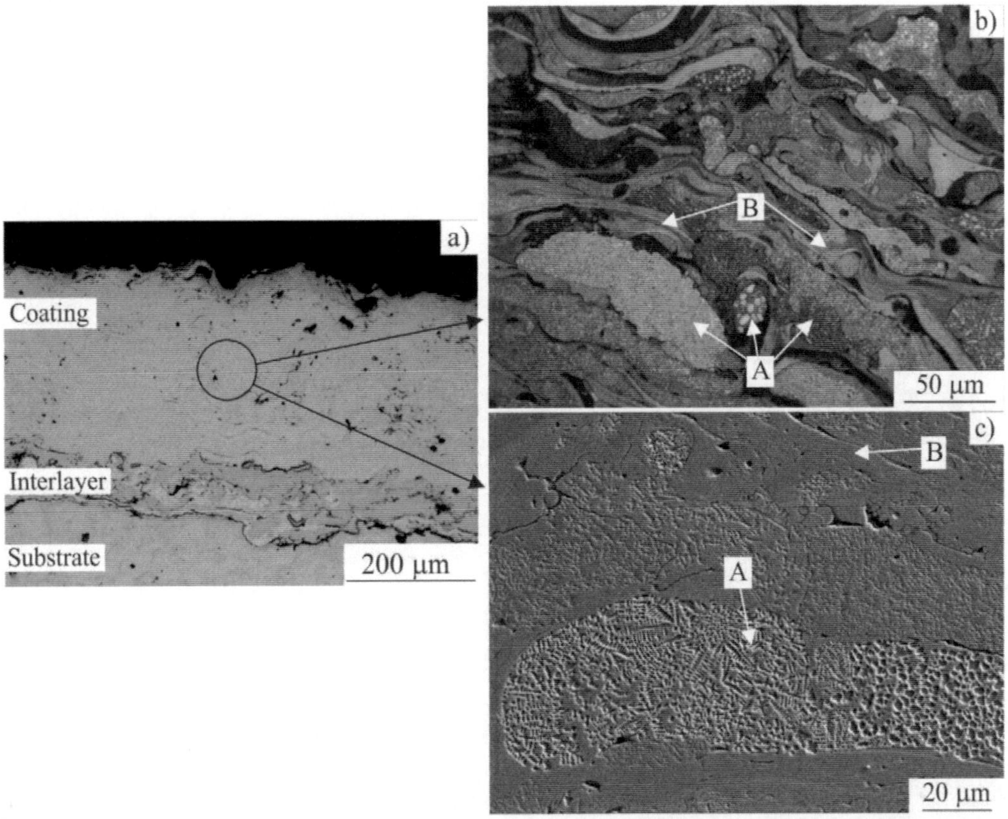

Figure 1. The structure of coatings: a) before etching, b, c) after etching; a, b) OM, c) SEM.

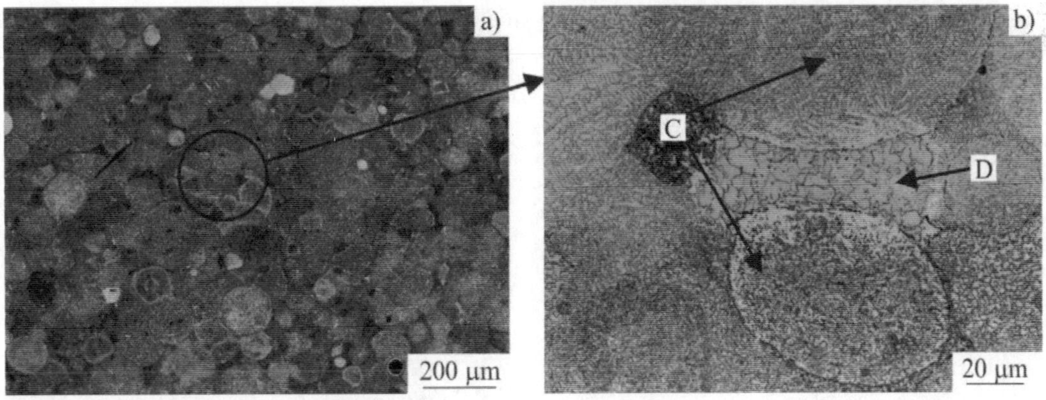

Figure 2. The structure of sintering material.

According to the data obtained by X-ray diffraction (Figure 3), the major phases of self-fluxing coatings as well as sintering materials are γ-Ni, Ni_3B, CrB and Cr_7C_3. It is noteworthy that the initial powder has the same phase composition.

Figures 1c and 2b show that the structure of A and C type particles is identical and constitutes eutectic of γ-Ni dendrite and Ni_3B phase. The structure of A type particles is more disperse which can be explained by the short time exposure of powder particles to high temperature. In this case any structural changes do not take place. The initial powder particles have the same disperse structure. The heating and cooling rates are less at SPS. The same conditions contribute to the transformation of material in a more stable state and lead to the changes that facilitate the increase of structural components.

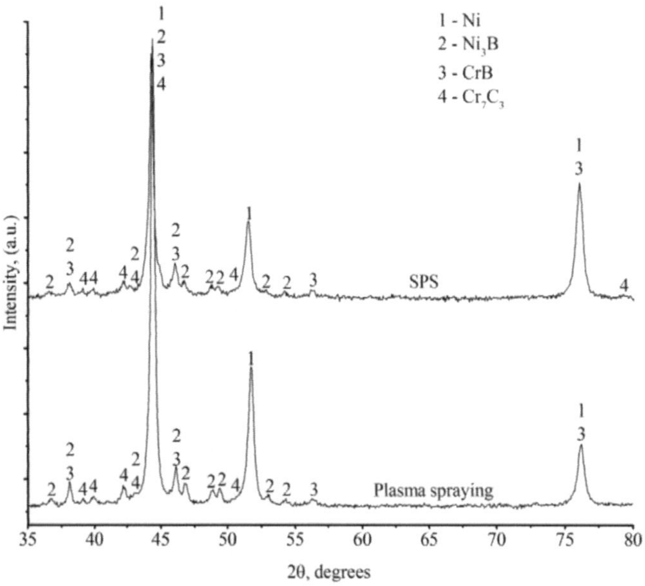

Figure 3. XRD spectra of the coating and sintering material.

According to the equilibrium diagram of Ni-C and Ni-B, C and B practically do not dissolve in Ni lattice (2.7 and 0.3 at. %, respectively) [33]. Chromium borides and carbides do not change their morphology in A type particles in comparison to the initial powder. Figure 4 obtained by SEM in regime backscattering electrons shows that the phases of these types are clearly seen. CrB is the dark-grey blocky particles, Cr_7C_3 is light-grey with the flower shape. The size of these inclusions of plasma coatings is smaller, than the one of the sintering material (Figure 4b). It can be explained by the difference of the technology parameters.

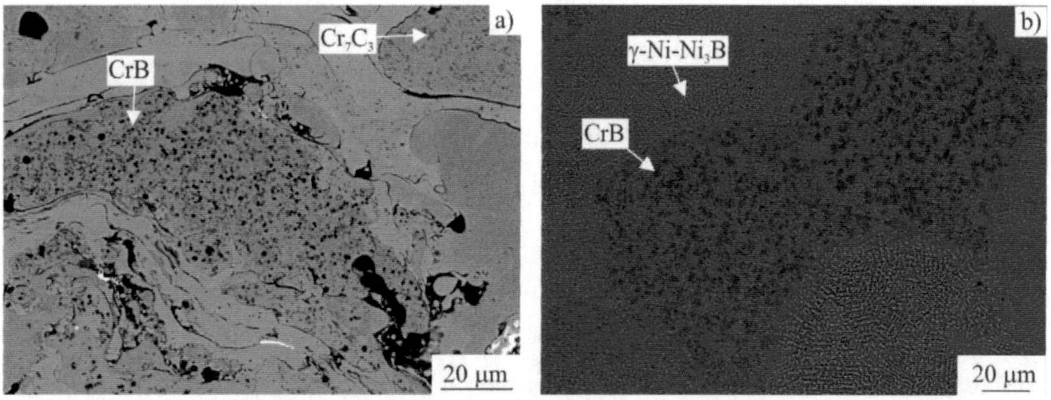

Figure 4. CrB and Cr7C3 inclusions of coatings (a) and CrB inclusions of sintering material.

The structure of powder particles which were melted (B and D type) is also different. Significant supercooling degrees of B type particles do not allow segregating all alloying elements from γ-Ni lattice. As a result the coating particles of the same type constitute the supersaturated solid solution of

Cr, Si and Fe in γ-Ni. TEM revealed that the size of subgrain in the particles of this type does not exceed 100 nm (Figure 5). The structure of D type particles of sintering material is showed on figure 2b. Mostly, these particles represent coarse Ni grains with size up to 10 μm in the particle core and γ-Ni-Ni_3B eutectic about the particle edges.

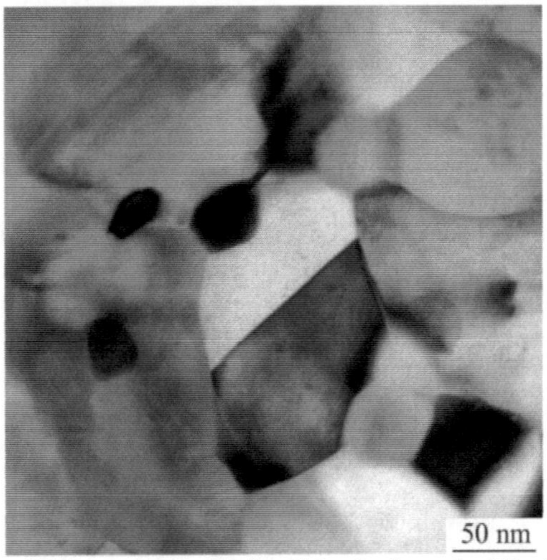

Figure 5. Bright-field image of the supersaturated solid solution of γ-Ni of coating.

It was revealed by microhardness measuring that the hardness of sintering material is less than coating (650 and 850 MPa, respectively). The study [34] by local microhardness measuring demonstrated that particles hardness changing depends on particles type. The slightly deformed particles have the smallest microhardness (600…700 HV). Figure 2a clearly shows that the sintering material is characterized by the presence of this type particles only. Besides long exposure of the material to high temperatures provides for the diffusion processes that further promote the decreases of the microhardness. The sintering material is denser (porosity less than 1 %) in comparison with plasma coatings (~2…3 %).

4. Conclusions

X-ray analysis showed that the phase composition of coatings and sintering materials are the same. The major phases of coatings and sintering materials are γ-Ni, Ni_3B, CrB and Cr_7C_3.

The OM and SEM revealed that the structure of plasma coatings and sintering materials are different. The unmelted particles in plasma jet or sintering constitute γ-Ni-Ni_3B eutectic and CrB and Cr_7C_3 inclusions. The prolonged exposure to high temperature of powder particles by sintering and low cooling rates provide the growth of structural components in comparison to material of plasma coatings. High cooling rates by plasma spraying of the melted particles contribute to formation of supersaturated solid solution of Cr, Si and Fe in γ-Ni. The structure of melted particles in sintering material has gradient composition: Ni grains with size up to 10 μm are in the core of the particles and γ-Ni-Ni_3B eutectic is about particles edges.

Microhardness measuring shows that sintering material has less hardness in comparison with plasma coatings (650 and 850 MPa, respectively).

Porosity measuring showed that sintering material is denser than coatings (porosity less than 1 % and about 2…3 %, respectively).

References

[1] Chaliampalias D, Vourlias G, Pavlidou E, Skolianos S, Chrissafis K, Stergioudis G 2009 *Applied Surface Science* **255** 3605–3612

[2] Hemmati I, Ocelik V and Hosson J Th M De 2013 *Physics Procedia* **41** 302-311
[3] Sudha C, Shankar P, Rao R V S, Thirumurugesan R, Vijayalakshmi M and Raj B 2008 *Surface & Coatings Technology* **202** 2103–2112
[4] Serresa N, Hlawka F, Costil S, Langlade C and Machi F 2011 *Applied Surface Science* **257** 5132–5137
[5] Serres N, Hlawka F, Costil S, Langlade C and Machi F 2011 *Journal of Thermal Spray Technology* **20** 336-343
[6] Gonzalez R, Garcıa M A, Penuelas I, Cadenas M, Fernandez M del R, Battez A H and Felgueroso D 2007 *Wear* **263** 619–624
[7] Zhang X C, Xu B S, Tu S T, Xuan F Z, Zhang Y K, Wang H D and Wu Y X 2009 *Fatigue & Fracture of Engineering Materials & Structures* **32** 84–96
[8] Zhao W, Wang Y, Dong L, Wu K and Xue J 2005 *Surface & Coatings Technology* **190** 293–298
[9] Lee C H and Min K O 2000 *Surface and Coatings Technology* **132** 49-57
[10] Zeng Z, Kuroda S and Era H 2009 *Surface & Coatings Technology* **204** 69-77
[11] Xuan H, Wang Q, Bai S, Liu Z, Sun H and Yan P 2014 *Surface & Coatings Technology* **244** 203–209
[12] Guoqing C, Xuesong F, Yanhui W and Wenlong Z 2013 *Surface & Coatings Technology* **228** 276-282
[13] Serres N, Hlawka F, Costil S, Langlade C and Machi F 2010 *Surface & Coatings Technology* **205** 1039–1046
[14] Zhanga Z, Wangb Z and Liang B 2009 *Journal of materials processing technology* **209** 1368–1374
[15] Pogrebnjak A D, Danilionok M M, Uglov V V, Erdybaeva N K, Kirik G V, Dub S N, Rusakov V S, Shypylenko A P, Zukovski P V and Tuleushev Y Zh 2009 *Vacuum* **83** 235–239
[16] Li R, Li Z, Huanga J, Zhanga P and Zhua Y 2011 *Applied Surface Science* **257** 3554–3557
[17] Wulina S, Echigoya J, Beidi Z, Changsheng X and Kun C 2001 *Surface and Coatings Technology* **138** 291-295
[18] Pogrebnjak A D, Danilionok M M, Uglov V V, Erdybaeva N K, Kirik G V, Dub S N, Rusakov V S, Shypylenko A P, Zukovski P V and Tuleushev Y Zh 2009 *Vacuum* **83** 235–239
[19] Chesov Yu S, Zverev E A and Plokhov A V 2010 *Obrabotka metallov* **1** 14-8
[20] Veselov S V et al 2010 *Obrabotka metallov* **4** 35-7
[21] Lin C M 2012 *Journal of Thermal Spray Technology* **21** 873-881
[22] Zhu H B, Li H and Li Z X 2013 *Surface & Coatings technology* **235** 620-627
[23] Chesov Yu S *et al* 2014 *Obrabotka metallov* **65** 8-11
[24] Tokita M 1993 *Journal of the Society of Powder Technology Japan* **30** 790–804
[25] Raychenko A I 1987 Osnovy protsessa spekaniya poroshkov propuskaniem elektricheskogo toka (Moscow: Metallurgiya)
[26] Kwon H, Jung S-A, Suh C-Y and Kim W 2016 *Ceramics International* **42** 8750–8753
[27] Ding L, Xiang D P, Pan Y L, Zhang T M and Wu Z Y 2016 *Journal of Alloys and Compounds* **661** 136-140
[28] Shongwe M B, Diouf S, Durowoju M O, Olubambi P A, Ramakokovhu M M and Obadeleb B A 2016 *International Journal of Refractory Metals and Hard Materials* **55** 16–23
[29] Sadat T, Dirras G, Tingaud D, Ota M, Chauveau T, Faurie D, Vajpai S and Ameyama K 2016 *Materials & Design* **89** 1181–1190
[30] Li Y, Katsui H, Goto T 2015 *Ceramics International* **41** 14258–14262
[31] Shevtsova L I 2014 *Obrabotka metallov* **65** *21-27*
[32] Kornienko E E, Nikulina A A, Drobyaz E A, Plotnikova N V, Lapushkina E Yu and Kuz'min V I 2015 *Applied Mechanics and Materials* **788** 252-258.
[33] Baker H 1998 *ASM Handbook Vol. 3: Phase Diagrams* (ASM International)
[34] Kornienko E E, Smirnov A I, Kuz'min V I 2015 *Applied Mechanics and Materials* **698** 405-410

Structure and properties of Ti-C-B coatings produced by non-vacuum electron beam cladding

O G Lenivtseva, N S Belousova, E A Lozhkina, T A Zimoglyadova, V V Samoylenko and L V Chuchkova

Department of Material Science in Mechanical Engineering, Novosibirsk State Technical University, 20 Prospekt K.Marksa, Novosibirsk, 630073, Russia

E-mail: lenivtseva_olga@mail.ru

Abstract. Cp-Ti/TiB+TiC wear-resistance coatings produced by non-vacuum electron beam cladding of boron carbide and titanium powders are studied in the paper. The X-ray phase analysis of the composite coatings microstructure showed that titanium carbide and boride reinforcing particles are evolved during the process. The obtained data are in good agreement with results of optical and electron microscopy. Undissolved particles of the initial boron carbide powder are detected in the coatings. The microhardness test as well as wear resistance test of materials under conditions of loose abrasive particles are conducted. It is established that the precipitation of reinforcing particles improves the tribological properties of the composite coatings.

Keywords: Cp-titanium, Titanium carbide, Titanium boride, Electron beam cladding, Microstructure, Microhardness, Wear resistance.

1. Introduction

Titanium and titanium-based alloys are attractive materials for many industries due to its unique physical, mechanical and biological properties. However, titanium alloys are characterized by low tribological properties, high and unstable friction coefficient, heavy adhesive wear and tendency to seizure [1-3]. Using the technology of surface hardening can significantly improve the wear resistance of the titanium. One of the most effective methods is to form a composite coating on the surface of titanium alloys. Carbides, borides, silicides and intermetallic compounds are used as the reinforcing components in the coatings of this type. Production of titanium with distributed above-mentioned particles is possible by introducing these particles into the molten bath or in the synthesis process (in-situ) [3-10].

For the moment one of the most researched technologies for the production of such coatings is the laser treatment. Using the laser cladding techniques can significantly increase the hardness and wear resistance of the titanium alloys as well as thermal stability of the alloys [11, 12]. [13].

Disadvantages of laser processing technologies associated with a metal high reflectivity and relatively low lasers efficiency. Special coatings are often applied to increase the absorptive capacity of the materials.

An alternative method of laser cladding technology is electron beam processing of powder materials [8-10]. Contrary to the laser, electron beam is a volume source of energy that penetrates deeply into the processed material. In this research the technology of electron beam cladding of powder mixtures in the air is used to improve the wear resistance of Cp-titanium. This technology is characterized by high performance and flexible control of processing parameters. Boron carbide powders are used as cladding components. Under electron beam influence on material, boron carbide particles are dissolved in the molten bath. In primary crystallization from supersaturated solution of carbon and boron precipitation of titanium carbide and boride crystals occurs. Ti-TiB-TiC composites combine high strength of carbides and borides as well as toughness and fracture resistance of the titanium substrate.

2. Materials and methods

Cp-titanium plates were used as substrates for the coating formation. The substrates dimensions were 100 mm x 50 mm x 12 mm. Titanium (30 wt.%) and boron carbide (20 wt.%) powders were used as the cladding materials. To protect the material being treated from the oxidation CaF_2 (40 wt.%) and LiF (10 wt.%) fluxes were added to the powder mixture during cladding. All components of the powder mixture were mixed together and uniformly applied to the surface of the titanium substrate. The density of the powder was 0.2 g/cm^3. Electron beam cladding was performed in Budker Institute of Nuclear Physics SB RAS using an electron accelerator ELV-6 type. The scheme of the electron accelerator is presented in [14, 15]. The scheme of electron beam cladding is presented in [16, 17].
Table 1 summarizes the main parameters of electron beam treatment. During the experiment three types of samples were formed: Sample 1 was obtained using 27 mA electron beam current, Sample 2 - 28 mA and Sample 3 - 29 mA.

Table 1. The modes of electron beam treatment.

The energy of the electron beam	1.4 MeV
Scanning frequency	50 Hz
The amplitude of the scanning	25 mm
The speed of the sample transfer	25 mm/s
The distance from the nozzle to the substrate	90 mm
The diameter of the beam	10 mm

To investigate the structural characteristics of the obtained materials an optical microscope Carl Zeiss Axio Observer A1m was used. Calculation of the volume fraction of the reinforcing particles was performed using software Image J image analysis. Chemical etching was carried out with a Kroll's solution. To identify structural features of the cladded layers a scanning electron microscope Carl Zeis EVO 50 XVP, equipped with microanalyzer Oxford Instruments X-Act EDX, was used. The phase composition of the material was examined using diffractometer ARL X'TRA. X-ray exposure was performed in step mode using Cu Kα radiation.

A microhardness distribution in depth of the cladded layer was evaluated by five tracks, using the Vickers Woulpert Group 402 MVD hardness tester. The load on the diamond indenter was 0.98 N.

Tribological test of materials obtained by electron beam cladding was carried out in accordance with GOST 23.208-79 (analogue ASTM G65-04) under conditions of loosely fixed abrasive particles. The test configurations are given in [18]. River sand was used as an abrasive. The sand was loaded into a hopper and fed into the friction zone where an experimental sample was contacted with a rotating rubber roller. The load during the test was 44 N. The worn material mass loss was taken as the magnitude of sample wear.

3. Results and discussion

When implementing non-vacuum electron beam cladding on the surface of titanium substrates, layer with a thickness of 1.2 ... 1.4 mm is formed. The cross section of Sample 2 is shown in Figure 1 a. The cladded layers have a high quality. No pores and cracks were detected. However, undissolved particles of the initial boron carbide powder are found in the coatings. These particles aggregation are clearly identified in the lower coating region and are shown by arrows in Figure 1 a. The presence of individual particles of this type in the coating, obtained by laser treatment has been shown in [7]. High

IOP Conf. Series: Materials Science and Engineering **156** (2016) 012021 doi:10.1088/1757-899X/156/1/012021

heating and cooling rates of the material are the features of the laser and electron beam treatments. Thus the boron carbide particles were not completely dissolve. The reaction zone is forming during the diffusion of boron and carbon atoms were observed around the boron carbide particles (Figure 1 b, g). Hence, boron carbide particles have a good contact with the titanium matrix and do not fall out under friction.

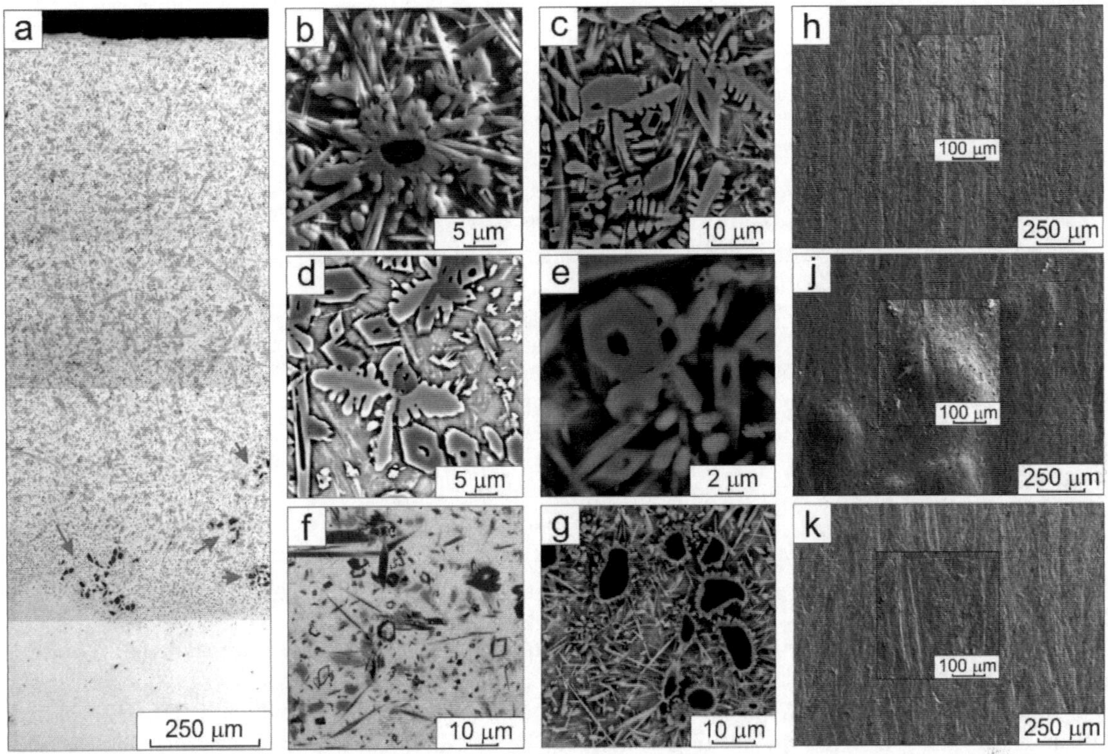

Figure 1. The structure of the Ti-B-C coatings, analyzed by optical (a) и scanning electron microscopy (b-g) and the surface topography after wear tests (h-j).

The diffraction pattern taken from the middle of the cladded layer of the Sample 2 is shown in Figure 2. The main phases, formed in the surface layer of titanium during the electron beam cladding are titanium carbide, titanium boride and alpha-titanium. No boron carbide phase was detected. This explained by preferential location of these particles close to the heat affected zone (HAZ) (Figure 1 a, g). The volume fraction of the particles in the cladded layer is less than 2 % and 3 % in the Samples 3 and 1, respectively.

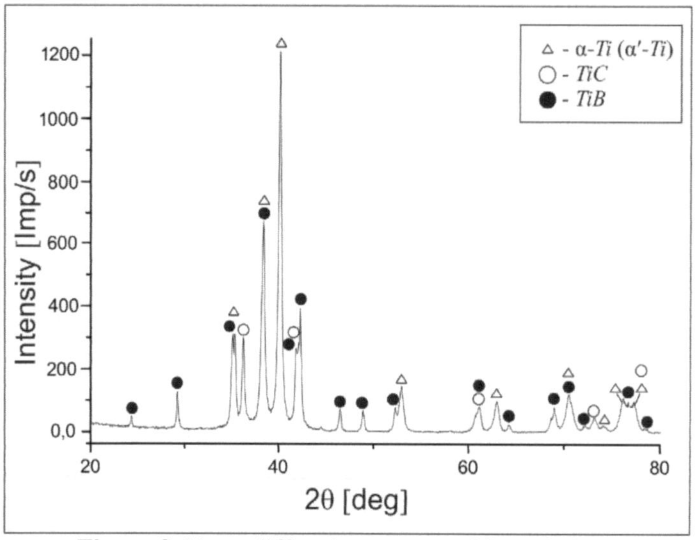

Figure 2. X-ray diffraction pattern of Sample 2

The volume fraction of the reinforcing phase in the cladded layers of the Samples 1, 2 and 3 has not been changed with increase of the beam current. This is due to the fact that in the process of cladding the base metal is exposed to a greater fusion penetration during the treatment at higher currents. increase of the power input leads to more boron carbide particles melting. Thus, the volume fraction of titanium boride and carbide crystals in the deposited layer is about ~ 32 %.

The microstructure of the samples is shown in Figure 1 b-g. There are a large number of initial particles of titanium boride precipitated in the shape of hollow hexagonal prisms (Figure 1 c-d). In [9] it is shown that under similar concentrations of elements in the Ti-B-C system the titanium boride crystals nucleate and grow firstly. These particles are characterized by the preferred crystal growth in one direction. When approaching the eutectic line, the formation of TiC-TiB eutectic takes place. Titanium carbide is evolved mainly in the shape of dendritic crystals. This is accompanied by the growth of the dendrites branches on the edges and inside the particles of titanium boride. The eutectic titanium boride has a fine needle-like morphology, while the needles are hollow as well. Near the heat affected zone due to lack of boron atoms thin defective crystals of titanium boride are formed (Figure 1 e).

Studies have shown that the maximum value of microhardness (~ 5900 MPa) is recorded in a Sample 1, obtained by cladding of the powder mixture at a beam current of 27 mA. Increase of the beam current leads to decrease of the average microhardness of the cladded layers. In the HAZ in all experimental samples a decrease in microhardness to 1600-1800 MPa is observed. The obtained data have a good correlation to tribological tests. Sample 1 has the lowest wear rate under conditions of loosely fixed abrasive particles (0.65 mg/min). When beam current is raised, a slight increase in the rate of wear is observed (0.74 and 0.66 mg/min for the Samples 2 and 3 respectively). The wear rate of commercially pure titanium is 2.79 mg/min. Figure 1 h - k shows SEM micrographs of the worn-out surface of the samples. On the surface of cp-titanium there is a typical adhesive wear surface characterized by deep grooves in the direction of wear (Figure 1 h). On the surface of the Samples 1 and 2 there are areas of selective material wear. Aggregations of undissolved particles of boron carbide are good barriers to the material intensive wear. According to the EDX analysis data, these particles have the following composition: 85.23 wt. % B, 12.28 wt. % C, 1.65 wt. % O, 0.11 wt. % Si and 0.73 wt. % Ti. In the Figure 1 j it can be seen that particles spalling is not observed in the process of wear. Wear of areas with no particles of boron carbide was more intensive. However, discontinuous grooves indicate on the presence of the reinforcing particles of titanium carbide and boride, preventing wear. Sample 3 has an almost flat wear surface (Figure 1 k). This is due to the absence of boron carbide particles aggregation in the middle zone and uniform wear of the coating.

4. Conclusions

By X-ray phase analysis and structural analysis of the coatings, produced by non-vacuum electron beam cladding, three phases were detected: titanium carbide, α-titanium and titanium boride. Highest wear resistance and microhardness were characterized for samples obtained by cladding with 27 mA beam current. The wear resistance of this sample is in ~ 4.4-times higher than that of commercially pure titanium.

Acknowledgement

The reported study was funded by RFBR, according to the research project No. 16-33-60066 mol_a_dk.

References

[1] Dong H 2010 *Tribological properties of titanium-based alloys* in *Surface Engineering of Light Alloys* (Woodhead Publishing)
[2] Leyens C and Peters M 2005 *Titanium and Titanium Alloys: Fundamentals and Applications* (Wiley-VCH Verlag GmbH & Co. KGaA)
[3] Baker T N 2010 *Laser surface modification of titanium alloys* in *Surface Engineering of Light Alloys* (Woodhead Publishing)
[4] Kondoh K 2015 T*itanium metal matrix composites by powder metallurgy (PM) routes* in *Titanium Powder Metallurgy: Science, Technology and Applications* (Elsevier Inc.)
[5] Banerjee R 2003 *Mater. Sci. Eng. A.* **358** 343-349
[6] Wang F 2007 *Mater. Sci. Eng., A.* **445–446** 461-466
[7] Zhang Y 2011 *Journal of Materials Processing Technology.* **211** 597-601
[8] Euh K 2001 *Scr. Mater.* **45** 1-6
[9] Lee C S 2003 *Metall. Mater. Trans. A.* **34** 1461-1471
[10] Oh J C 1998 *Scr. Mater.* **39** 1389-1394
[11] Candel J J 2010 *Surf. Coat. Technol.* **204** 3161-3166
[12] Kim I Y 2011*Wear.* **271** 1962-1965
[13] Guo C 2011 *Wear.* **270** 492-498.
[14] Bataev I A 2012 *Surf. Coat. Technol.* **207** 245-253
[15] Bataev I A 2013 *Appl. Surf. Sci.* **284** 472-481
[16] Bataev I A 2016 *Mater. Charact.* **112** 60-67
[17] Lenivtseva O G 2014 *Applied Mechanics and Materials* **682** 14-20
[18] Lenivtseva O G 2015 *Appl. Surf. Sci.* **355** 320-326

IOP Conf. Series: Materials Science and Engineering **156** (2016) 012022 doi:10.1088/1757-899X/156/1/012022

Formation of high-carbon abrasion-resistant surface layers when high-energy heating by high-frequency currents

N V Plotnikova[1], V Yu Skeeba[1], N V Martyushev[2], R A Miller[1] and N S Rubtsova[1]
[1] Faculty of Mechanical Engineering and Technologies, Novosibirsk State Technical University, 20 Prospekt K. Marksa, Novosibirsk, 630073, Russia
[2] Department of Materials Science and Technology of Materials, Tomsk Polytechnic University, 30, Lenina Av., Tomsk, 634050, Russia

E-mail: n.plotnikova@corp.nstu.ru

Abstract. The paper shows the possibility of carburization of low-carbon steel surface layers using high-frequency currents. The mathematical modeling of carburization using high-energy heating by high-frequency currents (HEH HFC) has been carried out, the temperature fields formed during the given processing have been calculated, as well as the structural changes in the surface layers have been simulated. The features of the structure formation in the surface layers of low-carbon steel after carburizing via HEH HFC have been determined by optical and scanning microscopy, which is confirmed by the computational models. The rational mode of fusion via HEH HFC has also been determined (power density of the source $q_s = (1.5 \ldots 4.0) \cdot 10^8$ W m^{-2}, (the relative travel speed of parts $V_p = 5 \ldots 100$ mm / sec), with forming the compressive retained stresses in the surface layer ($\sigma_{RS} \approx -300 \ldots -400$ MPa).

1. Introduction

The problem of improving the abrasion resistance of the surface layers of metallic materials for constructional purposes is relevant. This is due to the fact that a large amount of machine parts and structural elements are out of order due to their abrasion [1 - 10]. One of the most cost-effective methods for solving the problem is chemical and thermal processing of steel. The most common method today is the cementation process followed by quenching and tempering. The main disadvantage of this method is high labor intensity and duration of the process. The second problem is that in the standard processes of surface saturation of steel by reinforcing elements it is almost impossible to handle large-size items. An effective solution of the problem is the use of high-energy heating processing of the surface layer of the material for the purpose of alloying. In modern production such methods of surface hardening of materials as laser treatment [11, 12], and plasma arc welding [13 - 17], electron-beam treatment [18 - 22] are successfully used. These technologies can significantly accelerate the process of hardening of the surface layer. Such heat sources as laser beam, plasma and electron beam can be used for rapid heating or melting the surface in a short period of time during which the heat is not able to penetrate deep into the product, and therefore, will not change its properties.

There is little research devoted to surface carbon alloying of low-carbon steels when exposed to high-frequency currents [23, 24]. This is probably due to the fact that during this type of processing it is very difficult to secure the saturant (carbon) on the surface of the part. During HFC processing, when exposed to a strong magnetic field, the carbon powder "is blown" off the sample surface. To find the solution of this problem is also an urgent task.

The aim of this work was to develop a surface carbon alloying technology of low carbon steel using high-energy heating by high-frequency currents.

2. Materials and methods

2.1. Materials and methods of field experiments

The samples used for the surface hardening were the bars with the dimensions 10 × 10 × 100 mm made of steel 20 (0.19% C, 0.47% Mn, 0.20% Si, 0.009% P, 0.042% S, 0.15% Ni, 0.15% Cu), pre-treatment of which was carried out on the machining center DMC 635 and on the surface grinder 3G71. Graphite GL-1 (GOST 5279-74) was used as a saturant.

The bar surface was coated with graphite epoxy resin, after solidification of which the surface grinding was repeated. The saturant thickness was 0.1 ... 0.2 mm. The size control was carried on profilograph-profilometer Form Talysurf Series 2.

The samples with the hardened coating were processed on the contrivance, the main drive of which has stepless speed control in the range V_p = (5 ... 200) mm / sec. The energy source was the vacuum-tube generator HFG 6-60 / 0.44 with an operating frequency of the current 440 10^3 Hz. The heating process was carried out on the depth pattern (the thickness of the hardened layer did not exceed the penetration depth of the current into the hot metal - 0.6 ... 0.8 mm) in a continuous-sequential manner. When hardening, a loop-type inductor, equipped with a ferrite magnetic core N87, was used (Figure 1) [25 - 28]. HEH HFC of the experimental samples was carried out in the following modes: the power density of the source q_s = (1.5 … 4.0)·10^8 W m^{-2}, the travel speed of parts V_p = (5 ... 100) mm / sec. The active inductor wire width was R_s = 2 mm, the processing was carried out with a gap Δ = 0.1 ... 0.2 mm, followed by intensive water shower cooling of the surface (heat transfer coefficient α = 30·10^3 W / (m^2 ·°C) or by air cooling.

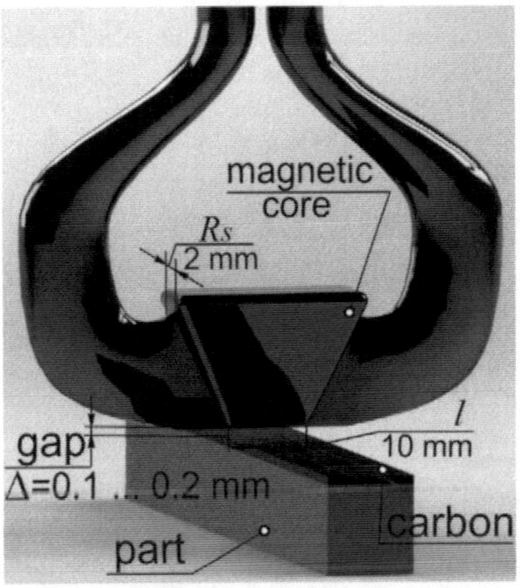

Figure 1. HEH HFC processing scheme

Structural studies of the samples were carried out using a light microscope Axio Observer A1m and scanning microscope EVO 50 XVP of the company «Carl Zeiss».

2.2. Mathematical modeling

Preparation of the finite element of the model was carried out in the program complex ANSYS. ANSYS Meshing Generator formed the hexahedral mesh using the following types of finite elements: Solid bodies - solids were simulated by the 8-node tetrahedrons SOLID 45; Surface bodies - surface bodies were simulated by the 4-node 4- carbon shell elements - SHELL 63; Line bodies - linear bodies were simulated by the 2-node linear elements LINK 8. The size of finite elements was 0.01 ... 1 mm. The total number of elements (Elements) was 108800.

The modeling of the HEH HFC process was carried out in SYSWELD system allowing by using the model of elastic and viscoplastic response of the material and the modern mathematical apparatus to calculate the temperature fields, distribution of the structural components, hardness, internal stress and strain [28 - 30].

The advantage of the mathematical apparatus of the heat conductivity theory can be taken only when describing the heat source at the point of action properly. [23]

The adequacy of the mathematical model was tested indirectly via the structural studies carried out beforehand and the determination of micro-hardness of the hardened layer.

3. Results and discussion

The calculations have shown that when HEH HFC heating we can achieve sufficiently high heating rates $V_{hr} = 5 \ldots 50 \cdot 10^3$ °C/sec and quenching rates (in the temperature range (700 ... 500) °C – $V_{qr\_700\text{-}500} = 3 \ldots 33 \cdot 10^3$ °C/sec, in the temperature range (400 ... 150) °C – $V_{qr\_400\text{-}150} = 200 \ldots 4100$ °C/sec). When the maximum surface temperature is 1700 ... 1750 °C, the formation of high-carbon layer is realized simultaneously with the surface fusion.

Such dynamics of the distribution of the thermal field in the cross section of the plate causes non-uniform structural and phase transformations in the material and occurrence of residual strains and stresses, the level of which on the surface may reach σ = -300 ... -400 MPa. When checking the adequacy of the mathematical model, the maximum error did not exceed 4 ... 8%.

Figure 2 shows the results of optical microscopy and modeling of structural and phase transformations for steel 20 with high-carbon surface layer formed after the air quenching. When HEH HFC processing the gradient material structure is formed: the structure of hypoeutectic cast iron is formed, and then the zone typical for hypereutectoid steel can be observed, which lengthens into the zone of the base metal (Figure 2). In this case the depth of the hardened layer was h = 0.23 mm, and the size of the transition zone was 0.04 mm. The microhardness of the surface layer was 6000 MPa, while the maximum compression residual stress on the surface σ = -340 MPa (calculated σ = -384 MPa). The feature of the transition zone in this case is the precipitation of excess cementite of widemanstatten type, indicating high heating temperatures and high quenching rate in that area (Figure 3a).

a b

Figure 2. Distribution of structural components of steel 20 with high-carbon surface layer: a) optical microscopy; b) the results of the simulation of structural and phase transformations.

Figure 4 shows the results of metallographic analysis and modeling of structural and phase transformations for steel 20 with high-carbon surface layer formed after water quenching. The formation of martensite is observed both in austenite forming a part of ledeburite fused layer (Figure 4a) and in the transition zone (Figure 3b). The formed structure has raised microhardness up to 7000-7500 MPa.

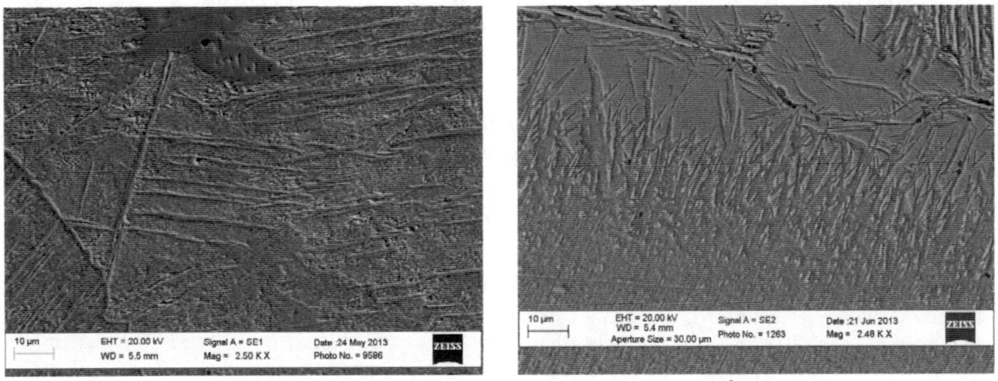

a b

Figure 3. The structural features of the transition zone: a) in air cooled samples; b) in water cooled samples

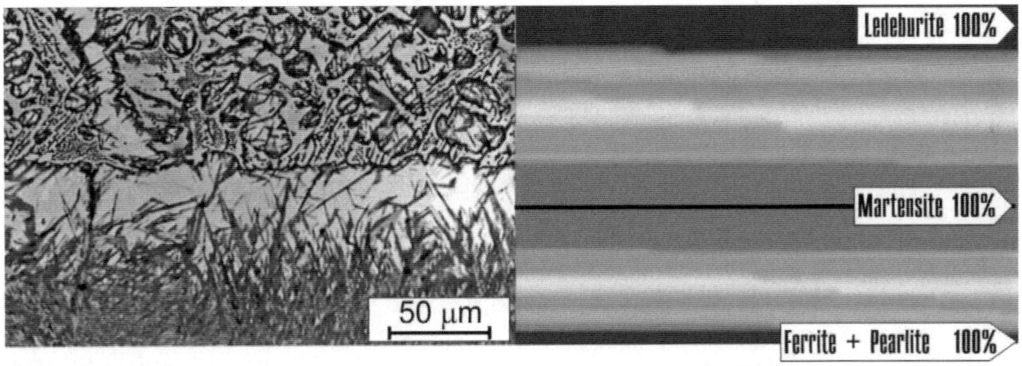

a b

Figure 4. Distribution of structural components of steel 20 with high-carbon surface layer near the transition zone: a) optical microscopy; b) the results of the simulation of structural and phase transformations.

4. Conclusions

1. The proposed technology of high-energy heating by high-frequency currents for producing high-carbon layers with increased hardness can compete with classic carburizing of the low-carbon steel. Due to instantaneous heating of the surface the carburizing layer is melted, mixed with the base and in the process of fast cooling the material forms a complex structure.

2. Via the optical and scanning microscopy it has been shown that in the process of HEH HFC heating followed by air quenching the hypoeutectic cast iron with high microhardness (6000-6500 MPa), and therefore high abrasion-resistance, is formed. If the processing finishes in water quenching, it will lead to the formation of the martensitic phase and, consequently, to an increase in the microhardness (7000-7500 MPa) and wear resistance.

3. By applying finite element method we have computationally solved the problem of modeling the stress and strain state of the material when surface high-energy hardening by high-frequency currents. It is shown that when high-energy heating by high-frequency currents the heating and quenching rates are, respectively: $V_{hr} = 5 \dots 50 \cdot 10^3$ °C/sec and $V_{qr\_700\text{-}500} = 3 \dots 33 \cdot 10^3$ °C/sec (in the temperature range 700 ... 500 °C). The level of residual compressive stresses on the surface of the sample is σ_{max} -300… -400 MPa. The developed mathematical model is recommended for optimization of technological

modes of processing, while ensuring large-sized hardened layers, comparable to the thickness of carburized layers produced by methods of classical carburizing technology (1-1.5 mm).

References

[1] Vodopyanov E M *et al* 2015 *Applied Mechanics and Materials* **756** 299-303
[2] Akarachkin S V *et al* 2015 *Applied Mechanics and Materials* **756** 286-293
[3] Skeeba V Yu *et al* 2016 *IOP Conf. Ser.: Mater. Sci. Eng.* **126** 012016
[4] Burov V G, Yanpolskiy V V and Rakhimyanov K Kh 2016 *IOP Conf. Ser.: Mater. Sci. Eng.* **126** 012016
[5] Skeeba V Yu 2015 *Actual Problems and Decisions in Machine Building* (Pfaffikon: Trans Tech Publ.)
[6] Tyurin A G *et al* 2008 *Obrabotka metallov* **1** 25-7
[7] Kornienko E *et al* 2015 *Applied Mechanics and Materials* **788** 252-8
[8] Skeeba V Yu *et al* 2016 *IOP Conf. Ser.: Mater. Sci. Eng.* **125** 012031
[9] Plotnikova N 2007 *Obrabotka metallov* **1** 35-6
[10] Skeeba V Yu, Pushnin V N, Kornev D Yu and Parts K A 2015 *Obrabotka metallov* **68** 31-41
[11] Abboud J H *et al* 2007 *Journal of Materials Processing Technology* **182** 427–31
[12] Das D K 1997 *Materials characterization* **38** 135-41
[13] Fadeev S N *et al* 2000 *Radiation Physics and Chemistry* **57** 653-5
[14] Bataev I A *et al* 2012 *Surface and Coatings Technology* **207** 245-53
[15] Skeeba V, Pushnin V and Kornev D 2015 *Applied Mechanics and Materials* **788** 88-94
[16] Hasan M, Wang J and Berndt C 2013 *Material and Manufacturing process* **28** 1279 – 85
[17] Chesov Yu S *et al* 2014 *Obrabotka metallov* **65** 11-8
[18] Lenivtseva O G *et al* 2015 *Applied Surface Science* **355** 320-6
[19] Ivancivsky V V, Bataev V A and Skeeba V Yu 2015 *Heat sources in the technological processes of machine parts manufacturing* (Novosibirsk: NSTU Publ.)
[20] Golkovsky M G *et al* 2013 *Obrabotka metallov* **61** 43-8
[21] Ivancivsky V V 2004 *News of the Tula State University* **5** 96-107
[22] Losinskaya A A *et al* 2013 *Obrabotka metallov* **61** 5-11
[23] Boonruang C *et al* 2012 *Chiang Mai J. Sci.* **39(2)** 254-62
[24] Plotnikova N *et al* 2015 *Applied Mechanics and Materials* **698** 351-4
[25] Ivancivsky V V, Skeeba V Yu and Pushnin V 2011 *Obrabotka metallov* **4** 19-25
[26] Ivancivsky V V, Skeeba V Yu and Stepanova N P 2005 *Obrabotka metallov* **3** 22-4
[27] Skeeba V Yu *et al* 2015 *Materials and Manufacturing Processes* **30** 1408-11
[28] Ivancivsky V, Parts K and Popov V 2015 *Applied Mechanics and Materials* **788** 129-35
[29] Skeeba V, Ivancivsky V and Pushnin V 2015 *Applied Mechanics and Materials* **698** 288-93
[30] Skeeba V Yu *et al* 2014 *Obrabotka metallov* **64** 90-102

IOP Conf. Series: Materials Science and Engineering **156** (2016) 012023 doi:10.1088/1757-899X/156/1/012023

Corrosion resistance of Ti-Ta-Zr coatings in the Boiling Acid Solutions

I A Polyakov[1], O G Lenivtseva[1], V V Samoylenko[1], M G Colkovski[2] and I S Ivanchik[2]

[1] Material Science in Mechanical Engineering Department, Novosibirsk State Technical University, 20 Prospekt K. Marksa, Novosibirsk, 630073, Russia

[2] Budker Institute of Nuclear Physics, Siberian Branch, Russian Academy of Sciences, 11 Prospekt Lavrentyev, Novosibirsk, 630090, Russia

E-mail: lenivtseva_olga@mail.ru

Abstract. In this study corrosion resistance of Ti-Ta-Zr coatings fabricated on VT14 titanium alloy workpieces using a high-energy electron beam injected in the atmosphere was investigated. Estimation of corrosion resistance of surface alloyed layers was carried out by the weight-change method. Boiling solution of 65 % nitric acid in water and 5 % of sulfuric acid in water were used as the corrosive environments. Investigation of samples after corrosion tests was carried out using a Carl Zeiss EVO 50 XVP scanning electron microscope.

Keywords: Titanium alloy, Zirconium, Tantalum, Electron beam cladding, Corrosion resistance.

1. Introduction

Titanium and its alloys possess a number of unique properties such as high specific strength, biocompatibility, heat resistance and excellent corrosion resistance [1, 2]. However, production of the constructions for critical applications in nuclear and chemical industries requires a higher corrosion resistance. Thus, tantalum-, zirconium- and niobium-based alloys are often used as structural materials in these areas. Meanwhile, the literature data give evidence of increase of titanium corrosion resistance by alloying it with tantalum and zirconium [3-5]. Taking into account a high price of aforementioned materials, it is more appropriate to fabricate the layers alloyed with tantalum and zirconium on the surfaces of titanium workpieces. Non-vacuum electron beam cladding can be considered as the efficient technology for fabrication of such materials [6-9].

In papers [6, 8-10] result of investigations of surface alloyed layers obtained on commercially pure (CP) titanium substrates are presented. Ultimate tensile strength of CP-titanium is about 300 MPa. Increase of strength of surface alloyed materials can be achieved by substitution of CP-titanium substrate by VT14 titanium alloy. Ultimate tensile strength of this material is 3-fold higher than that of CP-Titanium (VT1-0 grade) and equal to 980 MPa.

2. Materials and methods

VT14 titanium plates with dimensions of 12x50x100 mm were used as a base material in this study. A powder mixture used for cladding consisted of alloying elements and protective fluxes (Table 1).

Fluxes kept a molten bath from saturation with atmosphere gases and contributed in uniform distribution of alloying elements in the depth of cladded layers.

Table 1. Compositions of the powder mixtures and percentage of alloying components in the coating determined by EDX analysis, wt. %.

Sample label	Percentage of the components in the powder mixtures					Percentage of alloying components in the coating			
	Ta	Zr	Ti	CaF_2	LiF	Ta	Zr	Al	Mo
3Ta-4Zr	8.2	8.2	40	32.7	10.9	3.3	3.9	4.5	3.2
9Ta-3Zr	17	8.5	35	29.62	9.88	8.6	3.3	3.8	3.3

Before cladding, a powder mixture was homogeneously distributed on a surface of a treated workpiece. A surface density of the alloying powder placed to the workpiece surface before cladding was 0.3 g/cm^2 on the average; the total surface density of a powder mixture with flux was 0.45 g/cm^2.

Cladding was carried out in Budker Institute of Nuclear Physics SB RAS (Novosibirsk) using an ELV-6 electron accelerator. A beam current was 24 mA; a sample movement respectively to the electron beam was 10 mm per s; a scanning amplitude was 25 mm; a scanning frequency was 50 Hz.

An INCA X-ACT (Oxford Instruments) energy dispersive X-ray (EDX) analyzer was used for determining concentrations of alloying components in the surface alloyed layers.

Samples for corrosion tests were cut out of the cladded layers; their dimensions were 1x15x20 mm. These plates were hung on fluoroplastic filaments and immersed in 65 % nitric acid in water and 5 % of sulfuric acid in water. Diluted solution of sulfuric acid was chosen due to the aggressive action of this acid to titanium alloys. The temperatures of nitric and sulfuric acids were about 125° C and 100 °C respectively. At equal time intervals samples were removed, washed, dried and weighed. After that the experiment was continued. Summarized testing duration in nitric and sulfuric acids was 120 h and 50 min respectively. During the experiment a corrosion rate of cladded layers and a reference material (VT14 titanium alloy) was determined. Surfaces of samples after corrosion tests were analyzed by a Carl Zeiss EVO 50 XVP scanning electron microscope.

3. Results and discussion

Results of corrosion tests are presented in Table 2 and in Figure 1. It was found that the VT14 titanium substrate possessed the maximum dissolution velocity in both acids. Electron beam cladding of Ta-Zr powder mixtures lead to significant increase of corrosion resistance (Table 1, Figure 1 and Figure 2). The lowest corrosion rate (0.296 mm per year in nitric acid and 19.3 mm per year in sulfuric acid) was typical for samples alloyed with 9 % Ta and 3 % Zr. Corrosion rate of the alloys doped with 3 % Ta and 4 % Zr was 0.392 mm per year and 49.8 mm per year in nitric and sulfuric acids respectively.

Table 2. Corrosion rate of VT14 titanium alloy and Ti-Ta-Zr alloyed layers.

Sample label	Corrosion rate in the boiling concentrated solution of nitric acid in water, mm per year	Corrosion rate in the boiling concentrated solution of sulfur acids, mm per year
VT14	0.4768±0.016	53.7±3.6
3Ta-4Zr	0.392±0.036	49.8±6.4
9Ta-3Zr	0.296±0.008	19.3±2.5

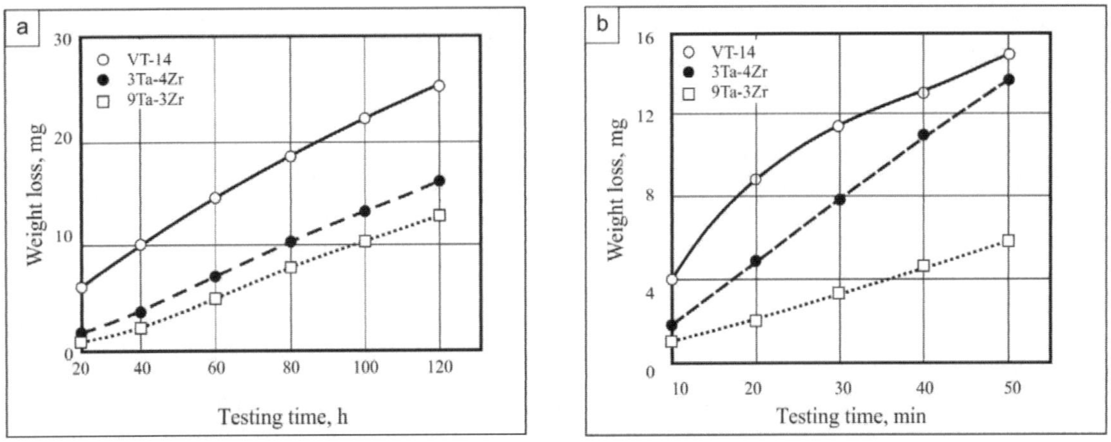

Figure 1. Relation between mass loss and test duration of samples in concentrated solutions of nitric (a) and sulfuric (b) acids.

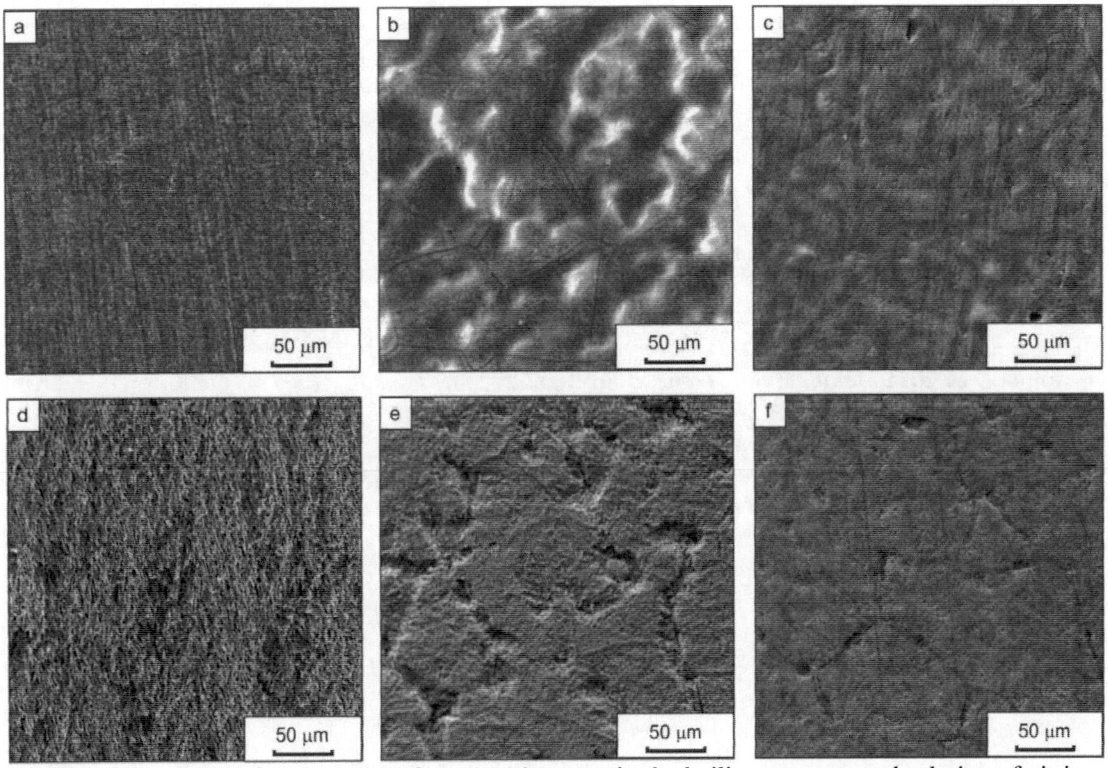

Figure 2. Surfaces of the samples after corrosion tests in the boiling concentrated solution of nitric acid (a-c) and sulfur acids (d-f): a, d – VT14 titanium alloy; b, e - 3Ta-4Z alloy; c, f –9Ta-3Zr alloy.

Surfaces of samples after corrosion tests in boiling solutions of acids were analyzed using a scanning electron microscope. Traces of preliminary mechanical treatment were observed on the surfaces of samples after testing in nitric acid (Figure 2, a-c). Damage of 3Ta-4Zr alloy exposed by boiling nitric acid occurred predominantly in the areas corresponding to the interdendritic space with formation of apparent relief which was not observed on the surfaces of other samples (Figure 2 a, c). The higher was a percentage of alloying components in the coatings the fewer interdendritic space was exposed to etching by a corrosive solution. Figure 2 c shows the contrast between dendritic axes and the interdendritic space.

In the boiling solution of 5 % sulfuric acid in water VT14 titanium alloy was demonstrated a tendency to general corrosion (Figure 2 d). Cladding of the Ti-Ta-Zr powder mixture led to transformation of the corrosion character. Thus, 3Ta-4Zr alloy fractured in the aggressive environment with formation of corrosion pits (Figure 2, e). It can be supposed that pits appeared in the area corresponded to the interdendritic space due to the lower percentage of alloying elements in them. Increase of Ta content in the cladded layer up to 9 % led to decrease of number of pits and their depth (Figure 2, f). There were no any corrosive pits on the surface of this alloy comparing to 3Ta-4Zr (Figure 2 e, f).

4. Conclusions

Coatings fabricated on the titanium substrate by non-vacuum electron beam cladding possessed a high corrosion resistance in the boiling solutions of nitric and sulfuric acids. Alloying titanium with 9 % Ta and 3 % Zr led to 1.6-fold and 2.78-fold increase of corrosion resistance in solutions of nitric and sulfuric acids respectively comparing to VT14 titanium alloy.

Acknowledgements

This work was carried out on the unique Experimental facility based on the ELV-6 industrial electron accelerator for the treatment of materials by a high-concentrated electron beam injected in the air atmosphere with financial support from the Russian Ministry of Education and Science. The unique ID of the project is RFMEFI60414X0135.

References

[1] Popov B N 2015 *Corrosion Engineering (*Amsterdam: Elsevier) 2015
[2] Leyens C and Peters M 2003 *Titanium and Titanium Alloys: Fundamentals and Applications* (Wiley-VCH Verlag GmbH & Co. KGaA)
[3] De Souza K A and Robin A 2003 *Materials Letters.* **57** 3010-3016
[4] Ionita D M 2011 *Metall. Mater. Trans. B.* **42** 1352-1357
[5] De Souza K A and Robin A 2007 *Mater. Chem. Phys.* **103** 351–360
[6] Golkovski M G 2013 *Metall. Mater. Trans. A.* **578** 310-317
[7] Golkovski M G 2011 *Advanced Manufacturing Technology.* **314-316** 23-27
[8] Ruktuev A A 2014 *Applied Mechanics and Materials.* **682** 100-103
[9] Samoylenko V V 2014 *IOP Conference Series: Materials Science and Engineering* **66** 012026
[10] Lenivtseva O G 2015 *Appl. Surf. Sci.* **355** 320-326

IOP Conf. Series: Materials Science and Engineering **156** (2016) 012024 doi:10.1088/1757-899X/156/1/012024

The study of the modes of Ta-Zr powder mixture non-vacuum electron-beam cladding on the surface of the cp-titanium plates

V V Samoylenko[1], E A Lozhkina[1], I A Polyakov[1], O G Lenivtseva[1], I S Ivanchik[2] and O E Matts[1]
[1]Novosibirsk State Technical University, 20 Prospekt K. Marksa, Novosibirsk, 630073, Russia
[2]Siberian State University of Water Transport, 33 Schetinkina st., Novosibirsk, 630099, Russia

E-mail: samoylenko.vitaliy@mail.ru

Abstract. The effect of the modes of non-vacuum electron-beam cladding of Ta-Zr powder mixtures on the structure and properties of the layers formed on the surface of cp-titanium were studied. The mode of the electron-beam alloying of titanium with zirconium and tantalum, which ensured the formation of a defect-free layer with a high content of alloying elements was selected. Metallographic examination indicated the presence of a dendritic- and plate-type structure of cladded layers. The microhardness of the layers, formed at the optimum mode, was not changed in the cross section and was equal to 450 HV.

1. Introduction

Modern development of industry puts high demands on the materials, working in particularly aggressive environments, such as boiling solutions of acids. These materials should provide high mechanical properties combined with high chemical resistance to corrosion. Stainless steel is one of the most common corrosion resistant materials. However, despite the favorable combination of mechanical properties the use of stainless steels is limited by the reduced level of corrosion resistance in boiling acids. Titanium alloys have higher corrosion resistance [1-3]. Titanium is almost not subjected to corrosion in oxidizing acid solutions over a wide range of concentrations and temperatures, as its mediums promote its passivation [3]. However, titanium strongly interacts with boiling solutions of reducing acids such as hydrochloric, phosphoric and sulfuric [1, 4, 5]. High resistance is observed only in dilute solutions [5].

One way to reduce the corrosion rate in boiling acids is alloying with the elements that contribute to passivation in these solutions (tantalum). The corrosion resistance of this alloy is increased due to the formation of a dense and stable passive film Ta_2O_5 [6]. It is indicated in the sources [6-8] that a significant effect is achieved upon alloying titanium with tantalum in an amount of 40% or higher. However, a significant drawback is the high density of tantalum, which is equal to 16.6 g/cm^3. Also this element is expensive as a consequence of its low prevalence in the crust. [7] The materials operation features should be also taken into account, since in some cases it's enough to protect only the surface of the item, which is subjected to aggressive action. In this regard, obtaining the workpieces sparingly alloyed with tantalum on the surface of less scarce material (titanium) is an important task. Such a solution can not only increase the corrosion resistance of the surface, provide savings of expensive element but significantly reduce contribution of the high density of tantalum into the total weight of the workpiece.

As the technology of production of tantalum-containing layers a method based on the partial melting of the alloying powder mixture by electron beams, injected in the air atmosphere is offered in the paper. This technology is implemented using an industrial electron accelerator ELV-6 at the Institute of Nuclear Physics SB RAS, which important advantage is the absence of a vacuum chamber. By doing so the remelted metal bath is protected from the atmosphere by means of the introduction of welding fluxes in the cladding mixture. It should be noted that in prior experiments powder mixture contained titanium powder in addition to tantalum and fluxes, which is added to wet the alloying component [9-11]. However, the melting point of tantalum and boiling temperature of titanium are close that may lead to evaporation of the wetting component and losses of alloying elements. To reduce the loss of tantalum it is rationally to replace the titanium powder to a refractory material with similar properties (zirconium). Experiments devoted to the cladding of Ta-Zr powder mixtures were conducted earlier [12, 13]. Nevertheless, optimization of the non-vacuum electron-beam treatment modes was not carried out.

In this regard, the aim of this study is to investigate the influence of the modes of cladding of the Ta-Zr powder mixture on the surface of titanium VT1-0 by electron beams, outputted in an air atmosphere.

2. Materials and methods

Cladding of the tantalum-zirconium layers was carried out on the plates made of commercially pure titanium VT1-0. Plates dimensions were 50x50x12 mm. Powder mixture for cladding consisted of the tantalum powder, the zirconium powder and the flux in an amount of 11.5 g of Ta, 4.7 g of Zr and 6.3 g of flux. Flux was the mixture of LiF and CaF_2 salts in the ratio 3:1 by weight.

The technological experiment in creation of the surface-alloyed layers consisted of several stages. At the preparatory stage, the alloying powder mixture was thoroughly mixed and applied to the greater face of a titanium plate. Then, the plate was placed on a table-manipulator so that the distance from the outlet window of the accelerator to the surface of the formed powder layer was 90 mm. The table progressively moved longitudinally with respect to the window with the rate of 10 mm/s. Thus the plate surface was scanned in the transverse direction by the electron beam with the peak-to-peak value of 50 mm. At the final stage, the mechanical removal of the slag crust from the surface of the cladded layer was performed. In order to study the effect of electron-beam cladding conditions on the structure and properties of the layers the beam current value varied in the range of 21 to 24 mA.

The structural features of cladded layers were studied by optical and scanning electron microscopy using Carl Ziess Axio Observer A1m and Carl Zeiss EVO 50 XVP microscopes respectively. Microsections for metallographic studies were prepared by the standard method. The samples were cut in the transverse direction of the surface-alloyed plates using a Struers Exotom-150 machine. Then cut samples were pressed into the polymeric material using a Buehler Simpli Met 1000 press. Further formed samples were ground and polished using a Buehler EcoMet 250/300 machine. The structure of the material was detected by chemical etching in a solution of 40% KOH, 30% H2O2 and H2O in the reactant ratio of 2:1:4 by volume. The solution was heated up to the temperature 70-80 °C. The microsections were immersed in the etchant for 60 seconds.

The chemical composition of the surface-alloyed layers was determined using an INCA X-ACT (Oxford Instruments) energy-dispersive analyser. To evaluate the level of mechanical properties microhardness tests were conducted using Wolpert Group Hardness 402 MVD. The measurements were carried out on polished thin sections. Indenter impressions were directed from the surface area of the cladded layer to the titanium base. The number of indenter impressions was 20. The distance between indenter impressions was 200 microns. The load on the diamond indenter corresponded to the value of 0.98 N.

3. Results and discussion

The structure of the layers cladded at different currents is shown in Figure 1. It can be seen that with an increase in electron beam current the thickness of the surface-alloyed layer increases. Thus, the electron-beam treatment of the tantalum-zirconium powder mixture at a current of 21 mA leads to the formation of a layer with the thickness of 1.1 mm (Figure 1 a). The thickness of the cladded material on the samples obtained at a current of 22 mA and 23 mA was 1.5 and 1.9 mm, respectively (Figure 1 b, c). Maximum thickness (2.1 mm) was fixed in the experiments for the layer formed at a current of 24 mA. (Figure 1 d).

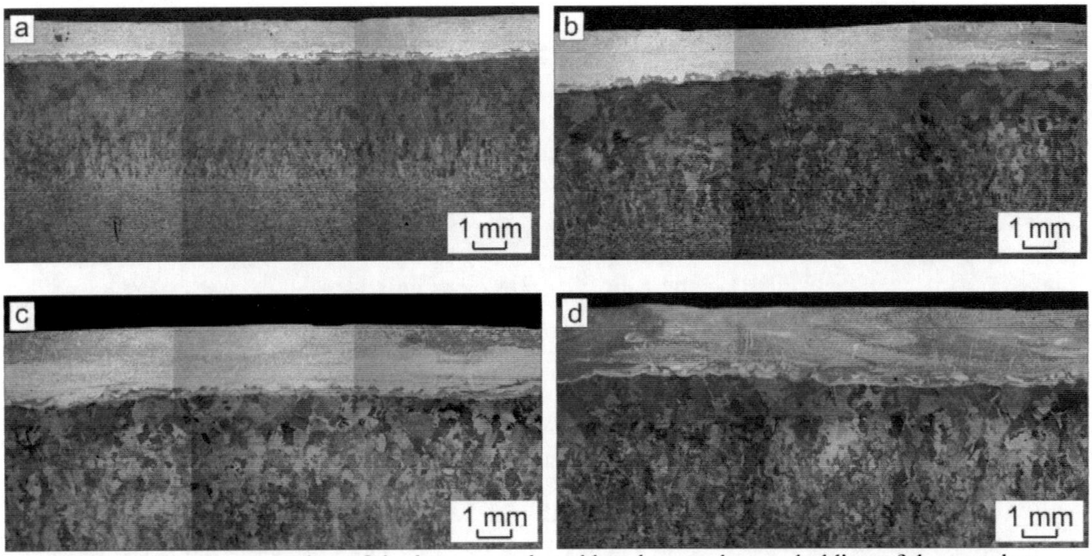

Figure 1. The cross section of the layers produced by electron-beam cladding of the tantalum-zirconium powder mixture at a current of: a - 21 mA, b - 22 mA, c - 23 mA, d - 24 mA

The structure of the layers formed at a current of 21 mA, has defects. The pores and the insoluble tantalum particles present in the material (Figure 2 a, b). Increasing the beam current by 1 mA helps eliminating pores in the cladded layer, but there is still a certain amount of insoluble particles of alloying components in cross section (Figure 2 c). Samples that were formed at a current of 23 mA and 24 mA are characterized by a homogeneous and defect-free structure (Figure 2 d, e). However, in the photomicrographs there is a significant contrast between the layers, obtained at the currents, which indirectly indicates the difference in the chemical composition. The sample that was cladded at a current of 23 mA is more alloyed with Ta and Zr due to less depth of the titanium fusion penetration compared to the sample formed at a current of 24 mA. This assumption was confirmed in [14] on the system of Ti-Ta. Thus, to obtain a defect-free surface layer with the highest concentration of alloying components electron beam treatment should be carried out at the beam current of 23 mA.

The structure of a layer formed at a current of 23 mA has a dendritic structure (Figure 2 d). Dendritic crystals grow preferentially in the direction from the surface layer to the melting zone of the base material that corresponds to the direction of the heat removal. At high magnifications thin-plate material structure can be seen (Figure 2 f).

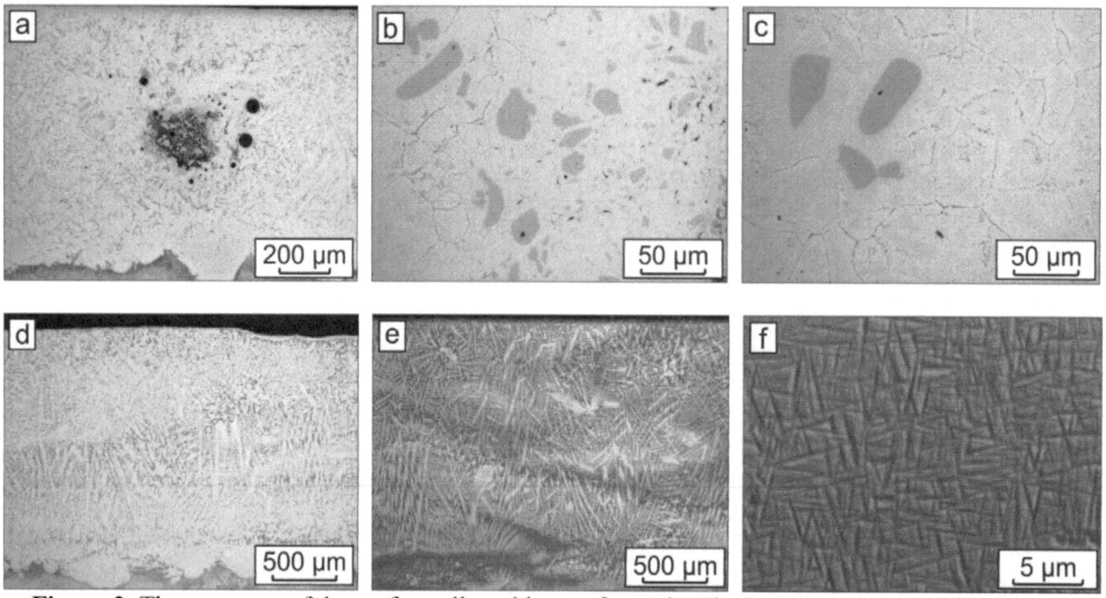

Figure 2. The structure of the surface-alloyed layers formed at the beam current of: a, b – 21 mA; c – 22 mA; d, f – 23 mA; e – 24 mA

The concentration of alloying elements in the layer formed on the optimal current amounts is 31.4 % (wt.) Ta and 11.74 % (wt.) Zr. Using EDX analysis it was established that there is a difference in chemical composition between the surface area and the central portion of the deposited layer (see Figure 3). The near-surface zone is less alloyed with tantalum and zirconium (spectrum 1), compared with the central part of the layer (spectrum 2). This is due to a large difference in the densities of alloying elements and the base material. Since during the electron beam treatment heavier components settle on the bottom of the molten metal bath, displacing the lighter.

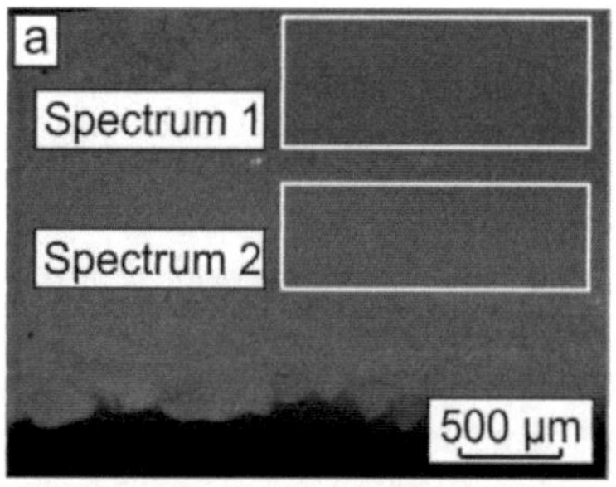

b

№	Content, wt %		
	Ta	Zr	Ti
Spectrum 1	28.5	10.9	60.6
Spectrum 2	32.5	12.6	54.9

Figure 3. EDX analysis of the Ti-Ta-Zr layer: a – the spectrum accumulation region; b – the chemical composition

To evaluate the mechanical properties, the microhardness tests were conducted. Distribution of the microhardness over the surface of the sample prepared at a current of 23 mA is shown in Figure 4. The average microhardness of the Ti-Ta-Zr layer does not vary throughout the thickness, and is about 450 HV. In the transition to the base metal is an abrupt drop in the level of microhardness up to 130 HV, which corresponds to the hardness of commercially pure titanium.

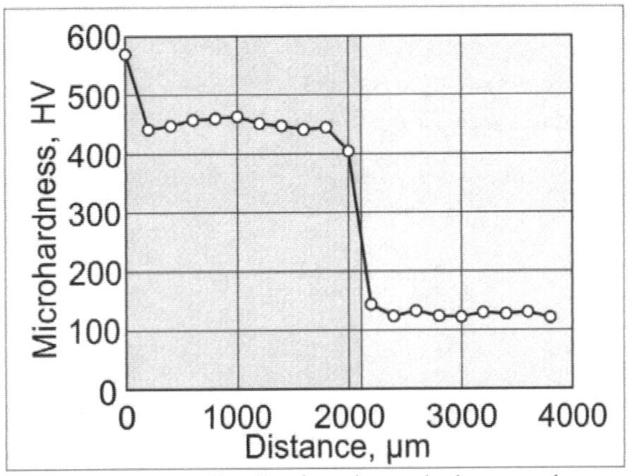

Figure 4. Microhardness distribution through the sample cross-section

4. Conclusions

The optimal mode of electron-beam cladding of the tantalum-zirconium powder mixture was selected by the work. The layers, formed at the optimal mode, had a homogeneous structure in which no defects such as pores and insoluble particles of alloying components were observed. It was found that the structure of cladded layers and the plate is presented a dendritic structure. The concentration of alloying elements in the surface layer according to the EDX analysis was 31.4 % (wt.) Ta and 11.74 % (wt.) Zr. The microhardness of the cladded material was 450 HV. The value of the microhardness did not change for the whole cross section of the formed layer, which indicated the homogeneity of mechanical properties in different parts.

Acknowledgments

The reported study was funded by RFBR according to the research project No. 16-33-00900 mol_a.

References

[1] Leyens C and Peters M 2003 *Titanium and Titanium Alloys Fundamentals and Applications* (Weinheim: WILEY-VCH Verlag GmbH & Co. KGaA) p 532

[2] Matthea J and Donachie Jr 2000 *Titanium: A Technical Guide* (ASM International, Materials Park, Ohio) p 381

[3] Shreir L L, Jarman R A and Burstein G T 1976 *Corrosion Metal/Environment Reactions* (London: Newnes-Butterworths)

[4] *Uhlig's Corrosion Handbook* 2011 ed Winston Revie R (Hoboken: John Wiley & Sons, Inc.) p 1288

[5] *Corrosion resistance of titanium. TIMET Corporation Brochure* 1997 (Denver: Titanium Metals Corporation)

[6] De Souza K A and Robin A 2007 *Mater. Chem. Phys.* **103** 351-60

[7] De Souza K A and Robin A 2003 *Mater. Lett.* **57** 3010-16

[8] Lyons L R 13 December 1960 Tantalum-titanium corrosion resistance alloy *U.S. Patent No* 2,964,399

[9] Golkovski M G, Bataev I A, Bataev A A, Ruktuev A A, Zhuravina T V, Kuksanov N K, Salimov R A and Bataev V A 2013 *Mater. Sci. Eng., A* **578** 310-17

[10] Ruktuev A, Golkovski M, Samoylenko V, Komarov P, Bataev I and Bataev A 2014 *Adv. Mater. Res.* **1040** 759-63

[11] Ruktuev A A, Samoylenko V V and Golkovski M G 2014 *Adv. Mater. Res.* **682** 100-103

[12] Samoylenko V V, Lazuenko D V, Lenivtsiva O G and Polyakov I A 2014 *IOP Conf. Series: Materials Science and Engineering* **66** 012026

[13] Samoylenko V V, Lazuenko D V, Lenivtsiva O G and Lozhkin V S 2015 *Appl. Mech. Mater.* **698** 424-29
[14] Samoylenko V V, Lenivtsiva O G, Polyakov I A, Laptev I S and Martyushev N V 2016 *IOP Conf. Series: Materials Science and Engineering* **124** 012117

IOP Conf. Series: Materials Science and Engineering **156** (2016) 012025 doi:10.1088/1757-899X/156/1/012025

The features of steel surface hardening with high energy heating by high frequency currents and shower cooling

V V Ivancivsky[1], V Yu Skeeba[1], I A Bataev[1], D V Lobanov[2], N V Martyushev[3], O V Sakha[1], I V Khlebova[1]

[1] Faculty of Mechanical Engineering and Technologies, Novosibirsk State Technical University, 20 Prospekt K. Marksa, Novosibirsk, 630073, Russia

[2] Bratsk State University, 40, Makarenko St., Irkutsk region, Bratsk, 665709, Russia

[3] Tomsk Polytechnic University, 30, Lenina Av., Tomsk, 634050, Russia

E-mail: skeeba_vadim@mail.ru

Abstract. The paper examines the process of surface hardening of steel 45 with the help of high energy heating by high frequency currents with simultaneous shower water cooling. We theoretically justified and experimentally proved a possibility of liquid phase forming in the course of heating not on the surface, but in the depth of the surface layer.

1. Introduction

Surface hardening of steel workpieces with concentrated energy sources is characterized by high rates of heating (tens of thousands of degrees a second) [1 - 7]. Under these conditions the heating of steel is carried out up to the melting temperature for the completion of the austenitizing process. While using the surface sources of heating (laser or plasma) the maximum values of temperatures are definitely observed on the surface of material proper [8, 9]. But for three-dimensional energy sources (electron beam, high frequency currents) this fact is not obvious. It is explained, first of all, by the physical nature of a three-dimensional source, i.e. by the law of energy distribution throughout the depth of a heated layer. A possibility of melted metal micro-volumes to form in the depth of material at heating with an electron beam in the atmosphere is shown in paper [10]. Ledeburite structure typical for the heat treatment of cast irons was registered in these areas during surface hardening of hypereutectoid steel.

At heating of steels in air medium by a concentrated electron beam the emitted energy distribution in material is similar to that of high energy heating by high frequency currents (HEH HFC) [11]. In this case during the surface hardening with HEH HFC one can also expect a possibility of liquid phase local volumes emergence in the depth of material.

The objective of this research is to determine the most heat-stressed layer during high energy heating of steel workpieces by high frequency currents with simultaneous shower cooling.

2. Materials and methods

HEH HFC allows implementing power density of around $3.0 \cdot 10^8$ W m^{-2} by a continuous method of heating. The processing scheme is presented in Figure 1 [12].

Typical features characteristic of HEH HFC consist in the following. For the maximum concentration of energy in the local material volume the heating is carried out at current frequency of 440000 Hz. To achieve this we used a flat loop inductor with minimum width of an active wire (b_{min} =1.2 mm) equipped by a ferrite magnet duct with high magnetic permeability. The treatment is carried

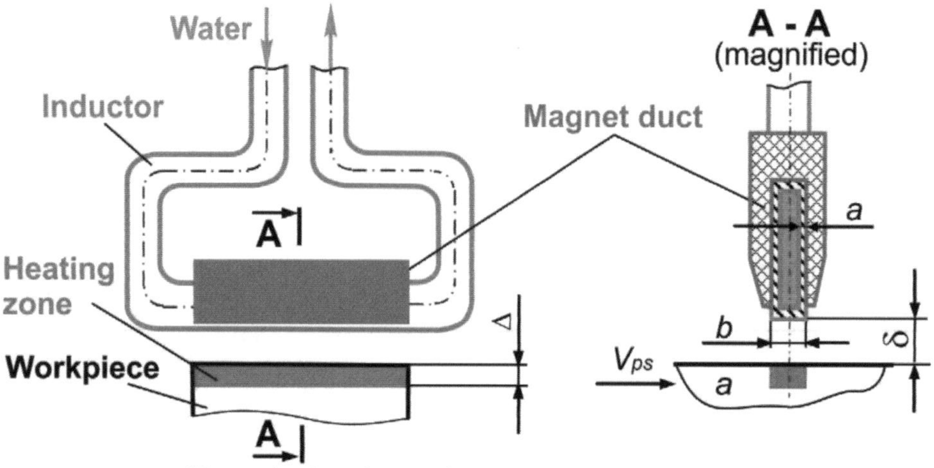

Figure 1. The scheme of processing with HEH HFC

out with minimum technological clearances of δ = 0.1... 0.5 mm. In order to prevent a possible break-down of an active wire and provide a reliable heat extraction the inductor wall thickness was chosen to be a = 0.12...0.15 mm [13, 14].

The source of energy emission at HFC heating is the eddy currents emerging in the material under the influence of alternating magnetic and electric fields. The value of specific heating power will be defined by the density of current J, whose change in the depth of metal is described by the dependence

$$J_Z / J_O = e^{-Z\cdot\sqrt{\frac{\pi\mu_0\mu f}{\rho_e}}}.$$

Here J_Z is the current density at the depth of Z; J_0 is current density on the surface; ρ_e is specific electric resistance; f is current frequency; μ_0 is the absolute magnetic permeability of vacuum, μ is a relative magnetic permeability of the material.

Specific electric resistance and magnetic permeability of steel change while heating, with specific resistance increasing up to the point of magnetic transformation, after which its growth slows. Magnetic permeability is hardly dependent on temperature up to approximately 650...700 °*C*, after which it rapidly decreases and approximately equals to the magnetic permeability of vacuum. Consequently, the energy distribution in the depth of the material is not constant.

The kinetic curves of workpiece surface heating by HFC have a bend in the interval of temperatures of 700...800 °*C*. The heating process is subdividing into the initial stage with great nearly constant speed of temperature increase and the stage of slower heating above temperature at which steel loses its magnetic properties. The redistribution of energy along section of a workpiece is the key reason of heating slowdown at the point of magnetic transformation. Indeed, there is always a temperature gradient during the heating on the section of a processed workpiece. The ρ_e and μ values depend on the temperature of the material. The spread of an electromagnetic process is thus occurring in an environment with ρ_e and μ variables. The paper [15] considers a case when material seemingly consists of two layers having different ρ_e and μ. If the first layer is heated above the temperature of 800 °C, and the second is not exposed to heating (20 °C), the distribution of eddy current would exactly correspond to a dependence presented in Figure 2. Therefore, under the condition that the top layer of material has lost ferromagnetic properties and the underlying layer is heated to temperature

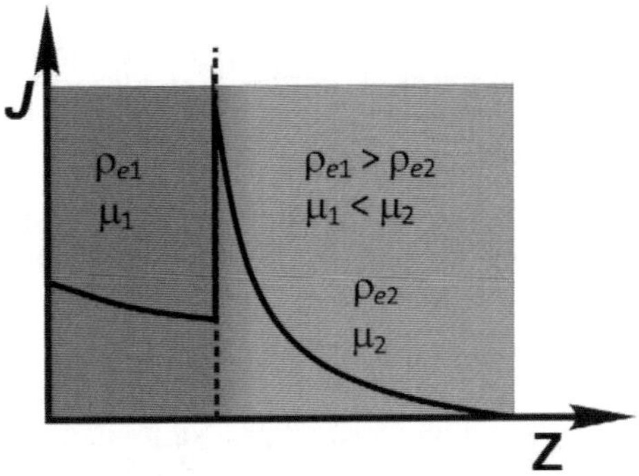

Figure 2. Current density distribution in the depth in a two-layer medium

not exceeding Curie temperature, there is a redistribution of current density. The maximum of energy emission shifts from the surface to the layer which did not lose ferromagnetic properties.

During modeling of temperature fields in steels under HEH HFC according to the algorithm presented in [11] we determined that at a certain combination of heating modes the temperature of the underlying layer could reach higher values than those on the surface (Figure 3). Thus temperature of the layer at the depth of Z = 0.2 mm from the surface reaches values of melting temperature of material while temperature of the surface layer is below this value.

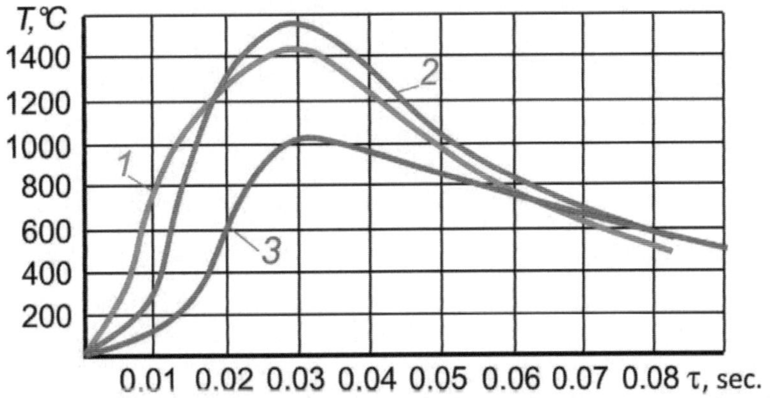

Figure 3. The heat cycles implemented on different depths of the surface layer with HEH HFC: the material is steel 45; specific heating power q = 200 MW/m^2, the part speed movement V_{ps} = 60 mm/s, b = 1.2 mm; *1* – Z = 0; *2* – Z = 0.2 mm; *3* – Z = 0.8 mm

It occurs due to the fact that at the heating of the surface layer to Curie T_k temperature, when steel loses ferromagnetic properties, a large part of energy is emitted to the underlying layer. Simultaneously, intensive heat removal takes place in the surface layer by cooling liquid being fed directly to the heating zone, while in the underlying layer heat removal is regulated by the material heat conductivity. Under these conditions the formation of liquid phase not on the surface, but in deeper layers of heated metal becomes possible.

We carried out the experiment in order to confirm the received modeling results of temperature fields with specific heating modes presented in Figure 3. While processing a cylindrical sample we registered the appearance of auto-oscillations of the inductor caused by periodic ejections of the melted metal to the surface (Figure 4).

Figure 4. The traces of meltdown on the hardened samples

3. Results and discussion

This phenomenon can be explained as follows. In these conditions the energy source is slowly moving, in other words, heat transfer speed is higher than the speed of a source motion. Therefore at the initial moment of heating no melting of metal is observed (Figure 3). At a certain moment of time the temperature at the depth of around 0.2 mm reaches the values of metal melting temperature (the curve 2, Figure 3). In this case a confined volume of melted metal forms in the near-surface layer in the zone located directly under the inductor. Thermal expansion of melt leads to the increase of pressure of this volume. The surface metal layer heated up to high temperatures (curve 1, Figure 3) turns plastic. It leads to dagger-shaped melting, in other words, melted metal is being thrown out of the underlying layer, leaving a crater (hole) on the surface of a sample, as well as beadings and drops of melted metal (Figure 5).

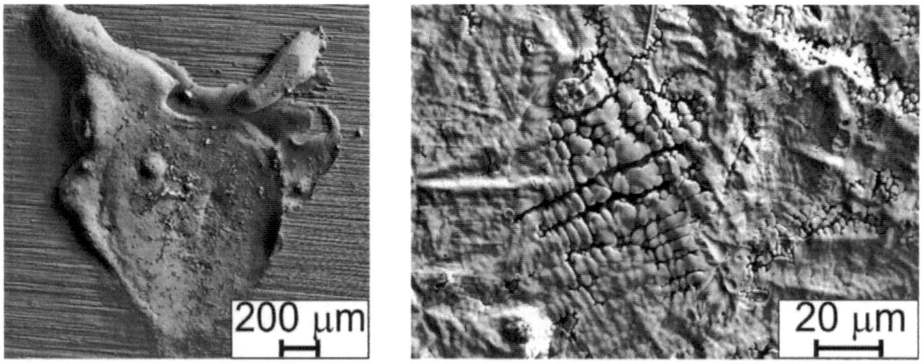

a b

Figure 5. The hole on the sample surface after the meltdown ejection (*a*) and a fragment inside the hole (*b*)

It is noteworthy that at certain combinations of surface hardening modes the melting of material can start on the surface layer. But, as can be seen in Figure 5, no traces of metal melting are observed up to the hole. The absence of melting traces after the hole is explained by the fact that there is a reduction of the clearance between the active inductor wire and processed surface in the course of an ejection of the melted metal, which leads to the increase of intensity of magnetic field between the inductor and the heated sample. This, in turn, increases a repellent specific mechanical effort between the inductor and the workpiece resulting in elastic deformation of conducting copper tubes and, as a consequence, the increase of the clearance and reduction of the specific heating density. After the passage of the melting zone, the value of the clearance is restored due to the inductor elasticity and the cycle of the workpiece surface heating repeats. The periodic formation of the metal meltdowns (Figure 4) is caused by this cyclic process. One important point is that there was no closure of the inductor active wire and the processed surface that could have led to the short circuit and melting of the copper in the inductor active wire and, as a result, its destruction. The X-ray microanalysis of the melted layer demonstrated no traces of copper in the meltdown (Figure 6), further supporting this idea.

As Figure 4 demonstrates, the ejection of the meltdown does not occur along the full width of the sample; instead it takes place in two locations. This is caused by the relatively labor consuming technology of the HEH HFC inductor production. The absolute linearity of the inductor active wire is difficult to provide. Its profile has linearity departures of about 0.02...0.04 mm that, undoubtedly, affect the level of specific heating density. The deepest meltdown occurs exactly in the two zones with minimum clearance between the inductor and the processed surface.

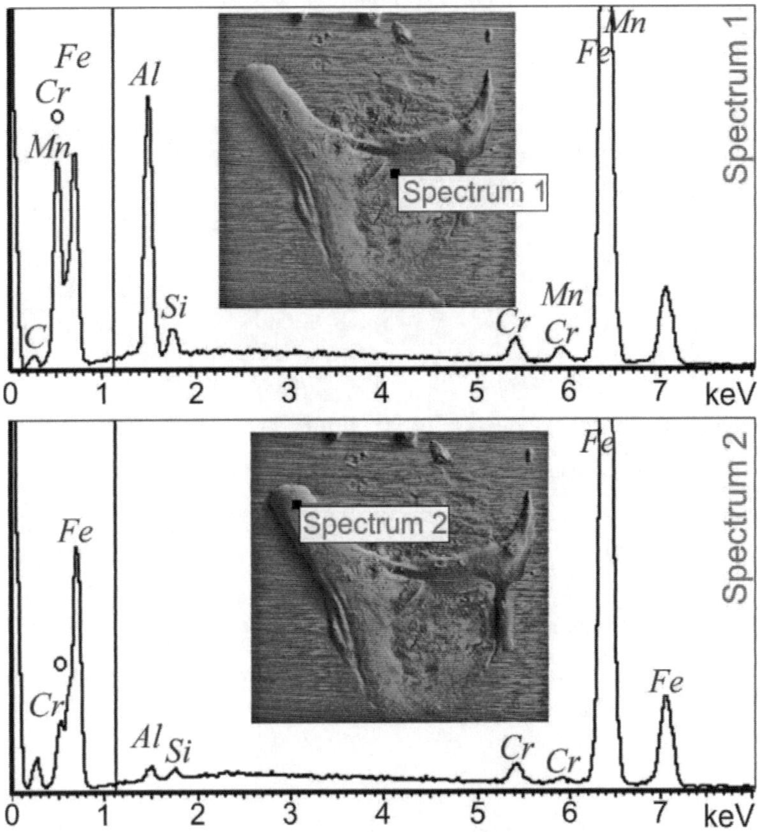

Figure 6. Diffraction patterns shot in two areas of local metal melting

Figure 7 demonstrates the surface fragments in the vicinity of the hole. The initial stage of metal melting is clearly visible here. In this zone metal melt formed at a smaller depth directly in the near-surface layer of the material. On the sample surface there are places of metal plastic deformation (swelling), as well as separate micro-craters of melted metal ejection.

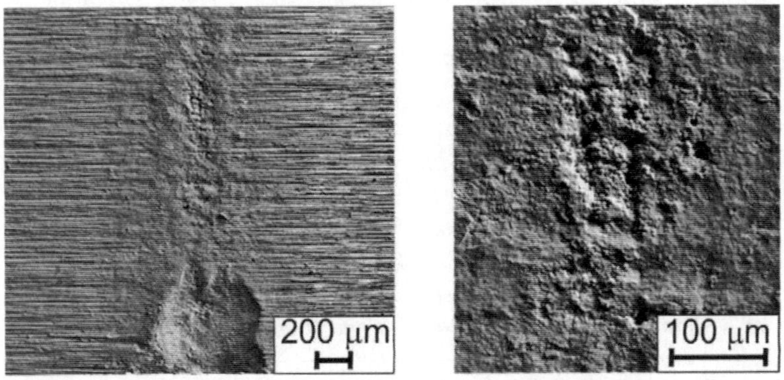

Figure 7. The fragments of the zone corresponding to the initial stage of metal melting

The images of different areas of the metal melting zone recorded on a raster *Carl Zeiss EVO50 XVP* electron microscope are presented in Figure 8. The presence of metal liquid phase, intensive cooling from the surface by feeding cooling liquid directly into the heating zone and intensive heat removal deep into the metal due to material conductivity lead to the conclusion about a possible receiving of an overcooled melt. The images demonstrate that dendrite structures form on the surface of the holes in the material. Metal crystallization occurred under the conditions of rapid cooling resulting in the emergence of significant magnitude gradients of stresses and, as a consequence, in the emergence of micro-cracks.

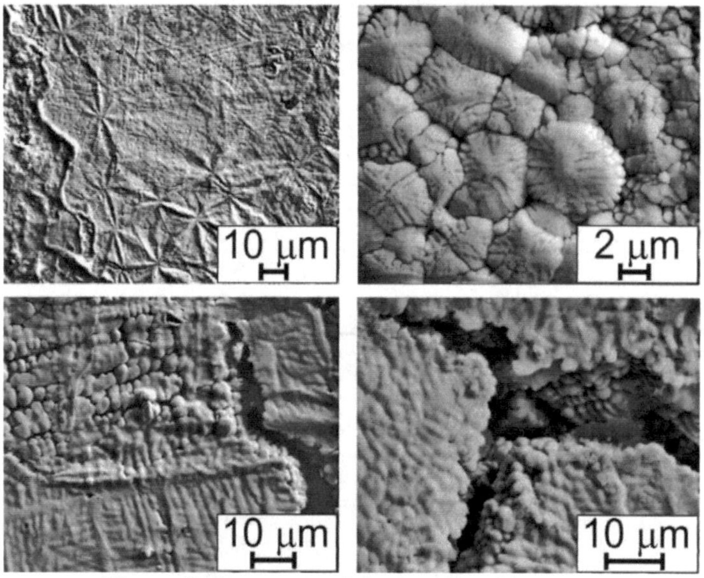

Figure 8. Fragments of the metal melting zone

4. Conclusions

During surface hardening of steels the presence of material melting is, as a rule, inadmissible. Thus, on the ground of the conducted research we came to the conclusion that the modes of surface hardening with HEH HFC and simultaneous shower cooling should be assigned on the basis of the most heat-stressed layer. For instance, for the steel 45 maximum temperature values are implemented at the depth of 0.15... 0.2 mm.

References

[1] Ion J C 2005 *Laser processing of engineering materials: Principles, procedure and industrial application* (Burlington: Elsevier Butterworth-Heinemann Publ.)
[2] Davis J R 2002 *Surface Hardening of Steels: Understanding the Basics* (Ohio: Materials Park, ASM International Publ.)
[3] Béjar M A and Henríquez R 2009 *Materials and Design* **5** 1726-28
[4] Skeeba V Yu *et al* 2014 *Obrabotka metallov* **64** 90-102
[5] Skeeba V Yu *et al* 2016 *IOP Conf. Ser.: Mater. Sci. Eng.* **126** 012016
[6] Skeeba V, Ivancivsky V and Pushnin V 2015 *Applied Mechanics and Materials* **698** 288-93
[7] Plotnikova N *et al* 2015 *Applied Mechanics and Materials* **698** 351-4
[8] Vodopyanov E M *et al* 2015 *Applied Mechanics and Materials* **756** 299-303
[9] Akarachkin S V *et al* 2015 *Applied Mechanics and Materials* **756** 286-293
[10] Bataev A A 2004 *Obrabotka metallov* **25** 18-9
[11] Ivancivsky V V 2004 Science Bulletin of Novosibirsk State Technical University **2** 161-72
[12] Ivantsivsky V V and Bataev V A 2005 *Polzunov Bulletin* **2-2** 104-12.
[13] Skeeba V *et al* 2015 Materials and Manufacturing Processes **30** 1408–11
[14] Skeeba V Yu *et al* 2016 *IOP Conf. Ser.: Mater. Sci. Eng.* **125** 012031
[15] Babat G I 1965 *Induction heating of metals and its industrial application* (Moscow: Energiya Publ.)

Comparison of mechanical and tribotechnical properties of UHMWPE reinforced with basalt fibers and particles

S V Panin[1,2], L A Kornienko[2], V O Alexenko[1,2], Huang Qitao[1] and L R Ivanova[2]

[1] National Research Tomsk Polytechnic University, 30 Lenina Avenue, Tomsk 634050 Russia
[2] Institute of Strength Physics and Materials Science of the Siberian Branch of the Russian Academy of Sciences, pr. Akademicheskii, 2/4, Tomsk, 634021, Russia

E-mail: vl.aleksenko@mail.ru

Abstract. Mechanical and tribotechnical properties of UHMWPE composites reinforced with basalt fibers and particles under dry sliding friction and abrasion were investigated. It is shown that adding of the basalt particles provides higher wear resistance under the dry sliding friction while at abrasion filling by the basalt fibers is more efficient since the wear resistance of the reinforced UHMWPE composites is by 3.7 times higher in contrast with the neat polymer. Wear mechanisms of the polymeric UHMWPE composites under various types of wear are discussed.

1. Introduction

Composites based on ultra-high molecular weight polyethylene (UHMWPE) are widely used in various industries. The application of composite materials with the UHMWPE matrix makes it possible to multiply increase the wear resistance of “metal-polymer” friction units. Recently, micro- and nanocomposites based on UHMWPE are actively developed [1-4].

At design of composite materials based on UHMWPE matrix their preferential operation conditions are to be taken into account: i) abrasion wear (e.g. liners of transporters or carriages); ii) dry sliding friction (gear wheels or gear transmission); iii) friction under boundary lubrication (polymer components of artificial joints).

During in-field exploitation the products made of UHMWPE matrix may experience all three above-mentioned types of wear. In this concern it is reasonable to use the fillers that can provide improvement of wear resistance to various types of wear, as well as to provide higher strength, lower friction coefficient, etc. In doing so, various particulate and fibrous materials are used to fill the polymers [5].

Note that composites filled with the basalt fibers are actively developed. This is related to their high mechanical properties, low cost and ecological compatibility, since the basalt is a natural mineral, whose main chemical components are SiO_2, Al_2O_3, CaO, MgO, Fe_2O_3 [6-9]. The melting temperature of the basalt makes 1500-1700 °C.

Regardless the fact that results on studying UHMWPE composite materials filled with the basalt fibers are reported in the literature the topic on the influence of filler size and shape are of particular importance in the sense structure formation as well as mechanical and tribotechnical properties.

This paper deals with the comparative analysis of basalt fibers and particles reinforced UHMWPE composites. The issues of structure formation as well as mechanical and tribotechnical properties under dry sliding friction and abrasive wear are studied.

2. Experimental

Ultra-high molecular weight polyethylene (UHMWPE) produced by Ticona (GUR-2122) with molecular weight 4.0 million and characteristic particle size of 5-15 μm, basalt particles with characteristic diameter of 7÷15 μm, basalt fiber with length of 260 μm and diameter of ~13 μm (aspect ratio 20:1) were used in the study. Specimens of polymeric composites were fabricated by hot isostatic pressing under specific pressure of 10 MPa and sintering temperature of 200 °C with subsequent cooling rate of 1.5°C/min.

Mechanical properties were determined with the help of the electromechanical testing machine Instron 5582 under uniaxial tension. Specimens were dog-bone shaped. The number of specimens of the each type was equal to 4. Friction coefficient for the specimens was measured by the "ball-on-disk" scheme with the help of tribometer "Tribotechnique" (France). Tests were carried out at the indenter sliding speed of 0.3 m/s (according to ASTM G99-95a and DIN 50324). The diameter of the indenter made of high strength steel was equal to 6 mm. The load makes 5 N.

Wear resistance under dry sliding friction was determined by "block-on-ring" scheme under loading of 68.8 N and disk rotation speed of 100 rev/min (according to ASTM G99). The SMT-1 wear testing machine was employed (sliding velocity – 0.32 m/s). Wear track surfaces of the specimens were examined by the optical profilometer New View 6200 (Zygo).

Abrasion testing was conducted with the help of MI-2 testing machine intended for the rubber abrasion (finger-on-disk). Abrasive wear resistance was evaluated under the load of 0.15 MPa. Sliding velocity of the steel disk regarding to the pair of polymeric specimens was equal to 17.0 m/min. Sand paper with fixed abrasive particles grade 240 P (grain size of 58.5 μm) was used. Volumetric abrasion was measured by weight loss for every 5 minutes. The test procedure meets the requirements of ASTM G99 and DIN 50324. Tribotechnical characteristics were evaluated by averaging the measuring data over four specimens of each type.

Permolecular structure investigations were carried out by the scanning electron microscope LEO EVO 50 at the accelerating voltage of 20 kV. Rupture surfaces were obtained by mechanical fracturing of notched specimens after preliminary exposure in liquid nitrogen.

3. Results and discussion

Tables 1 and 2 show the tribotechnical and mechanical properties of UHMWPE composites filled with the basalt fibers and particles. It is seen that with increasing basalt fillers weight fraction the density of the UHMWPE based polymeric material increases. If this takes place the Shore D hardness varies slightly as compared to neat UHMWPE. Tensile strength is reduced but slightly when adding more than 5 wt. % of the basalt fibers or particles. The value of elongation at failure for the composites "UHMWPE + n wt. % basalt filler" is also reduced. The friction coefficient *f* is increases from 0.120 to 0.152 when adding 20 wt. % of basalt fibers and up to 0.150 when adding 20 wt. %. of the basalt particles.

Table 1. Mechanical and tribotechnical characteristics of UHMWPE and composites "UHMWPE + n wt.% of basalt fibers"

Basalt fiber content, wt.%	Density ρ, g/cm^3	Shore hardness, *D*	Tensile strength σ_U, MPa	Elongation ε ,%	Friction coefficient ,*f*
0	0.934	57.7±0.6	32.3±1.7	485±24	0.120
5	0.960	58.75±0.7	33.7±1.9	415±27	0.134
10	0.990	55.34±0.7	31.5±1.9	409±59	0.126
20	1.054	60.1±0.6	29.1±1.9	345±55	0.152

Table 2. Mechanical and tribotechnical characteristics of UHMWPE and composites "UHMWPE + n wt.% of basalt particles"

Basalt particles content, wt.%	Density ρ, g/cm^3	Shore hardness, *D*	Tensile strength	Elongation ε ,%	Friction coefficient ,*f*

			σ_U, MPa		
0	0.934	57.7±0.6	32.3±1.7	485±24	0.120
5	0.962	55.6±0.9	32.5±2.2	439±40	0.112
10	0.991	56.4±0.8	30±1.7	410±36	0.136
20	1.063	56.9±0.9	31.5±2.8	401±26	0.150

In doing so, the strength properties of UHMWPE composites filled with the basalt fibers and particles are not substantially changed with increasing content of the fillers.

Tribotechnical properties of the UHMWPE composites filled with the basalt depend on the filler shape (fibers or particles). The diagrams of wear intensity of the wear track surface as well as its roughness for the UHMWPE reinforced with the basalt fibers and particles under dry sliding friction are shown at Figure 1 and 2 correspondingly. It can be seen that the wear resistance of the composites "UHMWPE + n wt. % of basalt fiber" is increased when adding 5 wt. % of the filler while further it sharply decreases. Similar trend is characteristic for the wear track surface roughness as the function of the fiber weight fraction.

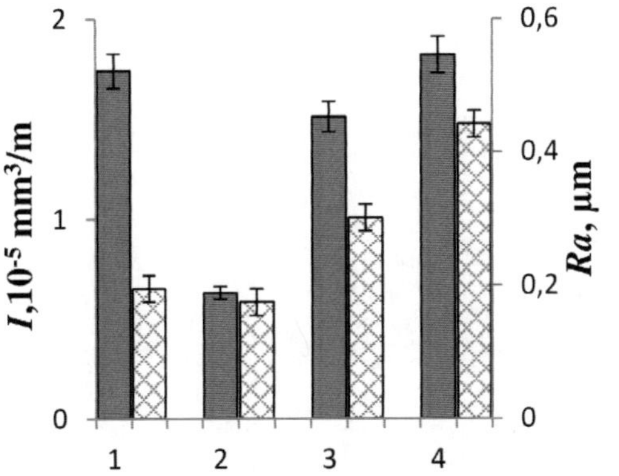

Figure 1. Wear rate (*I*) and surface roughness of the wear track (*Ra*) under dry sliding friction of UHMWPE composites filled with basalt fibers:(1) – UHMWPE, UHMWPE+ basalt fibers (2) – 5 wt. %, (3)– 10 wt. %, (4)– 20 wt. %

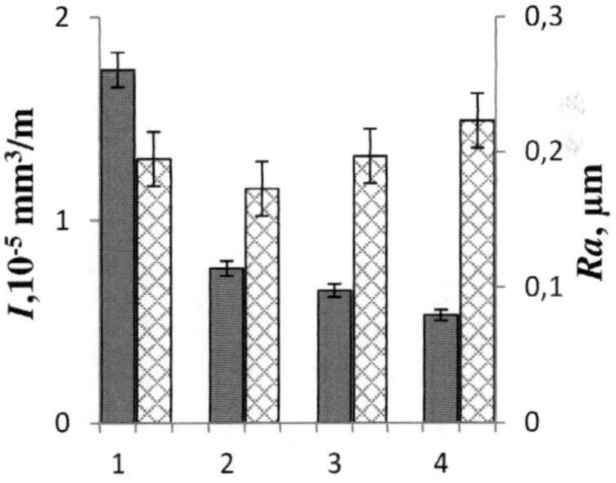

Figure 2. Wear rate (*I*) and wear track surface roughness (*Ra*) under dry sliding friction of UHMWPE composites filled with basalt particles: (1)– UHMWPE, UHMWPE+ basalt particles (2)– 5 wt. %, (3)– 10 wt. %, (4) – 20 wt. %

A quite different trend for the dependence of the wear rate versus the filler content is observed in the composites with basalt particles (Figure 2). It is seen that the wear resistance of the UHMWPE composites increases with enlarging particle weight fraction up to 20 wt. %. In doing so, the surface roughness of wear track surface in the composites is close to one in pure UHMWPE. Similar results were observed in our previous studies for UHMWPE composites filled with microparticles (aluminum oxide, aluminum oxyhydroxide, hydroxyapatite) [6].

The optical and SEM micrographs of the wear track surface morphology as well as permolecular structure of the composites with the basalt fibers and particles are shown in Figure 3 and Figure 4. It is seen that type and content of the filler affect them substantially. The obvious microscratches are formed on the friction surface. This pattern is made more manifested when the fiber content is enlarged. Most likely these scratches are resulted from destructive effects of the broken fibers that cut the surface of the softer polymeric matrix (Figure 3, c, d).

On the other hand, when adding the basalt particles into the UHMWPE matrix formation of spherulite permolecular structure is suppressed (Figure 3, c, d). Note that the basalt particles are uniformly distributed in the matrix. In doing so under the dry sliding friction they act as oscillating "strings" providing a sufficiently smooth wear surface of the composite with high level of wear resistance (Figure 4). Similar results were obtained in our previous studies of the UHMWPE composites filled with Al_2O_3 particles [6].

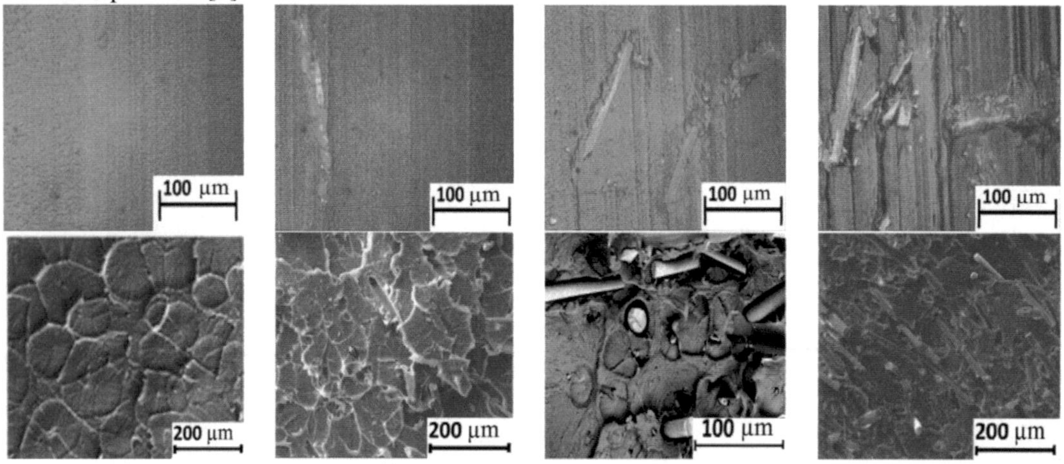

a) b) c) d)

Figure 3. Optical and SEM micrographs of wear track surfaces under dry sliding friction and permolecular structure: UHMWPE (a), UHMWPE + 5 wt. % of the basalt fiber (b), UHMWPE + 10 wt. % of the basalt fiber (c), UHMWPE + 20 wt. % of the basalt fiber (d)

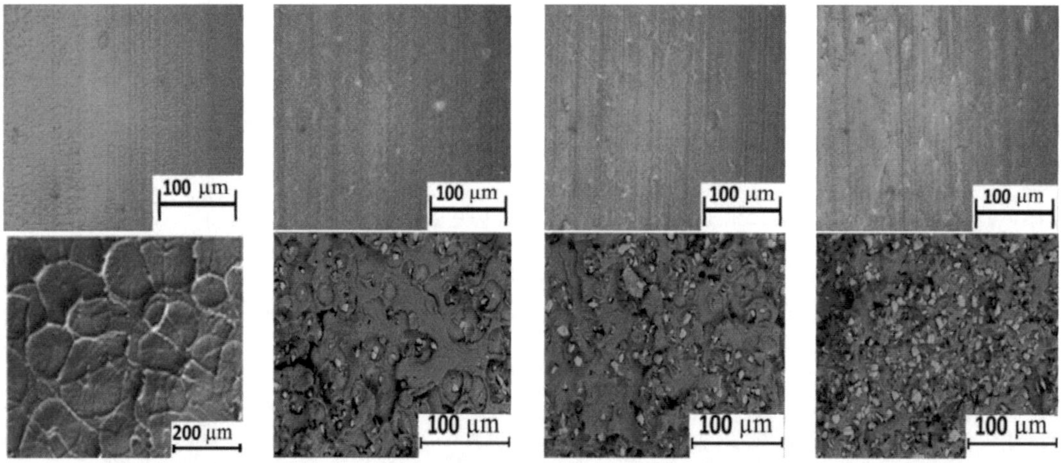

a) b) c) d)

Figure 4. Optical and SEM micrographs of wear track surfaces under dry sliding friction and permolecular structure: UHMWPE (a), UHMWPE + 5 wt. % of basalt particles (b), UHMWPE + 10 wt. % of basalt particles (c), UHMWPE + 20 wt. % of basalt particles (d)

At abrasive wear tests the role of the basalt fibers and particles in the formation of the tribotechnical properties of UHMWPE composites is reversed. Note, that in our previous studies it was shown that the size ratio of fixed abrasive particles and that of the filler plays a decisive role at abrasive wearing of polymeric composites [6,10].

The data on abrasive wear intensities of the UHMWPE composites filled with the basalt fibers and particles are shown at Figure 5 and 6. It is seen from figure 5 that increasing the basalt fiber content up to 10 wt. % results in increase of the abrasive wear resistance of the composites: at filling degree of 10 wt. % it rises by 2.8 times. With adding the basalt particles the opposite effect is evident: abrasive wear resistance is decreased: at filling degree of 10 and 20 wt. % it is nearly equal to that of the pure UHMWPE (Figure 6). Roughness of the wear track surfaces possesses the same trend.

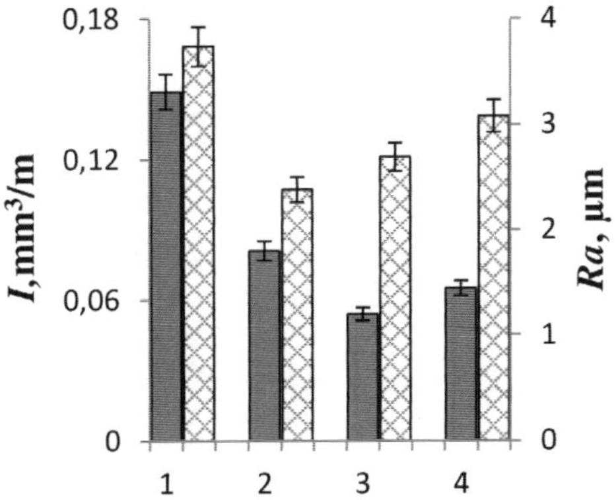

Figure 5. Wear rate (*I*) and surface roughness of wear track (*Ra*) under abrasive wear of UHMWPE composites with the basalt fibers: 1 – UHMWPE, UHMWPE + basalt fibers of 2) – 5 wt. %, 3) – 10 wt. %, 4) – 20 wt. %. P 240

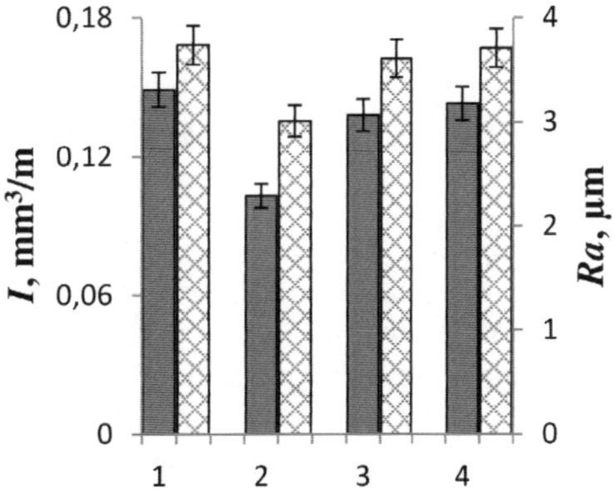

Figure 6. Wear rate (*I*) and surface roughness of wear track (*Ra*) under abrasive wear of UHMWPE composites with the basalt particles: 1 – UHMWPE, UHMWPE + basalt particles of 2) – 5 wt. %, 3) – 10 wt. %, 4) – 20 wt. %. P 240

Figure 7 shows the optical profilograms of the wear surface under abrasion of the composites with basalt fibers (a-d) and particles (e-h). Long basalt fibers resist the destructive action (cutting) of fixed abrasive particles (58.5 μm) and partly protect the matrix. At the same time the basalt particles has much smaller size (7 μm versus 58.5 μm). Therefore they are not able to protect the matrix from ploughing and cutting by less brittle abrasive particles of higher hardness.

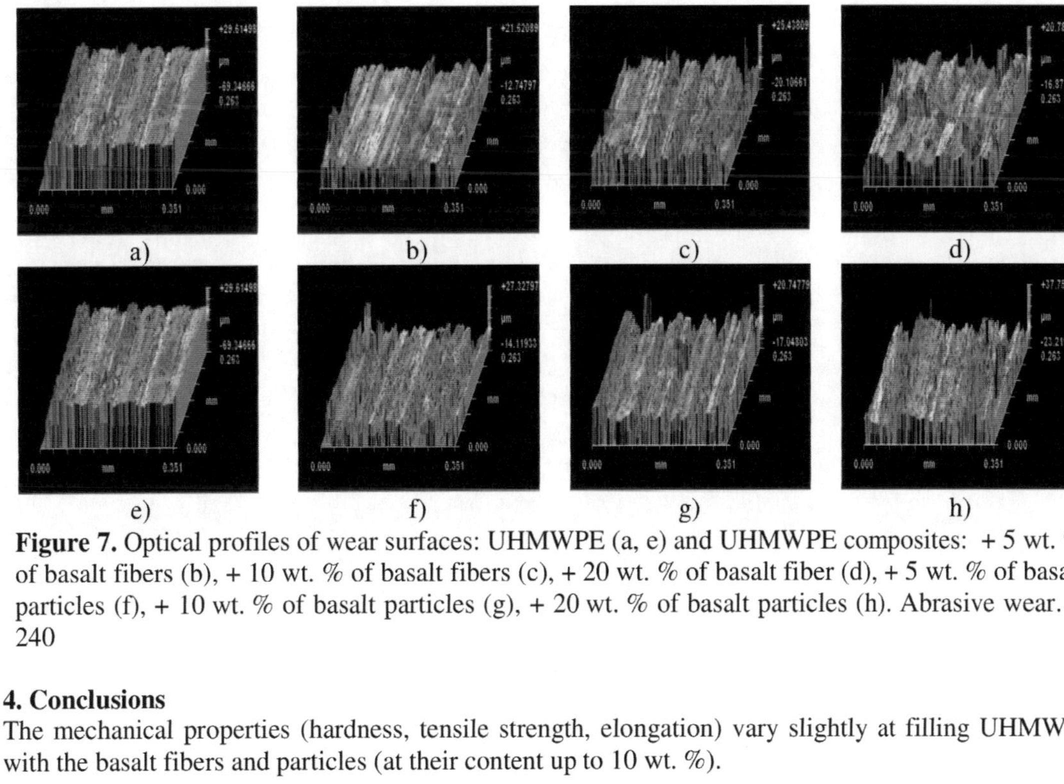

Figure 7. Optical profiles of wear surfaces: UHMWPE (a, e) and UHMWPE composites: + 5 wt. % of basalt fibers (b), + 10 wt. % of basalt fibers (c), + 20 wt. % of basalt fiber (d), + 5 wt. % of basalt particles (f), + 10 wt. % of basalt particles (g), + 20 wt. % of basalt particles (h). Abrasive wear. P 240

4. Conclusions

The mechanical properties (hardness, tensile strength, elongation) vary slightly at filling UHMWPE with the basalt fibers and particles (at their content up to 10 wt. %).

Formation of the spherulitic permolecular structure is suppressed when filling the UHMWPE matrix with basalt fibers and particles with weight fraction more than 5 wt. %.

The reinforcement of the UHMWPE-matrix by the basalt particles is effective under dry sliding friction: wear resistance of the composites increases 3-fold when 20 wt. % of the filler is added.

Adding basalt fibers into UHMWPE provides improvement of the abrasive wear resistance which is increased by 2.5 times when filler weight fraction is varied in the range of 10-20 wt. %.

References

[1] Harley L Stein 1999 *Engineered Materials Handbook* Vol.2: Engineering Plastics
[2] Galetz M C, Blar T, Ruckdaschel H, Sandler K W and Alstadt V 2007 *Journal of Applied Polymer Science* **104** 4173
[3] Jiansong Zhou and Fengyuan Yan 2005 *Journal of Applied Polymer Science* **96** 2336
[4] Wei Z, Ya-Pu Zhao, Ruan S L and Gao P 2006 *Surface and interface analysis* **38** 883
[5] Krasnov A P, Aderikha V N, Afonicheva O V, Mit V A, Tikhonov N N, Vasilkov A Yu, Said-Galiev A E, Naumkin A V and Nikolaev A Y 2010 *Friction and Wear* **31** **1** 93-108
[6] Ljukshin B A, Panin S V, Bochkareva S A etc. 2015 *Computer simulation and design of filled composites* (Novosibirsk, Izd-vo SB RAS) p 264
[7] Fiore V, Scalici T, Di Bella G and Valenza A 2015 *Composites Part B: Engineering* **74** 74
[8] Huaian Zhang, Yiming Yao, Deju Zhu, Barzin Mobasher and Liang Huang 2016 *Polymer Testing* **51** 29
[9] Gogoleva O V, Petrova P N, Popov S N and Okhlopkova A A 2015 *Journal of Friction and Wear* **36** **4** 301-303
[10] Panin S V, Kornienko L A, Nguyen Xuan T and Ivanova L R 2015 *Russian Physics Journal* **58** **6/2** 211-215

IOP Conf. Series: Materials Science and Engineering **156** (2016) 012027 doi:10.1088/1757-899X/156/1/012027

Wear Resistance of Friction Pair of Metal Composite/Copper under Electric Current

M I Aleutdinova[1,3], V V Fadin[1], V Ye Rubtsov[1,2] and K A Aleutdinov[2]

[1]Institute of Strength Physics and Materials Science, pr.Akademicheskii, 2/4, Tomsk, 634021, Russia

[2] National Research Tomsk Polytechnic University, Lenin avenue, 30, Tomsk, 634050, Russia

[3]Seversk Technological Institute - branch of State Autonomous Educational Institution of Higher Professional Education «National Research Nuclear University «MEPhI», pr.Kommunisticheskii, 65, Seversk, 636036, Russia

E-mail: aleut@ispms.ru

Abstract. Sliding of metal composites against copper counterbody under severe conditions (i.e. at the contact current density higher 50 A/cm^2 and at high roughness of counterbody) is carried out. It is shown that the composite of composition of Cu-30% of graphite shows low wear resistance in these conditions. Higher wear resistance is inherent in the composites containing lead and bearing steel. Impregnation of these composites by industrial oil hasn't led to noticeable increase in wear resistance.

1. Introduction

It is known that sliding with current collection is carried out against copper counterbody as a rule. Satisfactory wear resistance is reached in this case due to application of metal composites, low sliding surface roughness (R_a<0.8 μm), low contact current density (j<50 A/cm^2), low pressure (0,02 MPa), boundary lubrication, i.e. in the conditions of absence of high mechanical stresses in a surface layer [1-3]. Increase in these parameters has to cause decrease in wear resistance. The possibility of obtaining satisfactory contact characteristics under severe sliding is both of scientific and practical interest. One can assume that increase in wear resistance will increase the contact pressure. Wear resistance can also increase at change of metal initial structure. Studying such structures is always topical. Now composites of structures of Cu-30% Gr and 43% Cu-42% Fe-10% Pb-2% Zn-Gr (where Gr is graphite) are widely applied. These composites have structure of the pressed powder but show good characteristics at sliding with current density up to 50 A/cm^2.

The current collection materials containing a large amount of copper have rather high cost. Therefore, it is of interest to study a possibility of application of steel based composites with low copper content. One can suppose that the steel processed from grinding wastes of ball bearing production can serve as a basis of such composite. Determination of operability of this composite under sliding current collection can give an idea of prospect of application of this steel for creation of current collection materials. It is advisable to carry out the current collection sliding under severe conditions, i.e. at a high initial roughness of sliding surface of copper counterbody, at current density higher 50 A/cm^2.

The purpose of the present work is determination of operability of a composite based on recycled bearing steel in the conditions of current collection sliding against copper counterbody.

2. Experimental details

Model composites CM1 – CM3 had compositions of powder mixture as shown in the table 1 where BBS is recycled ball bearing steel. Commercial composites CM1 and CM2 had structures of the pressed powder product. Metallographic studying of a worn surface was carried out on an optical microscope Neophot-21. Hardness *HB* was determined on Brinell's hardness tester. Porosity *P* was obtained by Archimedes method. Bending strength was received by testing machine Instron-1185. Specific electric resistance ρ was determined by ammeter-voltmeter method. Initial parameters of a roughness (R_a=5 μm, R_z=31 μm) of a counterbody sliding surface were measured on the non-contact MICRO MEASURE 3D station profile-device. Copper (*HB*=0.7 GPa) served as counterbody.

Table 1. Mechanical and physical properties of metal composites.

Composition (vol.%) /properties	Hardness *HB* (MPa)	Bending strength σ(MPa)	Porosity *P* (%)	Specific electric resistance ρ (μΩ·m)
1. Cu-30% Gr (CM1)	94	14	28	0.12
2. Cu-42% Fe-7% Pb-2% Zn-11% Gr (CM2)	610	142	15	0.16
3. Cu-4% Gr-89% BBS (CM3)	1650	694	19	0.38

Wear tests of composites were carried out under the dry sliding electric contact and with lubricant at alternating current (50 Hz) under pressure of 0.02 MPa and 0.13 MPa, sliding velocity of 5 m/s using tribometer CMT-1. Tests were performed according to the scheme "pin-on-ring" similar to that represented elsewhere [4]. The sliding distance for each test was 9 km. Linear intensity of wear was found as $I_h=h/L$ where *h* is specimen height change for sliding distance *L*. The contact current density *j* is determined as $j=i/A_a$ where *i* is the current passing through the nominal contact area A_a=20 mm^2.

3. Results and discussion

Contact current density *j* is the main factor deteriorating a contact layer and limiting wear resistance. It is visible that increase of *j* causes increase in specific surface contact electric conductivity $r_s^{-1}=j/U$ (*U* is contact voltage drop) and also I_h (figure 1). But the surface layer of commercial composite CM1 quickly wears out under pressure of 0.02 MPa and catastrophically deteriorates under contact pressure 0.13 MPa at *j*=0 A/cm^2. One can see also that increasing of pressure leads to rise of conductivity r_s^{-1} and wear intensity I_h of composites CM2 and CM3. The composite CM2 shows the lowest I_h.

Distinction in tribotechnical behavior of composites is shown also in morphological details of a sliding surface. The composite CM1 does not form a transfer layer on a counterbody surface and thus isn't capable to create wear resistant contact at high roughness of counterbody. Therefore, it is not of interest for further studying. The composite of CM2 forms big sectors of copper on a working surface that is well shown on the optical image (figure 2, a-b). Besides, signs of formation of viscous liquid and absence of traces of adhesive interaction are observed in an explicit form. It should be noted that the composite CM2 forms a transfer layer on sliding surface of counterbody. The image of this transfer layer is visually close to the image of the layer presented in figure 2, a-b. Structural changes of surface layer of composite CM2 are not observed after sliding under pressure of 0.02 - 0.13 MPa and the cross section of surface layer has the image as in figure 2, e.

The sliding surface of composite CM3 contains traces of a plastic pushing off of surface layer by asperities of the interfaced counterbody surface and plastic shifts of big surface sectors are visible as a result of adhesive interaction with a counterbody (figure 2, c-d). Copper is not observed visually on a worn surface. The composite of CM3 forms a transfer layer on a counterbody only at sliding with current collection. This layer is a result of penetration of wear particles of a composite CM3 in copper surface. Structural changes in the CM3 surface layer are absent under pressure of 0.02 - 0.13 MPa and its cross section has an image as in figure 2,e.

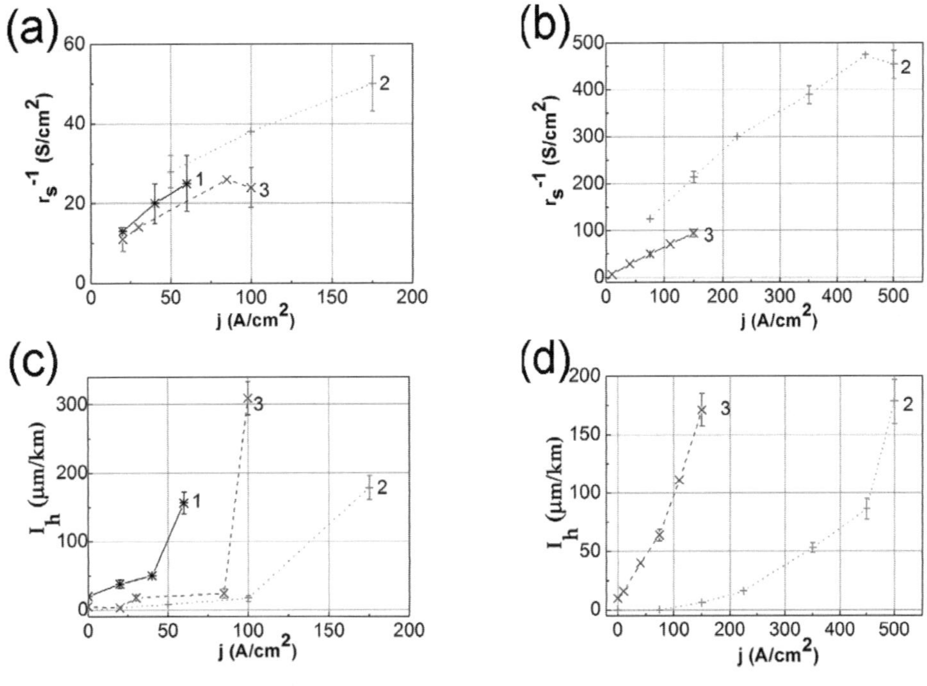

Figure 1. Current dependence of specific surface electric conductivity r_s^{-1} of contact and wear intensity I_h of composites CM1 (1), CM2 (2), CM3 (3) at contact pressures 0.02 MPa *(a,c)* and 0.13 MPa *(b,d)*

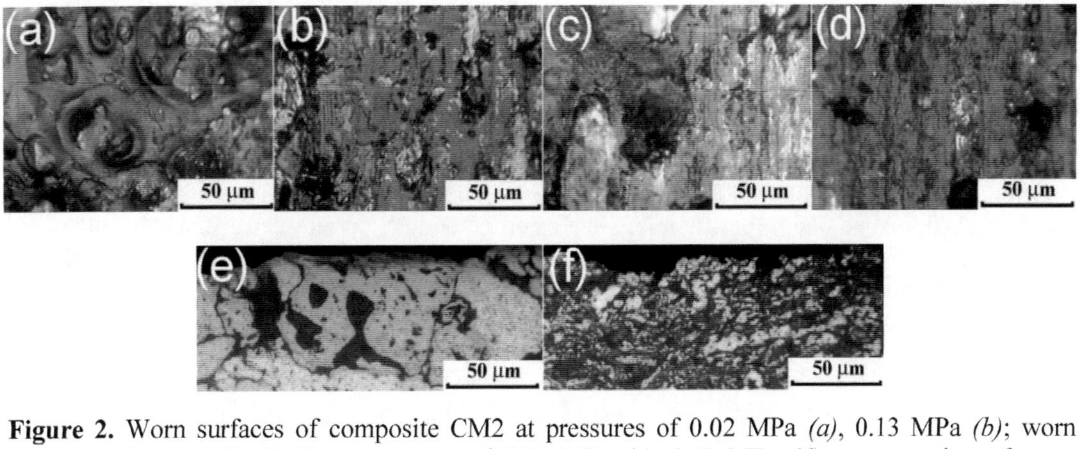

Figure 2. Worn surfaces of composite CM2 at pressures of 0.02 MPa *(a)*, 0.13 MPa *(b)*; worn surfaces of composite CM3 at pressures of 0.02 MPa *(c)*, 0.13 MPa *(d)*; cross section of worn surfaces of composite CM2 *(e)* and composite CM3 *(f)* at contact pressure of 0.02 MPa

Composites CM2 and CM3 have been impregnated with industrial oil at temperature of 100 °C for 1 hour and tested in similar conditions. Conductivity r_s^{-1} and wear intensity I_h of the impregnated composites don't differ considerably from the same contact characteristics of the composites presented in figure 1. One can note that wear of copper counterbody is not observed at presented sliding regimes.

The composite CM2 contains lead and carbon in the structure, forms a transfer layer on a counterbody and carries out sliding on this layer. It reduces adhesive interaction. Such type of self-organization of a sliding surface means that the initial roughness of a counterbody does not influence considerably on contact characteristics. Besides, the surface of a counterbody is not deformed plastically at sliding under pressure of 0.13 MPa. Therefore, the plastic flow can be realized only in transfer layer and in the surface layer of composite CM2. High plasticity of lead and copper on the interface of two bodies allows easily relax stresses due to plastic deformation. High concentration of copper in the structure of composite CM2 provides the good heat outflow from friction zone that leads to small increase of contact temperature and the plastic flow does not occur on a large scale. It expands possibilities of increasing of wear resistance of a composite CM2 at sliding against copper.

It should be noted that the working surface of a copper counterbody usually quickly deteriorates in the dry sliding contact with metals [5]. Low content of copper in structure of composite CM3 and low heat conductivity of its porous steel framework are the factors promoting increase of local surface temperature. Besides, strong adhesion in contact spots causes the high mechanical stresses which can not be relax due to local plastic micro-shears in surface layer owing to low plasticity of composite CM3. As a result, the pulling out of surface layer macro-fragments of a composite CM3 is performed and their following fixing on the counterbody sliding surface occurs. Thus the sliding surface of copper counterbody becomes protected from micro-cutting by asperities of a contact surface of a composite CM3 having higher hardness. Therefore, wear of a copper counterbody is absent but the CM3 composite surface layer deteriorates quickly without noticeable signs of self-organization (figure 2, f) that is observed as high wear.

Lubrication of a contact zone due to impregnation of a porous framework with industrial oil hasn't reduced adhesion in the sliding contact that was visible as absence of noticeable reduction of I_h. It is caused by impossibility to create a dividing film on contact spots owing to low viscosity of oil or its fast evaporation in a contact zone or its low ability to impregnation, etc. Probably the creation of the wear resistant sliding electric contact with use of recycled ball bearing steel can be carried out by a correcting of primary structure of composite or by increase in oil viscosity, i.e. due to use of greasing.

4. Summary

Dry sliding of metal composites against copper surface with high roughness under electric current can be accompanied by formation of a transfer layer on the interfaced surface. Absence of such layer leads to fast wear of Cu-30%Gr composite. The transfer layer in the form of wear particles penetrated in a sliding surface of a copper counterbody, prohibits from its wear, but does not prohibit from wear of the composite containing the recycled bearing steel. In this case catastrophic wear begins at the contact current density lower 100 A/cm^2. The composite containing lead demonstrates rather high wear resistance because it forms a plastic transfer layer on a counterbody surface that allows to relax the stress in a friction zone due to plastic micro-shears at the contact current density higher 100 A/cm^2. Impregnation of these composites with industrial oil (low viscosity) doesn't lead to significant changes of contact characteristics. It indicates expediency of use of grease. It should be noted that sliding against copper counterbody does not cause structural changes of surface layer of composites.

Acknowledgments

The work was carried out according to Project No. III.23.2.4 of SB RAS Fundamental Research Program III.23.2 and supported by Russian Foundation for Basic Researches project 16-48-700434

References

[1] Braunovich M, Konchits V V and Myshkin N K 2007 *Electrical contacts. Fundamentals, Applications and Technology* (CRC Press Taylor & Francis Group, New York)
[2] Kubota Y, Nagasaka S, Miyauchi T, Yamashita C, Kakishima H 2013 *Wear* **302** 1492–1498
[3] Dong L, Chen G X, Zhu M H, Zhou Z R 2007 *Wear* **263** 598–603
[4] Fadin V V and Aleutdinova M I 2009 *Russian Physics Journal.* **52 (6)** 607-611
[5] Kovalchenko A M, Blau P J, Qu J and Danyluk S 2011 *Wear* **271** 2707-3006

IOP Conf. Series: Materials Science and Engineering **156** (2016) 012028 doi:10.1088/1757-899X/156/1/012028

Experimental investigations of mechanical and electrical characteristics of a nanomodified epoxy resin DER-330

T A Brusentseva

Khristianovich Institute of Theoretical and Applied Mechanics SB RAS, Institutskay str.,4/1, Novosibirsk, Russia

E-mail: Brusentseva@itam.nsc.ru

Abstract. Technique preparation of a composite material based on epoxy resin DER-330 and 5 different nanopowders has been developed. The modifiers exert a positive effect on the elastic modulus, flexural strength, and dielectric permittivity of the epoxy-anhydride polymers. Weight content of nanoadditives in the matrix ranged from 0% to 4%. Dependences between mechanical properties of epoxy composites and concentration and nature of the filler were investigated in a three-point bending. Morphology of the fracture surface was studied. The complex analysis of nanofillers allowed us optimum fillers and concentrations when a significant increase of elastic and strength characteristics is achieved. It was showed that addition of silica fume nanoparticles A-380 to the matrix increases fracture stress on 30%. Composites with nanoparticles of aluminum nitride showed the best mechanical values in case of jointly investigated parameters. Growth of the elastic modulus is 12 % and increase of the fracture stress is 7 %. Adding alumina nanoparticles and nanofibers in the range of investigated concentration does not lead to significant changes in the mechanical characteristics of the composite

1. Introduction

The Epoxy-anhydride polymers are widely used as matrixes for modern composite electrical insulating materials [1]. These substances have great potential for the targeted modification of their properties and, therefore, are of an increasing interest both in scientific research and in engineering applications [2]. Not only the type of resin and the nature of a curing agent can be varied, but also a huge number of organic and inorganic modifiers and fillers can be added to improve the material characteristics. Modifications of the epoxy polymers are directed, first of all, at mechanical properties and dielectric properties. The reason for a significant improvement in the properties is presumably associated with a change in the structure of epoxy matrix near the filler surface [3, 4]. The dynamic development of industrial sectors is impossible without permanent improvement of materials. The use of composite materials (CMs) instead of the traditional ones allows significant reduction of the product weight without deteriorating the strength characteristics, which is particularly important for the aviation and aerospace industry. In addition, CMs may often have unique properties that are not inherent in traditional materials, which makes them indispensable for applications in certain areas of engineering.

This paper deals with the research of physical and mechanical characteristics of nano-modified epoxy composite materials, which are widely used in the construction industry as protective impregnations and coatings that increase the carrying capacity, chemical resistance, and durability of building structures. The goal of this work is to develop a method of epoxy resin binder modification by nanoparticles and a technology of obtaining the corresponding composite materials with improved performance.

2. Experimental

A polymeric matrix for the nanocomposite material was made from D.E.R. 330 epoxy resin (Dow Chemical Company, epoxy value 22.5), isomethyltetrahydrophthalic anhydride (IMTHPA) hardener (TU 8.103149 85), and benzyldimethylamine (BDMA) catalyst. Fillers used hydrophilic powders silica A380 (E1) and A200 (E2), differing particle sizes, powders of an aluminum oxide in the form of spherical particles (E3) and nanofibers (E4), and a powder of aluminum nitride(E5).

Dispersing powder in the resin was carried out using an ultrasonic generator Bandelin Sonopuls HD 3200. Stability of the suspension was monitored visually before and after curing. Curing was carried out in a stepwise mode, similar to that used in [5]. The completeness of the curing was monitored by IR spectroscopy.

Strength characteristics were studied in a three point bending mode. Experiments were conducted on a Zwick/Roell Z005 material testing machine according to ASTM D 790 testing standard. Rectangular samples ($4 \times 10 \times 80$ mm^3) were produced for the experimental using the technique described below. The loading (three point bending with unloading) was set by moving the mobile crosshead with a constant rate of 5%/min, deformation was measured by a crosshead motion sensor and the applied force was measured by an integrated force sensor. The distance between sup ports is 65 mm. The experiments were performed on up to 8% of deformation. The average test data of at least five samples are given. The error in the determination of the elastic modulus does not exceed 3% and that of tensile is no higher than 10%.

Figure 1 shows the concentration dependence of the Young's modulus of heterogeneous material with a variety of fillings. Judging schedule dependence of Young's modulus of the filler concentration for compositions with powders of silicon oxide (E1, E2) and aluminum oxide (E3) is the same. At a concentration of 0.2% to 1.2% modulus increases, reaching a maximum and then gradually decreases. Maximum increase of 7% is realized by using Al2O3 additive at a concentration of 0.8%. The dependence of the composition of the powder of aluminum nitride (E5) differs from others modulus monotonically increases and reaches a maximum at a concentration of 3.2% (E = 3890 MPa, an increase of 12% compared to baseline).

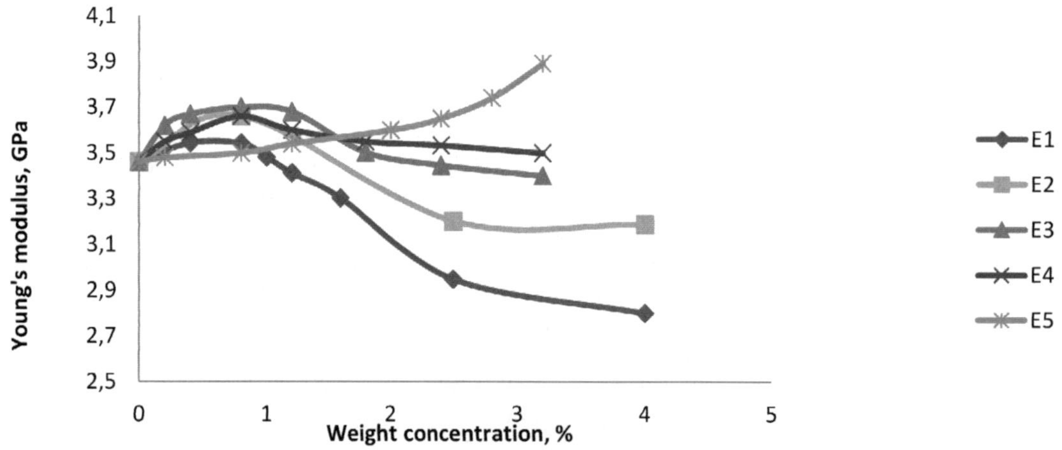

Figure 1. The dependence of the Young's modulus at the three-point bending from the weight concentration of fillers

Figure 2 shows graphs of the concentration of stress fracture powders. In a heterogeneous material containing particles of E1, E4, E5, realized higher stress at fracture over the whole range. The greatest positive influence nanosized powder silica A-380 (E1). fracture to stress 155 MPa base composition, adding 4% A-380 is increased by 30%. Powders E2 and type E3 have a strengthening effect on the epoxy composition at a concentration of 0.4- 1.2%.

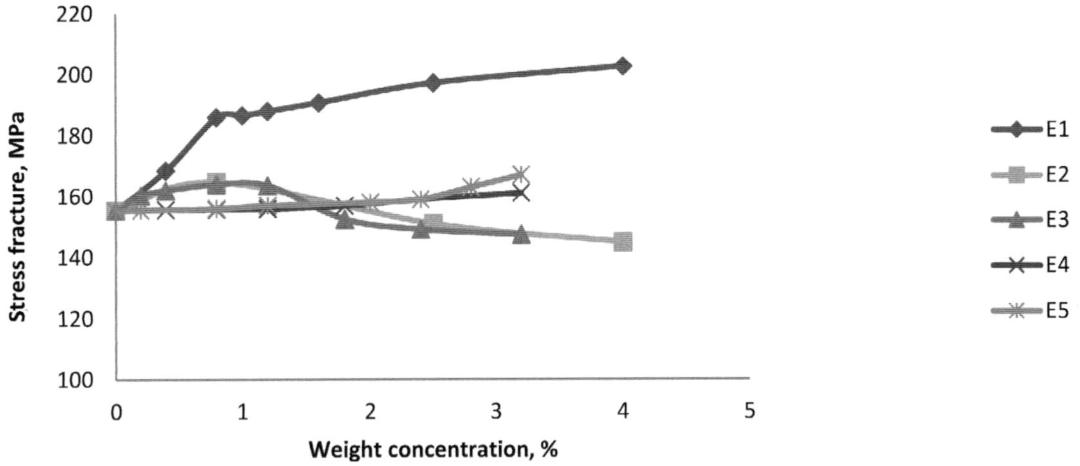

Figure 2. Dependence of fracture stress with three-point bending of the weight concentration of powders

2.1. The capacity and dielectric constant epoxy compositions

Compositions of epoxy resins with hardeners anhydrides are mainly used as a base insulating compounds. Determination of dielectric characteristics carried out for the samples containing alumina nanofibers at a concentration of 1.8 and 3.2%. Sample sizes - 5x5 cm, thickness 0.15 cm. Capacitance measurement carried out on the digital bridge R5058 at a frequency of 10 kHz. The measurement results are shown in Table 1.

Table 1. Capacity permittivity and epoxy compositions.

Parameters	Epoxy resin without filler	Epoxy resin with alumina nanofibers	
Concentration		1.8%	3.2%
Capacitance	45	44	42.2
The dielectric constant	3.05	2.98	2.86

It was established that the additive alumina nanofibers improves the capacitance and dielectric constant epoxy compositions. Previously it demonstrated that the filler does not impair the Young's modulus and fracture stress. Collectively, the data developed promising heterogeneous characterize materials based on epoxy resin for encapsulation of electrical equipment.

3. Conclusions

Epoxy nanocomposites based on polymer matrix DER330 + IMTHPA and five different nanofillers were obtained under the same conditions. The influence of the nature and amount of inorganic additives on the mechanical properties of epoxy anhydride nanocomposites was investigated. It has been established that the incorporation of silica nanoparticles at an amount of 0.2–1.2% of the weight of the binding agent leads to an increase in Young's modulus. Additives of silica A 380 with a particle size of 7 nm have the greatest effect on the fracture stress: a 30% increase over the base composition.

The addition of aluminum oxide nanoparticles and nanofibres does not lead to significant changes in the mechanical characteristics of the composite in the investigated concentration range. While alumina nanofibers additives improve the capacitance and dielectric constant epoxy compositions. Composites with aluminum nitride nanoparticles showed the best mechanical values on the strength of all investigated parameters: the growth in the elastic modulus is 12% upon an increase in the fracture stress by 7%.

Acknowledgement
The work was financially supported by the RFBR in the framework of a research project number 16-31-00135 mol_a

References

[1] A Li H, Zhang Z, Ma X, Hu M, Wang X, Fan P 2007 Synthesis and characterization of epoxy resin modified with nano-SiO2 and γ-glycidoxypropyltrimethoxy silane *Surf. Coat. Technol.* **9-11** 5269.

[2] Lee Y S., Cho T H, Lee B K, Rho J S, An K H, Lee Y H 2003 Surface properties of fluorinated single-walled carbon nanotubes *J. Fluorine Chem.* **120** 2 99-104.

[3] Pascault J P and Williams R J J 2010 *Epoxy Polymers: New Materials And Innovations* (Hardcover).

[4] Akatenkov R V, Aleksashin V N, Anoshin I V 2008 *The effect of small additions of functionalized multilayer carbon nanotubes onto strengthening kinetics and properties of epoxy composites* Proc. Int. Forum on Nanotechnologies *"Rusnanotech 08"* (Moscow) 410–12.

[5] Vaganova T A, Brusentseva T A, Filippov A A, Malykhin E V 2014 *Synthesis and characterization of epoxy-anhydride polymers modified by polyfluoroaromatic oligoimides* J. of Polymer Research **21** 11

Effects of transient thermal shock loadings on the structure of zirconia ceramics

E A Derkach[1,2], S P Buyakova[1,2,3] and S N Kulkov[1,2,3]

[1] National Research Tomsk Polytechnic University, 30 Lenin ave., 634050, Tomsk, Russia
[2] Institute of Strength Physics and Materials Science SB RAS, 2/4 pr. Akademicheskii, 634055, Tomsk, Russia
[3] National Research Tomsk State University, 36 Lenin ave., 634050, Tomsk, Russia

E-mail: eadtomsk@yandex.ru

Abstract. In this paper the influence of thermal shock loadings on the phase composition and microstructure of $ZrO_2(Y_2O_3)$ and $ZrO_2(MgO)$ ceramics was studied. It was found that thermal shock strains were no effect on phase composition of $ZrO_2(Y_2O_3)$ ceramics. Reducing the concentration of high-temperature tetragonal t-ZrO_2 and cubic c-ZrO_2 modifications and rising the content of low-temperature monoclinic m-ZrO2 crystal system with increasing the number of thermal shock strains were observed. The values of region coherent X-ray scattering of $ZrO_2(Y_2O_3)$ ceramics didn't changed, while the crystallite size of $ZrO_2(MgO)$ ceramics decreased. The formation of block structure in all studied ceramics was observed, the sizes of the blocks formed in the $ZrO_2(Y_2O_3)$ and $ZrO_2(MgO)$ ceramics were slightly different. The formation of the blocks in $ZrO_2(Y_2O_3)$ ceramics occurred on the grain boundaries, while the crystallites in $ZrO_2(MgO)$ ceramics were crushed due to phase transformation.

1. Introduction

The impetuous progress of high temperature technics needs to develop new materials for providing effective thermal insulation, both stable heat flux and under temperature extremes. In terms of thermal insulation characteristics, special attention is given to the zirconia based ceramics, which have an abnormally low for ceramic materials thermal conductivity ($\lambda = 1.7 - 2.0$ W / (m · K)), and a high melting point (mp = 2715 °C). Resistance to thermal shock strains is not a fundamental characteristic and it largely depends on the structure and composition of the product. Zirconia exists in three polymorphic modifications: the low-temperature stable monoclinic m-ZrO_2, the high-temperature tetragonal t-ZrO_2 and the high-temperature cubic c-ZrO_2. In terms of exploitation the main attention is given to two high-temperature modifications. They are obtained by creating an interstitial solid solution of oxide materials such as MgO, Y_2O_3, CaO, CeO_2 and others, which prevent a phase transformation during cooling process. Stabilized by different modifying additives, ceramics have differences not only in chemical composition, but also in mechanical and physical properties. Though the investigation of thermal shock resistance of zirconia-based ceramics began many years ago [1 - 7], but today the effect of thermal shock strains on the microstructure of the material was not fully defined. Gaining knowledge about the influence of the thermal shock loadings on the ceramic's structure enables to control its thermal shock resistance at the micro scale. Thereby, the aim of this work is to study the effect of thermal shock strains on the behavior of zirconia-based ceramics stabilized 5 mol. % yttria and 8.6 mol. % magnesia.

2. Materials and methods

As samples to study we used ceramic materials obtained by sintering zirconia partially stabilized 5 mol. % yttria $ZrO_2(Y_2O_3)$ and 8.6 mol. % magnesia $ZrO_2(MgO)$ powders. Studied samples had a cylindric form. Thermal shock loading carried on cooling specimens in water from 1000 °C to 20 °C. Totally we had nine thermal shock strains. Analysis of the microstructure of the ceramics was carried out on the polished surfaces of samples by optical microscopy. Phase composition was studied using XRD analysis in the diffraction angle range from 25 to 100°, obtained with the filtered CuKα radiation.

Calculation of the size of coherent X-ray scattering was carried out with broadenings of X-ray reflections by:

$$D = \frac{\lambda_p}{\Delta B \cos\theta},$$

where λ_r – X-ray wavelength, ΔB – width of the X-ray reflections at half height, θ – angle of X-ray diffraction. Measuring of block sizes was carried out using random cross-sections method, the number of measurements was at least 300.

3. Results and discussion

An x-ray analysis (Figure 1) showed that the of $ZrO_2(MgO)$ ceramics was represented by two high-temperature structural modifications: tetragonal t-ZrO_2 and cubic c-ZrO_2. Phase composition of ZrO_2

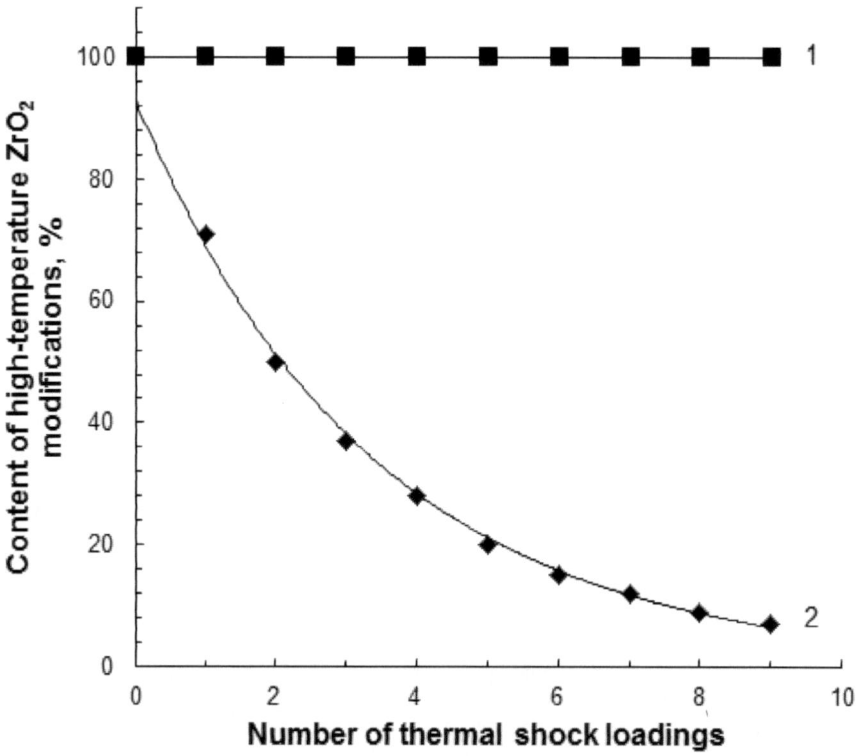

Figure 1. Dependencies of contents of high-temperature modifications: 1) t-ZrO_2 in the $ZrO_2(Y_2O_3)$ ceramics and 2) t-ZrO_2 and c-ZrO_2 in the $ZrO_2(MgO)$ ceramics vs. the number of thermal shock loadings.

(Y_2O_3) ceramics was represented only by the tetragonal t-ZrO_2 modification. Other X-ray diffraction reflections were not observed.

An analysis of the phase composition of the studied samples after thermal shock strains showed that phase composition of $ZrO_2(Y_2O_3)$ ceramics was unchanged, regardless of the number of thermal shock loadings. The X-ray reflections showed only tetragonal t-ZrO_2 zirconia. However, in the $ZrO_2(MgO)$ ceramics occurred decrease in the proportion of high-temperature t-ZrO_2 tetragonal and cubic-ZrO_2 phases and increasing proportion of monoclinic m-ZrO_2 phase with increasing number of thermal shock loadings. One of the reasons of increasing in the proportion of monoclinic m-ZrO_2 modification may be the disintegration of the solid solution ZrO_2-MgO, accompanied by the release of Mg^{2+} cations during the process of thermal stresses.

Calculating of the size of coherent diffracting domain (CDD) by the broadening of X-ray reflections (Figure 2) showed that this value for $ZrO_2(Y_2O_3)$ ceramics was 63 ± 3 nm and remained unchanged

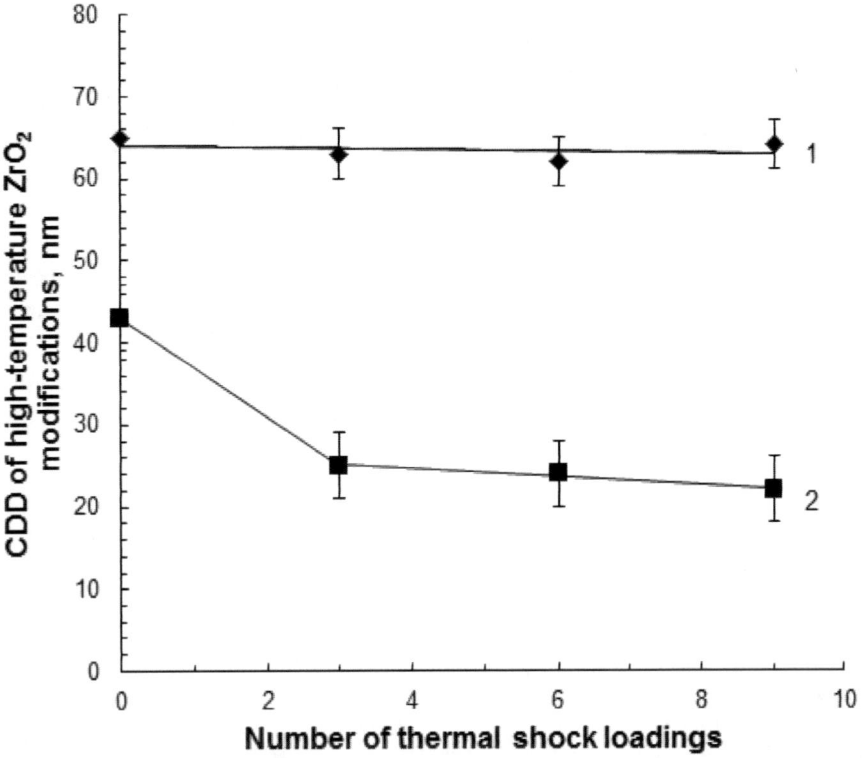

Figure 2. Dependencies of CDD of high-temperature: 1) t-ZrO_2 in the $ZrO_2(Y_2O_3)$ ceramics and 2) c-ZrO_2 in the $ZrO_2(MgO)$ ceramics vs. the number of thermal shock loadings.

irrespective of the number of thermal shock loadings. CDD values of $ZrO_2(MgO)$ ceramics before thermal shock loadings were 43 ± 5 nm and after three thermal shocks we observed a decrease in size up to 25 ± 5 nm. Resizing can be associated with increasing in the proportion of monoclinic m-ZrO_2 phase. As is known the phase transformation from the high-temperature cubic c-ZrO_2 modification to the low-temperature monoclinic m-ZrO2 phase is accompanied by significant compressive stresses due to increasing of the unit cell volume.

Figure 3 shows images of the $ZrO_2(Y_2O_3)$ sample's surfaces before and after nine thermal shocks. It was detected the formation of a block structure, which is generated in the process of thermal shock loadings as a consequence of the formation of cracks due to the interaction of compressive and tensile stresses during the quench. This effect occurred in studied specimens in different ways. Barely visible cracks appeared on the surface of the $ZrO_2(Y_2O_3)$ ceramics after the first thermal stress; after the third thermal shock we detected a clear boundaries of the blocks. The subsequent thermal shock loadings led to the production of new cracks in the already formed blocks: the structure was further fragmented. Growth and extension of cracks occurred on the grain boundaries. After nine thermal stresses the average block size decreased from 314 ± 26 to 268 ± 23 μm (Figure 4). Formation of the block structure in ZrO2(MgO) ceramics occurred in differ way: fragmentary structure with average size 289

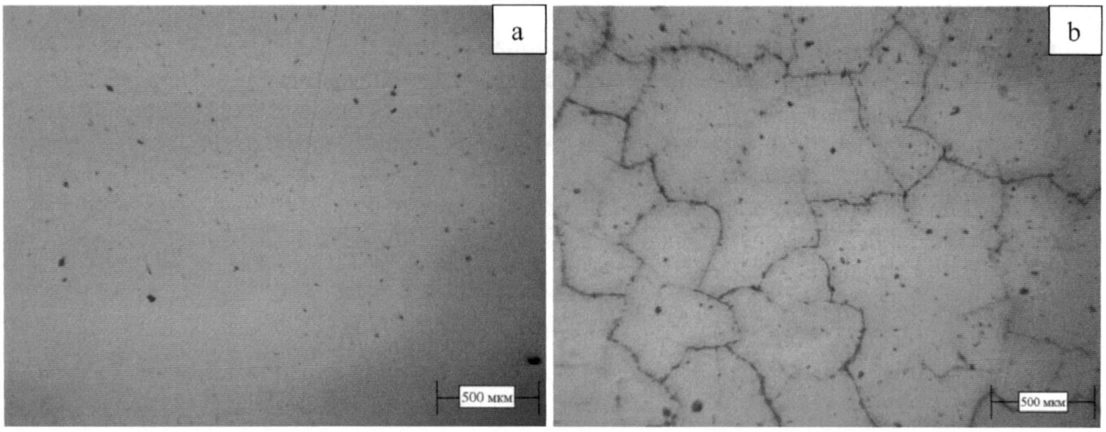

Figure 3. The images of the surfaces of $ZrO_2(Y_2O_3)$ ceramics (a) before and (b) after thermal shock loadings.

± 29 μm was detected after the first thermal shock. Subsequent thermal loading was no result in further fragmentation of the surface. Moreover the samples were not destroyed despite the appearance of cracks.

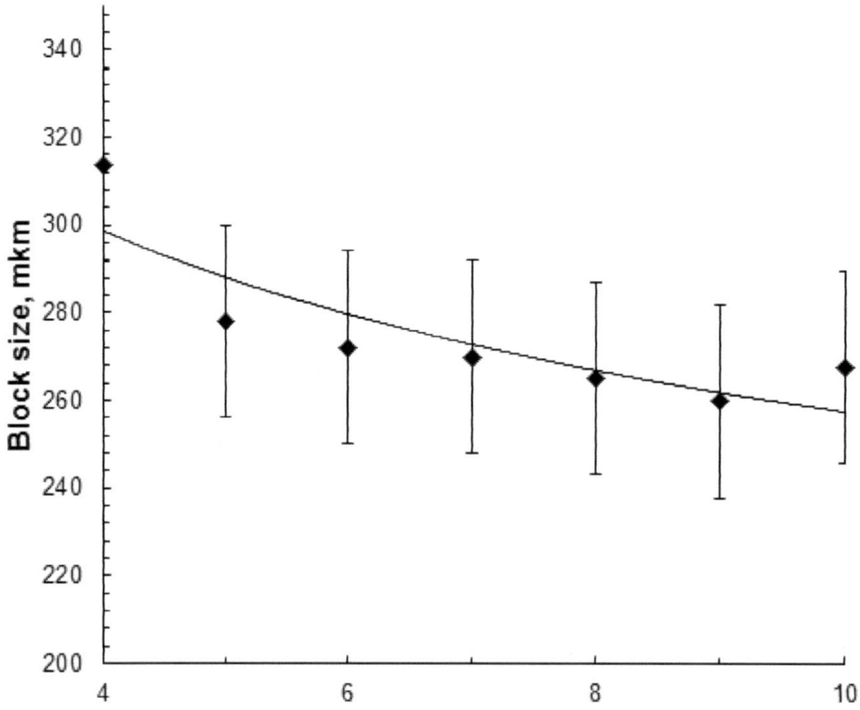

Figure 4. Dependencies of block size in $ZrO_2(Y_2O_3)$ ceramics vs. the number of thermal shock loadings.

4. Conclusion

Thus, it was found that $ZrO_2(Y_2O_3)$ ceramics retained the tetragonal crystal system t-ZrO2 during all thermal shock loadings rather than $ZrO_2(MgO)$ ceramics. In $ZrO_2(MgO)$ ceramics we detected a gradual decline in concentration of high-temperature tetragonal t-ZrO_2 and cubic c-ZrO_2 phases at the same time with appearance and increment of low-temperature m-ZrO_2 phase with increasing the number of thermal shock strains. For $ZrO_2(Y_2O_3)$ CDD values remained unchanged, while for $ZrO_2(MgO)$ ceramics crystallite size decreased. Formation of the block structure was observed in all studied ceramics, the sizes of the formed blocks in ZrO_2(Y2O_3) and $ZrO_2(MgO)$ ceramics were slightly different. The appearance of blocks in $ZrO_2(Y_2O_3)$ ceramics occurred on the grain boundary, while in $ZrO_2(MgO)$ ceramics crystallites were crushed due to phase transformations.

Acknowledgment
The research was conducted with partial financial support within the agreement with the Ministry of Education and Science of the Russian Federation № RFMEFI60714X0056.

References

[1] Fargas G, Casellas D, Llanes L and Anglada M 2003 *Journal of the European Ceramic Society* **23** 107
[2] Angle J P, Steppan J J, Thompson P M and Mecartney M L 2014 *Journal of the European Ceramic Society* **34** 432
[3] Pia G, Casnedi L and Sanna U 2016 *Ceramics International – 201* **42** 5802
[4] Wang R, Li W, Li D, Shen X, Zhang X, Li B, Yao J, Wu X and Fang D 2015 *Journal of Alloys and Compounds* **626** 56
[5] Promakhov V, Buyakova S and Kulkov S 2011 *Fundamental problems of advanced materials* **8** 11
[6] Promakhov V, Buyakova S, Illavszky V, Kulkov S and Gomze L 2014 *Journal of Silicate Based and Composite Materials* **3** 81
[7] Dyatlova E M, Barantseva S E, Kakoshko E S and Kononovich E S 2005 *Glass and Ceramics* **8** 10

IOP Conf. Series: Materials Science and Engineering **156** (2016) 012030 doi:10.1088/1757-899X/156/1/012030

Zen law and features of liquidus–solidus curves in binary state diagrams based on elements VIIIA and IB of the periodic table

A I Potekaev[1], A A Kondratyuk[2], S A Porobova[3], A A Klopotov[1,3], T N Markova[3], Yu A Kakushkin[3] and V D Klopotov[2]

[1]Tomsk State University, Tomsk, Russia
[2]Tomsk Polytechnic University, Tomsk, Russia
[3]Tomsk State University of Architecture and Building, Tomsk, Russia

E-mail: klopotovaa@tsuab.ru

Abstract. The paper presents the analysis of binary state diagrams based on elements VIIIA and IB of the periodic table and crystal geometry parameters of solid solutions and intermetallic compositions. The analysis shows an explicit correlation between the type of the evolution of phase diagrams classified by Lebedev depending on the nature of atomic volume deviations observed in solid solutions and intermetallic compounds from Zen law.

1. Introduction

An increased interest to state diagrams of binary and ternary systems from elements VIIIA and IB of the D.I. Mendeleyev's periodic table is caused by the presence in these systems of solid solutions and intermetallic compounds with unique combinations of physical and mechanical properties as well as important magnetic and optical characteristics. The principal distinctive feature of these functional materials is a weak stability of the crystalline lattice in temperature ranges preceding structural phase transitions (SFT). The weak stability and instability of the crystalline lattice in advance of SFT manifest themselves in features of temperature dependences of physical and mechanical properties (the so-called subtransient phenomena [1, 2]). The development of ideas on the nature of the stability of crystalline structures of solid solutions and intermetallic compounds is unthinkable without the research on clarification of features of their crystalline structure [3-9]. Up to the present moment, a purposeful study of the correlation of state diagrams of systems from elements VIIIA and IB of the D.I. Mendeleev's periodic table with crystal-geometry parameters of solid solutions and intermetallic compounds has not been carried out.

The aim of the work is to carry out a search for common patterns and correlations between the structure of state diagrams of binary systems from elements VIIIA and IB of the D.I. Mendeleev's periodic table and the deviation of atomic volumes in solid solutions and intermetallic compounds from Zen law.

2. Materials and methods

In [3-8] it is shown that different types of state diagrams of alloys, depending on the alloyability of elements, have certain regularities in their structure (size factor, electronic structure of atoms) that depend on the ratio of values of interatomic interactions of alloying elements.

There is a wide variety of state diagrams depending on the value of the mixing energy W for a binary alloy: in the approximation of the interaction among nearest neighbors

$$W = U_{AB} - \frac{U_{AA} + U_{BB}}{2}, \quad (1)$$

where U_{AB}, U_{AA}, and U_{BB} are potentials of interatomic interaction among atoms of different kinds *AB* and one *AA* and *BB*, respectively [9]. The necessary condition for the formation of a solid solution is the proximity of the value W to zero. To fulfill this condition, two alloying elements in binary alloys must have the structure of electron atom shells, atomic radii, and the energy of chemical bonds close to each other. In this case, the atomic substitution of one another is not associated with difficulties of the structural or the energy nature, thus, continuous solid solutions appear. The mutual solubility of components decreases when W becomes positive. Chemical compounds in alloys are formed when the numerical value of the mixing energy W becomes significantly negative. In the case when an intermetallic compound has a narrow homogeneity region and a significant negative mixing energy W, the location of stability regions of crystalline structures depends on the size factor and the external incomplete electron atom shell of the alloying component.

The classification proposed by T.A. Lebedev is promising if to consider the relationship of individual diagrams within a single type [10]. In this classification, it is proposed to consider the evolution of diagrams not only within a single type, but bonds of individual types of diagrams as well. Based on this classification of state diagrams we have presented, with some simplifications, the real state diagrams of binary systems from elements VIIIA and IB of the D.I. Mendeleev's periodic table [11], the order and the arrangement of which allow to clearly classify them by types (Figure 1). The evolution of alloy state diagrams corresponds to the change in the ratio of potential energies of interaction of different types of atoms AB and the same type AA and BB.

The diagram of the system Ag-Co is basic for the first type of phase diagrams under the classification of T.A. Lebedev. For this system, there is a limiting case for diagrams of the type I (Figure 1), in which the miscibility is so small that it can be taken as close to zero. An increase in the miscibility for diagrams of the type I leads to the evolution: the following sequence of diagrams for systems under consideration:

(Ag-Co)→(Co-Cu)→(Au-Pt)→(Cu-Ni)→(Co-Ni).

Such evolution of phase diagrams of the type I shows that there is an increase in the energy of the interatomic interaction of different types of atoms *AB*, i.e., U_{AB}, relative to energies of the interatomic interaction of the same type of atoms U_{AA} and U_{BB}. This is reflected in the change in the evolution of concentration dependences of atomic volume deviations in solid solutions and in ordered phases from Zen law in systems formed by elements VIIIA and IB of the D.I. Mendeleev's periodic table (Figure 2). It shall be noted that the deviation in concentration dependences of atomic volumes from Zen law is negative, which indicates a decrease in the atomic volume during the formation of solid solutions in alloys of systems under consideration.

In the investigated binary systems the following set of crystallo-geometrical parameters was used [13]: the atomic volume of elements Ω; the value of deviation from the linear dependence of the atomic volume $\Delta\Omega/\Omega^{EXP}=(\Omega^{EXP}-\Omega_Z)/\Omega^{EXP}$ on the concentration (Ω^{EXP} is the atomic volume in the alloy determined from the experimental values of parameters of elementary cells of the alloy, Ω_Z is the atomic volume determined by Zen's law.

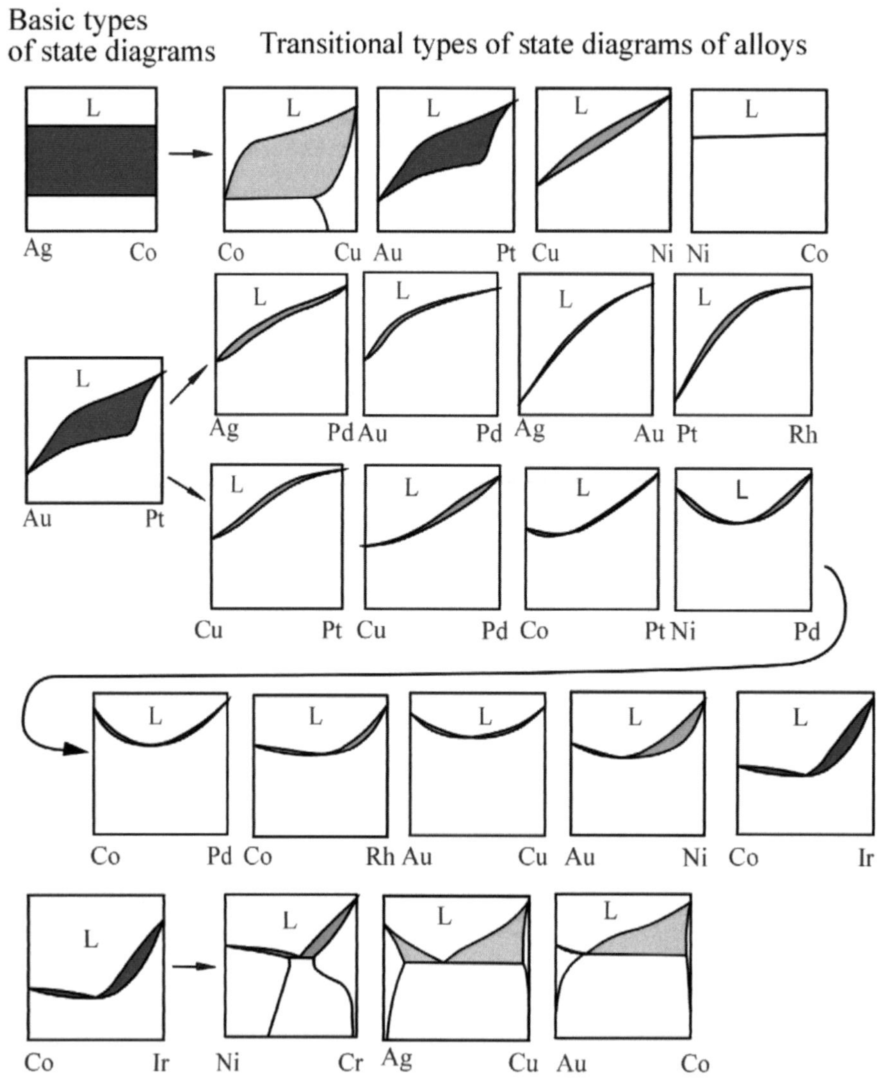

Figure 1. The schematic of the evolution of phase diagrams formed by elements VIIIA and IB of the D.I. Mendeleev's periodic table based on the T.A. Lebedev's classification [10]

The second type of phase diagrams, for which the system Au-Pt is basic, is characterized by modification of phase diagrams under two scenarios. The first scenario proposes modification of phase diagrams due to the fact that the liquidus line goes to a rapprochement with the solidus line. Then, the sequence of diagrams is as follows:

(Au-Pt)→(Ag-Pd)→(Au-Pd)→(Pt-Rh)→(Ag-Au).

The second scenario proposes that the solidus line goes on rapprochement with the liquidus line. In this case, the sequence of diagrams is as follows:

(Au-Pt)→(Cu-Pt)→(Cu-Pd)→(Co-Pt)→(Ni-Pd)→(Co-Pd)→(Co-Rh)→(Au-Cu) →(Au-Ni)→(Co-Ir).

IOP Conf. Series: Materials Science and Engineering **156** (2016) 012030 doi:10.1088/1757-899X/156/1/012030

Such evolution of state diagrams leads to an appearance of minima points on solidus and liquidus lines (Figure 1). It should be noted that this sequence is the most numerous among all of the considered phase diagrams formed from elements VIIIA and IB of the periodic table.

The third type of state diagrams is presented by a number of diagrams in which the breakdown degree of liquid solutions increases to produce, in the end result, a mechanical mixture of pure components (Figure 1)

(Co-Ir)→(Ni-Cr)→(Ag-Cu)→(Au-Co).

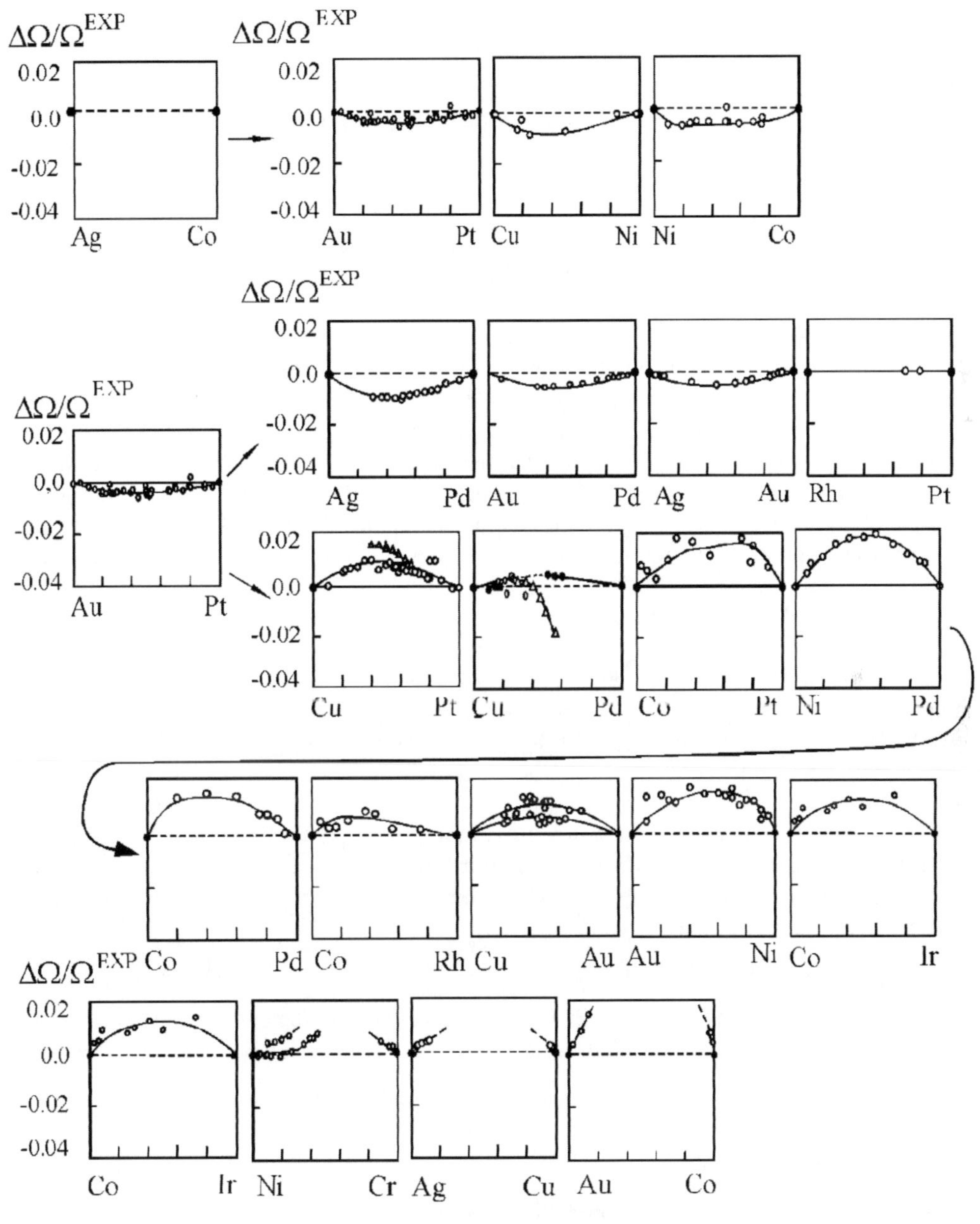

Figure 2. The schematic of the evolution of concentration dependences of atomic volume deviations in solid solutions and in ordered phases from Zen law in systems formed by elements VIIIA and IB of the periodic table

It is important that the minimum on the diagram of continuous solid solutions (Co-Ir, Figure 1) cannot be mixed with the eutectic point. The fact that at the beginning of the given series of diagrams with the eutectic point contains the diagram with the minimum point on solidus and liquidus curves does not give grounds to identify the minimum point on this diagram with the eutectic point. There is a tendency that reflects the process of a possible eutectic formation. This phenomenon is associated with the corresponding prerequisites that a eutectic transformation in the subsequent systems may occur at the place of a weakly-stable or an unstable compound.

Atomic volumes in solid solutions and intermetallic compounds (according to lattice parameters) were calculated for all above-mentioned binary phase diagrams [12]). The obtained values of atomic volumes have allowed determining the deviation of concentration dependences of atomic volumes from Zen law [1, 2].

According to the above mentioned distribution of phase diagrams, depending on the features of liquidus and solidus curves, a detailed arrangement of concentration dependences of atomic volume deviations in solid solutions and in ordered phases from Zen law in systems formed by elements VIIIA and IB of the D.I. Mendeleev's periodic table has been carried out (Figure 2). This arrangement has allowed revealing the correlation between liquidus (solidus) curves and concentration dependences of atomic volume deviations in solid solutions from Zen law in systems under consideration.

In diagrams of the type I under the classification of T.A. Lebedev only negative deviation of concentration dependences of atomic volumes from Zen law is observed (Figure 2).

Phase diagrams of the type II under the classification of T.A. Lebedev, for which the diagram of the system Au-Pt is basic, are characterized by both positive and negative deviation of atomic volumes in solid solutions and in ordered phases from Zen law, depending on the scenario of the evolution of phase diagrams (Figure 2). A positive deviation from Zen law leads to modification of phase diagrams due to the fact that the liquidus line goes on rapprochement with the solidus line. The sequence of diagrams of binary systems with a negative deviation of atomic volumes from Zen law results in a sequence of phase diagrams in which the solidus line goes on rapprochement with the liquidus line. Such evolution of state diagrams on the solidus and liquidus line leads to an appearance of the tendency to formation of the minimum (Figure 1).

Phase diagrams of the type III, for which the diagram of the system Co-Ir is basic, are characterized by a positive deviation of atomic volumes in solid solutions from Zen law in places of their existence on diagrams. At the same time, a significant increase in deviations of atomic volumes from Zen law in regions preceding solubility ruptures on phase diagrams is observed (Figure 2).

3. Conclusion

Thus, a clear correlation between the type of the evolution of phase diagrams under the classification of T.A. Lebedev depending on the nature of atomic volume deviations in solid solutions and in intermetallic compounds from Zen law in systems formed from elements VIIIA and IB of the D.I. Mendeleev's periodic table has been revealed.

Acknowledgement

The paper includes the results obtained in the course of the implementation of the project (No. 8.1.42.2015) within the framework of the Program "Scientific Foundation named after D.I. Mendeleev of Tomsk State University" in 2015-2016.

References

[1] Potekaev A I, Klopotov A A, Kozlov E V, et al. 2004 *Pretransitional weak-stable structure in NiTi* (Tomsk: NTL) p 296 (in Russian)

[2] Potekaev A I, Klopotov A A, Morozov M M, et al. 2014 *Structural features of binary systems with weak-stable* (Tomsk: NTL) p 304 (in Russian)

[3] Kornilov I I and Budberg P B 1961 *State diagrams of binary and ternary titanium systems* (Moscow: Nauka) pp 40-41 (in Russian)

IOP Conf. Series: Materials Science and Engineering **156** (2016) 012030 doi:10.1088/1757-899X/156/1/012030

[4] Vozdvizhinsky V M 1972 *Forecast of binary state diagrams. According to statistical criteria* (Moscow: Metallurgiya) p 326 (in Russian)
[5] Kornilov I I 1974 *Equilibrium metalloid systems//intermetallic compounds* (Moscow: Metallurgiya) pp. 244-279 (in Russian)
[6] Kaufman L 1970 *Stability of crystalline lattices in transition metals//Phase stability in metals* (Moscow: Mir) pp. 134-161 (in Russian)
[7] Goodman D, Bennett L and Watson R 1986 *Tendencies of mutual solubility in alloys of transition metals//Phase diagrams in alloys* (Moscow: Mir) pp. 25-35 (in Russian)
[8] Watson R and Bennett L 1986 *Structural maps and parameters determining the phase stability in alloys//Phase diagrams in alloys* (Moscow: Mir) pp. 36-44 (in Russian)
[9] Coreno-Alonso O and Coreno-Alonso J 2004 Volume size factor and lattice parameter in cubic intermetallics with $L1_2$ or B2 structure derived from the ''Macroscopic Atom'' model *Intermetallics* **12** 117-122
[10] Lebedev T A 1951 *Some questions in the general theory of alloys* (Leningrad: newspaper and magazine book) p 136 (in Russian)
[11] Liakishev N P 1996-2000 *State diagrams of binary metallic systems* (Moscow: Mashinostroenie) **1 – 3** (in Russian)
[12] ViUars P and Calvert L D 1985 *Pearson's Handbook of Crystallography Data for Intermetallic Phases* (Ohio: Metals Park) **1** 1332
[13] Zen E J 1934 Validaty of «Vegard Law», *MSA*. **41** 523-524

Selective laser melting of titanium alloy: investigation of mechanical properties and microstructure

A V Agapovichev, V V Kokareva, V G Smelov and A V Sotov

Department of engines manufacturing technology, Samara University, Samara, Russia

E-mail: charming_carrot@mail.ru

Abstract. This article presents the mechanical properties and microstructure of titanium alloy after selective laser melting (SLM). Titanium alloys are ideal material for selective laser melting (SLM), because they are expensive and difficult to machinery using traditional technologies. The application of SLM in the biomedical area has been slow due to the stringent performance criteria and concerns related to personification and part quality. In this article we focused on the manufacture by SLM and determination of microstructure and mechanical properties of titanium alloy (Ti Grade 2 Powder) using tensile tests and X-ray diffraction. The results reveal that the alloy exhibits a pronounced the homogeneous microstructure and high mechanical strength.

1. Introduction

Today titanium alloys with their favorable biocompatibility play an important role for implant applications. The titanium alloys offer good biomaterial properties such as a very high biocompatibility, a density as low as that of bone, high mechanical strength and fatigue resistance, low elastic modulus and good wear resistance. Commercially pure titanium is used preferentially for endosseous implant applications. There are currently four Ti grades and one titanium alloy specially made for dental implant applications. These metals are specified according to ASTM as grades 1 to 5. Grades 1 to 4 are unalloyed, while grade 5, with 6% aluminum and 4% vanadium is the strongest [1]. In Russia, we use commercially pure titanium grades VT1-0 for manufacturing implants. The selection of VT1-0 for medical implantation is determined by a combination of most favorable characteristics including immunity to corrosion, bio-compatibility, strength, low modulus and density and the capacity for joining with bone and other tissue - osseointegration. The mechanical and physical properties of titanium alloys combine to provide implants which are highly damage tolerant. The human anatomy naturally limits the shape and allowable volume of implants. Creating a biomechanical interaction of the implant with human bone tissue is current multidimensional problem, based on the development of technological methods of implants production. It should be taken into account the personalized approach in this direction, since the shape of any individual human bones is a unique.

Selective Laser Melting (SLM), an additive manufacturing technique makes it possible to produce parts with intricate shapes [2-4]. Selective laser melting method allows getting in a short time, metal products with complex geometric form, with a uniform distribution of density and homogeneous structure, taking into account the specific physiological human features. This technology provides an opportunity to produce ready-to-use parts with minimal post processing requirements. Moreover, the high cooling rates are resulting in parts with superior properties. However, SLM widespread application is largely constrained by the lack of own production powders. In medicine, the introduction of additive technology is very promising, as the cost of implants, especially individual, very high, and the decrease in production terms is of great interest in the development and manufacture of new types of implants [5].

The goal of this work is to study the influence of mode SLM process on the structure and mechanical properties of produced samples.

2. Materials and methods

The raw materials for the SLM are powders which are subject to number of requirements [6]: the spherical form of particles, dispersion structure, phase and chemical composition as microcrystalline structure and a particle size distribution, processing characteristics (high flowability, packed density, etc.).

In this paper we used the titanium powder VT1-0 as the raw material. For the elemental composition analysis, we used electron probe microanalysis. Powder structure was investigated in a scanning electron microscope Tescan Vega.

The microstructure of samples, which were obtained by the SLM, depends essentially on the original powder characteristics and processing conditions [7, 8]. The main parameters of laser melting are the laser power (W), scan speed (mm/s), layer thickness (mm), the distance between tracks (mm), the protective atmosphere. The quality of sintered layers is characterized by maximum achievable accuracy, uniform density, the maximum and minimum thickness of the processed layer.

Laser melting was carried out on the installation SLM 280HL with the size of the working chamber 280h280h350 mm. The laser source is YLR-100-SM single-mode CW ytterbium fiber laser (1064–1100 nm). The spot size and maximum laser power are 40 μm and 400 W. The titanium powder particles floated on a titanium substrate.

The study of mechanical properties of the samples was performed on a floor system for fatigue testing model INSTRON 8802. Metallurgical studies have been conducted using a microscope Metam LP-32.

3. Results and discussion

Mass concentration of chemical elements corresponds to reference titanium powder, the powder has a sufficiently high purity, does not contain impurities (Table 1). The maximum particle size of the initial powder was 77.5 μm, particles have an equiaxial regular geometric shape and an average particle size is 50 μm. The flowability of titanium powder was 6 seconds/100g. Figure 1 shows the surface morphology of the titanium initial powder particles VT1-0.

Table 1. Chemical composition of the initial powder

Element	Mass concentration of chemical elements (%)	Sample (%)
Ti (base)	99.68	99.5 – 99.99
Other impurities	0.32	0.01-0.5

SLM offers several advantages compared to conventional production techniques, such as reduction of production steps, a high level of flexibility, a high material use efficiency and a near net shape production. Furthermore, hard materials or materials with a high melting point can be processed with SLM. Most important, because of the layer-wise building, SLM enables the production of parts with a high geometrical complexity. However, the unique conditions during the SLM process give rise to some problems. Because of the short interaction times and accompanying highly localized heat input, large thermal gradients exist during the process. These lead to the build-up of thermal stresses, while the rapid solidification leads to segregation phenomena and the development of non-equilibrium phases. Moreover, non-optimal scan parameters may cause melt pool instabilities during the process, which leads to an increased porosity and a higher surface roughness. To ensure optimal building

IOP Conf. Series: Materials Science and Engineering **156** (2016) 012031 doi:10.1088/1757-899X/156/1/012031

conditions, the influence of the different process parameters, such as layer thickness, scan spacing, scan strategy, laser power and scan speed on the microstructure and mechanical properties and other properties such as the density and surface quality have been investigated.

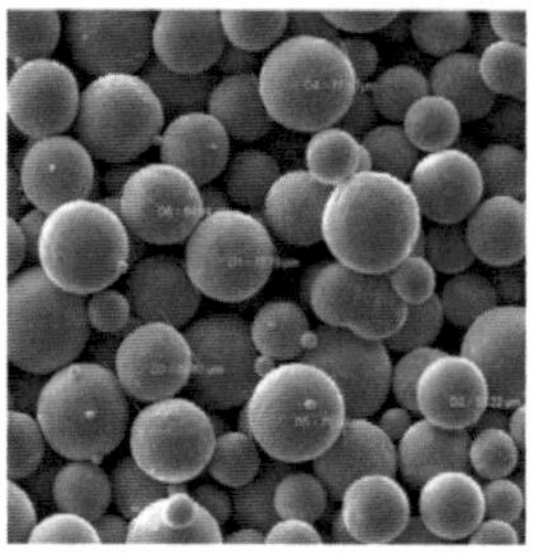

Figure 1. The surface morphology of the titanium powder VT1-0.

As a result of working out the SLM optimal processing conditions we manufactured rectangular cross section samples, the plate with a length 70 mm, width 2 mm and height 10 mm. The processing parameters under consideration included laser power (150, 165 W), scanning speed (100-500 mm/s). The amount of transmitted energy was evaluated by using the linear energy density (LET) radiation power equal to the ratio (P) to a scanning speed V (LET = P/V).

The range of the parameter LET amounted 0.30 ... 1.65 J / mm, the scanning speed was determined in increments of 50 mm / s. The main parameters of the SLM are given in Table 2.

Table 2. Parameters of Selective Laser Melting

No. Processing conditions	Scanning speed (mm/s)	Laser power (W)	Linear energy density (J / mm)
1	100		1.65
2	150	165	1.10
3	200		0.83
4	400		0.38
5	450	150	0.33
6	500		0.30

The Figure 2 shows a process of selective laser melting titanium of powder on SLM 280HL machine. The growing of samples was carried out in of argon, the thickness of the layer construction of 50 μm. Time of samples manufacturing was 2 h. Figure 3 shows the samples produced. Also we made individual prototypes of small joint implants; 3D models are produced as a result of treatment CT scans of individual patients in the Samara State Medical University (Figure 3). Due to the high values of the grown samples roughness (Rz 20), surface requires subsequent machining of the contact surfaces of implants small joints.

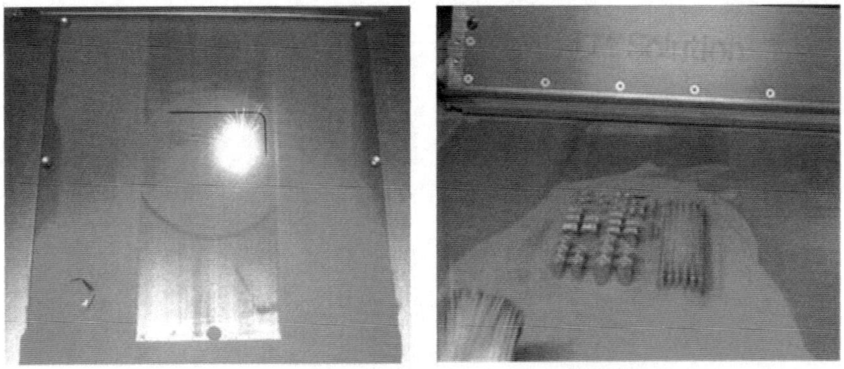

Figure 2. SLM process of titanium powder on the SLM 280HL machine.

Figure 3. Samples were produced by SLM process.

All tensile test samples of the SLM material were tested for each treatment to determine the mechanical properties in accordance with ISO 5832-2-2014. Results of mechanical tensile tests on the samples produced by the SLM are given in Table 3. The test results show that ultimate stress limits of SLM-alloy exceed the ultimate stress limit of the titanium alloy obtained by traditional technologies, reference material (200-680 MPa, ISO 5832-2-2014). The SLM material is much stronger than the reference material. This is true for all material processed by SLM because the rapid cooling conditions always lead to a fine microstructure.

Table 3. Mechanical properties of samples.

No. Processing conditions	Ultimate stress limit, σ (MPa)	Relative extension ε (%)
1	717	21.7
2	732	19.0
3	750	16.7
4	740	19.0
5	708	17.3
6	702	12.5

The dependence of ultimate stress limit from the technological selective laser melting parameters is shown in Figure 4.

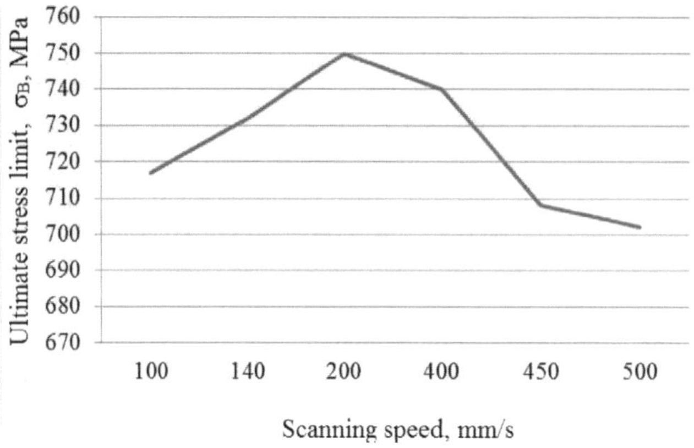

Figure 4. The dependence of the mechanical properties of titanium alloy samples VT1-0 from the technological SLM parameters.

The graph shows that the best value of the ultimate stress limit of material reaches at a laser scanning speed of 200 mm/s. In this paper we show the material microstructure obtained by titanium powder melting at six different processing conditions (Table 2). Metallographic studies showed homogeneity of the structure in the presence of small pores. The pore size for the adopted processing conditions of melting observed in the range of 10-50 m (Figure 5).

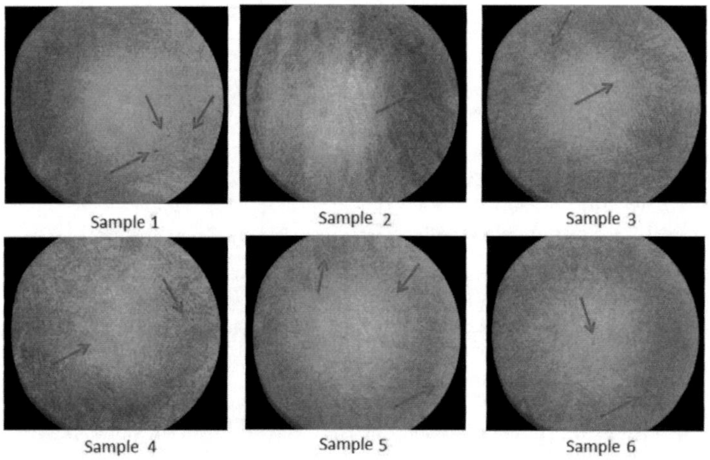

Figure 5. The microstructure of the samples produced by SLM

4. Summary

The SLM technology is a successful method of melting titanium directly by consolidating powder mixtures of commercially pure metals. In this research the change in surface morphology of SLM titanium with process parameters, such as laser power, scan speed was investigated to control the pore structure. Following results were obtained:

1) The powder is characterized by high purity, does not contain harmful impurities, which is an important requirement of the medical industry. The presence of fraction 35-50 μm does not prevent normal formation layer and does not reduce the flowability of the powder.
2) Titanium samples and prototypes of small joint implants. The quality of sintered layers is characterized by uniform density, the strength of 750 MPa, a spherical particle shape. SLM technology provides the high mechanical properties of samples, and has good prospects for further receiving implants of titanium alloys.

The results presented in the paper are the basis of research and development aimed at improving the SLS process and enhances the performance properties of the implants produced. [9] Next important step is the selection of the optimal parameters of selective laser melting and subsequent processing as by experimentation and mathematical modeling techniques that will improve and accelerate the process of creating a product with the specified requirements, including the desired roughness of samples [10]. To improve the quality of titanium surfaces samples we choose the suitable post-processing by methods of electrochemical machining [11].

Then we plan to carry out complex research aimed at the formation of biocompatible coatings on the samples in order to improve the functional characteristics of the implant, and experiments on the biocompatibility of implants manufactured prototypes.

Acknowledgments
The work described in this article is made at the equipment of Collective Use Center of CAM-technologies (a unique cipher RFMEFI59314X0003 agreement).

References

[1] Abdel-Hady Gepreel M and Niinomi M 2013 Biocompatibility of Ti-alloys for long-term implantation *Journal of the Mechanical Behavior of Biomedical Materials* **20** 407-15

[2] Attar H, Löber L, Funk A, Calin M, Zhang L, Prashanth K, Scudino S, Zhang Y and Eckert J 2015 Mechanical behavior of porous commercially pure Ti and Ti-TiB composite materials manufactured by selective laser melting *Materials Science and Engineering* **625** 350-56

[3] Popovich A, Sufiiarov V, Borisov E and Polozov I 2015 Microstructure and mechanical properties of Ti-6AL-4V manufactured by SLM *Key Engineering Materials* **651-653** 677-82

[4] Sufiiarov V, Popovich A, Borisov E and Polozov I 2015 Selective laser melting of titanium alloy and manufacturing of gas-turbine engine part blanks *Tsvetnye Metally* **2015, 8** 76-80

[5] Frazier W E 2014 Metal additive manufacturing: A review *Journal of Materials Engineering and Performance* **23** 1917-28

[6] Sallica-Leva E, Jardini A and Fogagnolo J 2013 Microstructure and mechanical behavior of porous Ti-6Al-4V parts obtained by selective laser melting *Journal of the Mechanical Behavior of Biomedical Materials* **26** 98-108

[7] Smelov V, Sotov A and Murzin S 2016 Particularly selective sintering of metal powders by pulsed laser radiation *Key Engineering Materials* **685** 403-07

[8] Khaimovich A and Balaykin A 2014 Analysis of titanium alloys plastic properties under severe deformation conditions in machining *ARPN Journal of Engineering and Applied Sciences* **9 10** 1828-33

[9] Agapovichev A, Balaykin A and Smelov V 2015 Production technology of the internal combustion engine crankcase using additive technologies *Modern Applied Science* **9 4** 335-43

[10] Bose S, Vahabzadeh S and Bandyopadhyay A 2013 Bone tissue engineering using 3D printing *Materials Today* **16, 12** 496-504

[11] Nekhoroshev M, Pronichev N and Smirnov G 2014 Computer simulation of high-speed anodic dissolution processes of geometrically-complex surfaces of GTE details *Open Mechanical Engineering Journal* **8** 436-40

IOP Conf. Series: Materials Science and Engineering **156** (2016) 012032 doi:10.1088/1757-899X/156/1/012032

Investigation of possibility to fabricate Si_3N_4-TiN ceramic nanocomposite powder by azide SHS method

L A Kondratieva[1], I A Kerson, A Yu Illarionov, A P Amosov and G V Bichurov

Samara State Technical University, 244 Molodogvardeyskaya Street, Samara 443100, Russia

E-mail: schiglou@yandex.ru

Abstract. The process of self-propagating high-temperature synthesis with use of a powder of sodium azide NaN_3 as a nitriding agent (the SHS-Az method) was applied to fabricate a nanocomposite powder Si_3N_4-TiN. Combustion of the initial mixtures of NaN_3 only with precursors that are halides of silicon and titanium: Na_2SiF_6, $(NH_4)_2SiF_6$, Na_2TiF_6, $(NH_4)_2TiF_6$ did not allow us to synthesize the composite powder of Si_3N_4-TiN, as the phases of silicon nitride Si_3N_4 were not formed. After water washing, the ultrafine powdered product of combustion consisted of one target phase of titanium nitride only and a large amount of impurity of side phases. Replacement of the halide salt of one of the elements (Si or Ti) by the powder of this element in the initial mixture of SHS-Az system resulted in formation of silicon nitride together with the titanium nitride and impurities. The least amount of impurities was obtained by burning the initial mixtures of $xSi + y(NH_4)_2TiF_6 + zNaN_3$ system. Only the initial mixture of $9Si + (NH_4)_2TiF_6 + 6NaN_3$ allowed us to obtain as a result of the SHS-Az process the nanostructured composite powder of Si_3N_4-TiN without impurities.

1. Introduction

Among nonoxide ceramics, the ceramics made of silicon nitride Si_3N_4 stands out because of its good combination of properties that, along with small weight, high values of hardness, stiffness, heat resistance, high-temperature strength, and corrosion resistance, has a high fracture toughness and is used for the manufacture of high-speed ball bearings of gas turbine engines, parts of automotive engines, and cutting tools [1]. However, the production of components made of very hard ceramic Si_3N_4 is only possible using diamond tools, which greatly complicates and increases the cost of manufacturing of components, especially of complex shape. A simple and cheap method of electrical discharge machining is not applicable here due to a very low conductivity of Si_3N_4 ceramics, so much attention is paid to obtaining a conductive composite ceramics based on Si_3N_4, primarily, the Si_3N_4-TiN ceramics [2, 3]. Titanium nitride TiN has not only high electrical conductivity but a high melting temperature, hardness, wear resistance, and corrosion resistance, it is compatible with Si_3N_4, improves the sintering and properties of Si_3N_4 ceramics [4]. In this case the structure of Si_3N_4-TiN composite ceramics is of great importance. Ceramics of Si_3N_4 - 30 vol. % TiN, obtained by spark plasma sintering of Si_3N_4 and TiN nanoparticles, has the wear resistance which is 3 times higher than that of composite ceramics of this composition, sintered by hot pressing of commercially available powders of micron size [5]. The composite ceramics with hard TiN nanoparticles damages a counterface by friction much less than ceramics with TiN particles of micron size, that is important e.g. for hybrid bearings in which the balls are made of ceramics, and races are made of metal [6].

High quality composite nanostructured ceramics can be obtained by sintering only a homogeneous mixture of nanopowders, which is hard to achieve because of the great tendency of nanopowders to agglomeration [1, 7]. The nanoparticles stick together, forming a porous, fairly strong aggregates, poorly distributed among the nanoparticles of the other phase, making it difficult to compact. Nanopowder mixture of Si_3N_4-TiN can be obtained as ex-situ by mechanical mixing of Si_3N_4 and TiN nanopowders, prepared in advance, and in-situ by formation of nanoparticles of Si_3N_4 and/or TiN by chemical reactions from precursors in the powder mixture [7]. However, it is not always possible to achieve uniform distribution of nanoparticles through simple mechanical mixing, and after sintering, the material can be anisotropic, porous, with low strength. More uniform distribution is achieved in the case of in-situ methods for forming nanoparticles through chemical reactions from precursors inside the body of the composite. Another important advantage of in-situ methods is the significantly lower cost of powders of the precursors in comparison with the high cost of nanopowders prepared in advance and used in ex-situ methods [8].

One of the most promising in-situ methods is the method of self-propagating high-temperature synthesis (SHS), also known as combustion synthesis, which is attractive for its simplicity and efficiency and is used for producing various ceramic powders including Si_3N_4-TiN composite [9-11]. Composite powder of Si_3N_4-TiN was synthesized by burning the mixture of elemental powders of Si (6 μm) and Ti (74 μm), diluted with submicron Si_3N_4 powder (40-50 wt. %), in the atmosphere of nitrogen at 5 MPa [9]. In another work, various titanium silicides (TiSi, Ti_5Si_3, $TiSi_2$) were used as the source of raw materials (precursors of Si and Ti), which also were burned in a nitrogen atmosphere at a pressure of 5 MPa [10]. The use of Ti_5Si_3 was optimal which had generated a two-phase composite powder of Si_3N_4- (30-35 wt. %) TiN with a particle size of 5-10 microns without impurity of free silicon. Research of regularities of combustion of samples, pressed from a mixture of powders of Si_3N_4/Ti = 1/3, in the gaseous nitrogen at the pressure in the range from 0.45 to 1.82 MPa showed that the increase in pressure contributes to the degree of nitriding of Ti; complete nitriding of titanium was achieved at 1.48 MPa in the absence of its reaction with Si_3N_4, and Si_3N_4-TiN composite formed which was free from impurities [11].

However, the above composite powders of Si_3N_4-TiN obtained in-situ by SHS are not nanoscale. Of great interest is to apply the SHS method for synthesis of nanopowder composition of Si_3N_4-TiN. This possibility can be provided by the SHS method using a powder of sodium azide NaN_3 as a nitriding reagent (the SHS-Az method) whereby nanostructural composite powders of TiN-BN and AlN-BN were synthesized on the base of using powders of halide salts of Ti, B, and Al elements to be nitrided as precursors instead of powders of the pure elements of Ti, B, and Al in the composition of the reagents [12, 13]. The aim of this work was to investigate the possibility of obtaining the composite nanopowder of Si_3N_4-TiN by the SHS-Az method using halide salts of the elements of Ti and Si as precursors of these elements.

2. Materials and methods

In the present work, powders of silicon, titanium, sodium azide and halide salts of silicon and titanium, which are produced by industry in Russia, were used as initial reagents: silicon of Kr1 brand (purity 98.0 wt. %; particle size less than 10 μm); titanium of PTM brand (99.0 %; less than 45 μm); NaN_3 (98.71 %); Na_2SiF_6 (99.1 %); $(NH_4)_2SiF_6$ (99.1 %); Na_2TiF_6 (98.0 %); $(NH_4)_2TiF_6$ (99.0 %).

All initial powders were sifted on the vibrating screen of LE-203/I type through a set of standard sieves. If necessary, the powders were dried in a vacuum drying box at a temperature of 80 °C within 1-2 hours. A mixture of the initial powders, taken in a predetermined ratio, was prepared manually in a porcelain mortar for 5-10 minutes until a uniform distribution of components. The prepared mixture of powders (the charge) was filled in a tracing paper cup with a diameter of 30 mm and a height of 45 mm and was placed in the filter assembly made from carbonic cloth. The relative density of the powder mixture was 0,34. The assembled structure was mounted on the subject shelf of a azide SHS laboratory reactor of a constant pressure with a volume of 4.5 liters. The pressure of 4 MPa of gaseous nitrogen was created in the reactor. The arrangement of this reactor and methods of work with the measurements of the linear velocity and the maximum combustion temperature by thermocouple method using tungsten-rhenium thermocouples TR 5/20 was described in detail in [14]. Recording the temperature and rate of combustion was carried out using an analog-to-digital converter connected to a computer.

The cooled sample was removed from the reactor and was destroyed by hand in a porcelain mortar to obtain a free-flowing powder. Then water washing the powder product of combustion was carried out to remove by-product residue, which, as a rule, was soluble in water sodium halide (its main mass escaped from the sample at high combustion temperatures). The powder was diluted with distilled water to ratio of 1:10, the resulting suspension was shaken and filtrated on a vacuum funnel. Rinsing was monitored by measuring the acid-base balance of the wash water (pH).

Shooting X-ray spectra was carried out on an automated diffractometer ARL X trA-138 using Cu-radiation with continuous scanning in the range of angles 2θ from 20 to 80 degrees with a speed of 2 deg/min. A quantitative estimate of the content of different phases was carried out by measuring the ratio of intensities of analytical lines. Morphology and particle size of the powder compositions was carried out using a scanning electron microscope (SEM) Jeol JSM-6390A.

3. Results and discussion

The masses of the components of the charge in the SHS-Az systems were calculated taking into account the stoichiometric ratio of the components, wherein the amount of halogen in the halide salts was bound to fix the sodium released by the decomposition of sodium azide NaN_3 in a neutral compound: chloride or fluoride of sodium. The calculation was carried out at the following ratio of Si_3N_4 and TiN in the composite powder: Si_3N_4-TiN, $3Si_3N_4$-TiN, Si_3N_4-3TiN. For example, for the system 1: "$xNa_2SiF_6 + yNa_2TiF_6 + zNaN_3$", where x, y, z are numbers of moles:

$3Na_2SiF_6+Na_2TiF_6+16NaN_3=Si_3N_4\text{-}TiN+24NaF+21{,}5N_2$; (x=3, y=1, z=16);

the increase of silicon halide salt Na_2SiF_6 in the initial mixture:

$9Na_2SiF_6+Na_2TiF_6+40NaN_3=3Si_3N_4\text{-}TiN+60NaF+53{,}5N_2$; (x=9, y=1, z=40);

the increase of titanium halide salt Na_2TiF_6 in the initial mixture:

$3Na_2SiF_6+3Na_2TiF_6+24NaN_3+=Si_3N_4\text{-}3TiN+36NaF+32{,}5N_2$; (x=3, y=3, z=24).

The reactions for the other investigated SHS-Az systems were similarly recorded, in which only the charge compositions are given for the sake of brevity:

System 2: "$x(NH_4)_2SiF_6 + y(NH_4)_2TiF_6 +zNaN_3$", (3, 1, 24; 9, 1, 60; 3, 3, 36).
System 3: "$xNa_2SiF_6 + y(NH_4)_2TiF_6 + zNaN_3$", (3, 1, 18; 9, 1, 42; 3, 3, 30).
System 4: "$x(NH_4)_2SiF_6 + yNa_2TiF_6 + zNaN_3$", (3; 1; 22; 9; 1; 58; 3; 3; 30).
System 5: "$xTi + yNa_2SiF_6 + zNaN_3$", (1; 3; 12; 3, 3, 12; 1, 9, 36).
System 6: "$xTi + y(NH_4)_2SiF_6 + zNaN_3$", (1, 3, 18; 3, 3, 18; 1, 9, 54).
System 7: "$xSi + yNa_2TiF_6 + zNaN_3$", (3, 1, 4; 9, 1, 4; 3, 3, 12).
System 8: "$xSi + y(NH_4)_2TiF_6 + zNaN_3$", (3, 1, 6; 9, 1, 6; 3, 3, 18).

All 24 of the SHS-Az investigated charges were capable of self-combustion, but their temperature and speed of combustion could be very different: from 600 to 1800°C and from 0.15 to 1.7 cm/s, respectively. The phase composition of the combustion products also varies greatly in different systems.

The combustion products of system 1 include a large amount of the NaF salt (from 24 to 80 wt. %) as the combustion products of this system were not subjected to water washing. This salt was not detected in the products of combustion of other systems, because NaF is highly soluble in water and was removed by washing. Titanium nitride TiN (5-18%) is produced in combustion of powder mixtures of system 1, but silicon nitride Si_3N_4 is not formed, titanium silicide $TiSi_2$ (12-18%) and free silicon Si (0-46%) are also formed in this case. Such powdery product is agglomerates of particles of equiaxed or spherical shape with an average particle size of 200-300 nm.

In systems 2 and 3, only TiN and Si are formed, and the more titanium nitride (from 47 to 80%), the greater halide salt of titanium (NH4)2TiF6, and the more free silicon (from 20 to 53%), the greater silicon-containing halide salt. In the system 4, free titanium Ti (10-39%) is added to the phases of TiN and Si in the combustion products; and in the case of charge composition x=3; y=3; z=30, a bit of silicon nitride of α-Si_3N_4 (7%) and sufficient amount of water-insoluble salt Na_2TiF_6 (22%) are added. This case is the only charge composition among the 12 formulas for systems 1-4, whose products of combustion contain Si_3N_4 along with TiN. Thus, in the SHS-Az systems 1-4 consisting of sodium azide and only precursors which are reviewed halide salts of both elements Si and Ti, the composite powder of Si_3N_4-TiN is not synthesized, with the exception of formula "$9(NH_4)_2SiF_6$ + Na_2TiF_6 + $58NaN_3$", whose combustion products contain only 25% TiN and 7% α-Si_3N_4, and much more (68%) by-products of Si, Ti, Na_2TiF_6. The size of particles of different shape of the powdered product of combustion of systems 1-4 is in the range of 100-500 nm, that is, this product is ultrafine powder consisting of titanium nitride TiN, free silicon Si, and titanium silicide $TiSi_2$ (in the case of system 1) or free titanium Ti (in the case of system 4).

To continue the study of possibility of synthesis of nanostructured composite powder of Si_3N_4-TiN in the SHS-Az systems, the precursor that is a halide salt of one of the elements to be nitrided was replaced by powder of this pure element: it was the titanium powder Ti in systems 5 and 6, and it was the powder of silicon Si in systems 7 and 8. Such changes led to the fact that during the combustion of almost all charge compositions of systems 5-8, the silicon nitride was formed along with the titanium nitride (with the exception of initial mixture "Ti + $9Na_2SiF_6$ + $36NaN_3$" of system 5 and mixture "3Si + $3Na_2TiF_6$ + $12NaN_3$" of system 7, whose combustion did not cause the synthesis of silicon nitride). For systems 5 and 6 with the titanium powder, the content of silicon nitride in the product of combustion was small: from 13 to 19%, and in this case it was a mixture of α-Si_3N_4 and β-Si_3N_4 for system 5, but only α-Si_3N_4 for system 6. For systems 7 and 8 with the powder of silicon, the content of silicon nitride in the product of combustion could be much more: up to 41-72% of a mixture of α-Si_3N_4 and β-Si_3N_4. It is interesting to note that the combustion products of system 6: "xTi + $y(NH_4)_2SiF_6$ + $zNaN_3$" contain a great quantity of $TiN_{0.30}$ phase (38 to 65%) along with the TiN phase (16-21%). It should also be noted that for almost all charge compositions of system 5-8, the washed combustion products contained side phases of impurities (along with the target phases of TiN and Si_3N_4): Ti (0-15%), Si (13-72%), and Ti_5Si_3 (0-9%) for system 5; a water-insoluble salt Na_2SiF_6 (0-10%) or Si (0-23%) for system 6; Si (6-17%) for system 7; also Si (0-10%) for system 8. And the only initial mixture "9Si + $(NH_4)_2TiF_6$ + $6NaN_3$" of the system 8 made it possible to obtain a pure mixture of target phases: 28% TiN, 11% α-Si_3N_4, 61% β-Si_3N_4 without any impurities which corresponds approximately to the Si_3N_4-TiN composition of the composite powder. The diffraction pattern of the phase composition and the morphology of particles of this composite powder are shown in figures 1 (b) and 2 (b). Figure 2 (b) demonstrates the typical relatively large columnar crystals of β-Si_3N_4 with a transverse size of 200-600 nm and more thin fibers and small equiaxed particles of α-Si_3N_4 and TiN with a size of 100-120 nm. This mixture of ultrafine and nanosized particles can be named as nanostructured composite powder of Si_3N_4-TiN. The great content of β-Si_3N_4 in the form of large columnar crystals without impurities of free silicon can be explained by the relatively high temperature of the combustion 1800°C [15]. Much lower combustion temperature (900°C) of neighbouring formula "3Si + $(NH_4)_2TiF_6$ + $6NaN_3$" of the same system 8 leads to the formation of significantly smaller particles TiN and Si_3N_4 (70-100 nm), but it is accompanied by partial nitriding the silicon (10% free silicon remains in the product of combustion), as can be seen from figures 1 (a) and 2 (a). A similar explanation can be given to results of performing the SHS-Az method with another adjacent initial mixture "3Si + $3(NH_4)_2TiF_6$ + $18NaN_3$" with the combustion temperature 700°C and 5% free silicon in the final product with a particle size of 80-130 nm. The combustion products of these two neighboring initial mixtures can be assigned to the nanopowders, but they contain impurities of free silicon (5-10%) and much more TiN (57-81%) compared to Si_3N_4 (14-33%), which corresponds in moles approximately to the composition of (4-13)TiN-Si_3N_4-(1,5-1,8)Si.

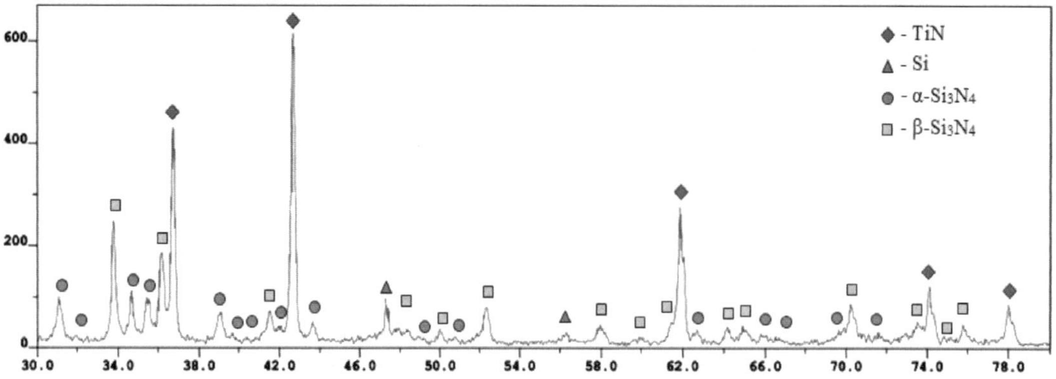

(a)

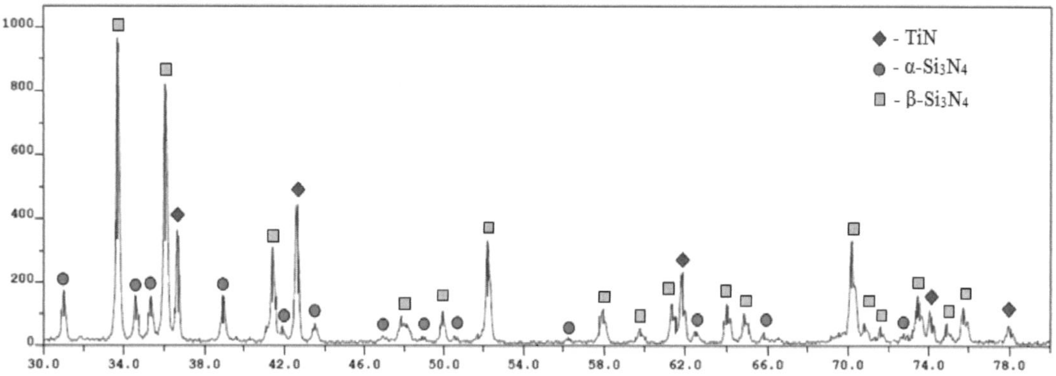

(b)

Figure 1. XRD patterns of the washed products of combustion for system 8: «$xSi + y(NH_4)_2TiF_6 + zNaN_3$»: (a) x=3; y=1; z=6; (b) x=9; y=1; z=6.

(a) (b)

Figure 2. SEM image of particles of washed products of combustion for system 8: "$xSi + y(NH_4)_2TiF_6 + zNaN_3$": (a) x=3; y=1; z=6; (b) x=9; y=1; z=6.

4. Conclusion

The obtained results show that the combustion of powder mixtures of sodium azide NaN_3 only with precursors, which are halide salts of silicon and titanium: Na_2SiF_6, $(NH_4)_2SiF_6$, Na_2TiF_6, $(NH_4)_2TiF_6$, does not allow producing a composite powder of Si_3N_4-TiN, because the phase of silicon nitride Si_3N_4 is not formed, and after water washing, ultrafine powdered product of combustion consists of only one target phase of titanium nitride TiN and a large amount of impurities of side phases. Replacement of the halide salt of one of the elements (Si or Ti) on the powder of this element in the initial mixture of SHS-Az system leads to the formation of silicon nitride together with titanium nitride. In this case with Ti powder, the content of silicon nitride in the product of combustion is small: from 13 to 19%, but with Si powder, the content of silicon nitride is much larger: can go up to 41-72% of a mixture of α-Si_3N_4 and β-Si_3N_4. The least amount of impurities (from 0 to 10% of free silicon) was obtained in combustion of initial mixtures of system 8: "xSi + y$(NH_4)_2TiF_6$ + zNaN_3". Burning two mixtures of this system (x=3; y=1; z=6 and x=3; y=3; z=18) occurred at not high temperature of 700-900°C and led to the formation of nanopowder composites: (57-81%)TiN-(14-33%)Si_3N_4-(5-10%)Si. And the only initial mixture "9Si + $(NH_4)_2TiF_6$ + 6NaN_3", that burned at high temperature 1800°C, allowed us to obtain a pure mixture of target phases: 28% TiN, 11% α-Si_3N_4, and 61% β-Si_3N_4 without impurities, which can be described as a nanostructured composite powder of Si_3N_4-TiN consisting of ultrafine columnar crystals of β-Si_3N_4 with a transverse size of 200-600 nm and nano-sized fibers and equiaxed particles of α-Si_3N_4 and TiN with a size of 100-120 nm.

Acknowledgements

The authors thank V A Novikov for capturing diffraction patterns for X-ray phase analysis of the products of SHS-Az. The work is performed at financial support of the Ministry of education and science of the Russian Federation within the basic part of State task for Samara State Technical University to perform research (code 1583).

References

[1] Basu B and Balani K 2011 *Advanced Structural Ceramics* (Hoboken, New Jersey: Wiley)
[2] Guo Z, Blugan G, Kirchner R, Reece M, Graule T and Kuebler J 2007 *Ceram. Int.* **33** 1223-9
[3] Evdokimov A A, Sivkov A A and Gerasimov D Yu 2016 *Glass Ceram.* **72** 381-6
[4] Huang J L and Nayak P K 2012 *Nanocomposites - New Trends and Developments* Ed F Ebrahimi (Rijeka: InTech) p 421-36
[5] Yoshimura M, Komura O and Yamakawa A 2001 *Scripta Mater.* **44** 1517-21
[6] Tatami J, Kodama E, Watanabe H, Nakano H, Wakihara T, Komeya K, Meguro T and Azushima A 2008 *J. Ceram. Soc. Japan* **116** 749-54
[7] Palmero P 2015 *Nanomater.* **5** 656-96
[8] *Nanomaterials and Related Products: Catalog & Price-List* URL: http:// www.plasmachem.com (accessed: 19.04.2016)
[9] He W, Zhang B, Zhuang H and Li W 2004 *Ceram. Int.* **30** 2211–4
[10] Manukyan K V, Kharatyan S L, Blugan G and Kuebler J 2007 *Ceram. Int.* **33** 379–83
[11] Yeh C L and Chen C Y 2009 *J. Alloys Comp.* **486** 853–8
[12] Amosov A P, Shiganova L A, Bichurov G V and Kerson I A 2015 *Mod. Appl. Sci.* **9** 133-44
[13] Shiganova L, Bichurov G, Kerson I, Novikov V and Ermoshkin A 2016 *Key Eng. Mater.* **685** 578
[14] Shiganova L A, Bichurov G V, Amosov A P, Titova Y V, Ermoshkin A A and Bichurova P G 2011 *Rus. J. Non-Ferr. Met.* **52** 91-5
[15] Amosov A P, Borovinskaya I P, Merzhanov A G and Sytchev A E 2005 *J. Self-Prop. High-Temp. Synth.* 2005 **14** 165-86

Development of feedstock of tungsten-nickel-iron-polyformaldehyde for MIM technology

D V Kostin[1,3], A V Parkhomenko[1], A P Amosov[1], A R Samboruk[1] and A V Chemashkin[2]

[1]Samara State Technical University, 244 Molodogvardeyskaya Street, Samara 443100, Russia
[2]Federal Research and Production Center "Production Association "Start"named after M V Protsenko", 1Mira Prospekt, Zarechny, Penzenskaya oblast 442960, Russia

[3]E-mail: deman777_91@mail.ru

Abstract. The article presents the results of the research and development of technology and formulation of the feedstock from domestic metal powders and polymers to fabricate complex-shaped components from heavy alloy of VNZh 7-3 brand (90 wt. % tungsten - 7% nickel - 3% iron) by Metal Injection Molding (MIM technology). The metal part of the feedstock is composed of powders of tungsten, nickel and iron, and the polymer part is composed of polyformaldehyde with the addition of low-density polyethylene and beeswax. The modes of mixing the components and the influence of the composition of the feedstock on the melt flow rate and the homogeneity of the feedstock were investigated. The optimal formulation of the feedstock was determined. Microstructure, density and hardness of control samples fabricated by MIM technology from the developed feedstock, correspond to, and in some respects are superior to the samples of VNZh 7-3 alloy fabricated by technology of traditional powder metallurgy.

Keywords: heavy alloy, metal powders, MIM technology, feedstock, polymers.

1. Introduction

Heavy composite pseudoalloys of tungsten-nickel-iron system (alloys of VNZh type in Russia) with a high specific weight and improved mechanical properties are well known and are used for manufacturing components and structures with special properties [1]. They are produced only by powder metallurgy methods: compaction of the powder mixture followed by sintering, hot pressing, infiltration of porous tungsten skeleton by molten metal binder [1]. However, these traditional methods of powder metallurgy allow the production of components of the simplest shapes only. In this connection, of interest is the use of MIM technology for the production of complex-shaped components from alloys of VNZh type. Technology of Metal Injection Molding (MIM) is a technology for injection molding of blanks of complex-shaped components from mixtures of melts of polymers with metal powders followed by removal of the polymer binder and sintering of metal powders [2]. This technology combines the capabilities of the technology of injection molding of thermoplastics by automatic molding machines and the technology of powder metallurgy and is widely used abroad for large-scale production of components for various purposes. The raw material

IOP Conf. Series: Materials Science and Engineering **156** (2016) 012033 doi:10.1088/1757-899X/156/1/012033

for the MIM-technology is a feedstock consisting of a mixture of metal powder and polymer binder. The most common binder for feedstocks is polyformaldehyde or polyacetal. The introduction of MIM technology at the factories of Russia is hampered by lack of feedstocks from domestic components, the development of which has just started [3, 4]. Therefore, it is important to develop the domestic MIM- feedstocks and for heavy alloys of VNZh type.

The aim of this work is to develop a MIM-feedstock of VNZh 7-3 heavy alloy from domestic metal powders and polyformaldehyde as a binder.

2. Materials and methods

The chemical composition of the VNZh 7-3 heavy alloy is the following: 6.5-7.5 wt. % nickel (Ni); 2.5-3.5% iron (Fe); tungsten (W) is the base [1]. The most suitable for MIM technology are metallic powders of spherical shape, which are not larger than 20 μm with an average size of about 5-10 μm [2]. Therefore, for development of metallic part of the feedstock, the following powders of domestic production were selected: the powder of carbonyl nickel of PNK-1L8 brand, consisting of rounded particles with a size up to 1 micron; the powder of carbonyl iron of VM brand, which is a mixture of spherical particles of size from 0.1 μm to 10 μm; the powder of tungsten of PV-2 brand, consisting of rounded particles in the size range 1 to 5 microns. For development of polymeric part of the feedstock, the following materials were used: polyformaldehyde with processing additives of "Tekhnaset A-110" brand (main binder); low-density polyethylene of 15813-020 brand (additional binder to hold the structure of the porous skeleton after removal of the main binder); beeswax according to GOST 21179-2000 (surfactant).

Particle-size distribution of powders was determined by laser particle-size analyzer Analysette 22 Compact. Microstructure of the materials was investigated by scanning electron microscope (SEM) Jeol JSM-6390A, and phase composition of materials was determined by X-ray diffractometer (XRD) ARL X'TRA.

The modes of preparation of the mixture of initial metal powders, or a powder batch, were worked out in a ball mill with metallic grinding bodies in the form of balls with a diameter from 5 to 10 mm and rollers with a diameter of 5 to 8 mm and a length of 8-10 mm. The powder batch load ranged from 300 to 500 g. For the preparation of feedstock, a laboratory installation was used consisting of two main parts heated by ultrathermostat: a batch blade mixer of polymer and powder components with a volume of 0.2 dm^3 and a hydraulic press with a die for extrusion a cord with a diameter of 5 mm followed by cutting it into granules 5 mm in length [3].

The quality of the feedstock was controlled for the following characteristics: a melt flow index of the feedstock on plastometer MELT INDEXER MI-2 at a temperature of 190°C, with inner diameter of capillary of 2 mm and a load of 21.6 kg; the homogeneity (uniformity) of the feedstock on a helium pycnometer AccuPyc II 1340.

Blanks of control samples for testing were molded in injection molding machine MicroPower 5/7.5 (obtaining green blanks). The removal of the polymer part from the feedstock (obtaining brown blanks) was conducted in a stepwise mode, close to that adopted for feedstocks on polyformaldehyde base. Sintering of the control samples was conducted in the environment of hydrogen with temperature up to 1440°C.

The hardness of the sintered samples of the VNZh 7-3 alloy was determined by the method of Brinell, the density was determined on a helium pycnometer AccuPyc II 1340.

3. Results and discussion

A study of particle-size distribution of the initial metal powders showed that the powders of nickel and tungsten are composed of particles with size of 1-5 μm, but these particles are combined into agglomerates with size of 40-70 μm. To prepare a homogeneous mixture of these powders, it is necessary to achieve the destruction of the agglomerates. To determine the optimal preparing time of the powder batch, the experiment was conducted on the dynamics of grinding components of the batch for four hours with sampling every half hour. The results of this experiment are presented in table 1 in the form of the change of wt. % different size fractions with time.

Table 1. The change of particle-size distribution with time of grinding, wt. %.

Grinding time, hour	Size fraction of powder mixture, µm							
	0-40	0-30	0-20	0-10	0-5	0-2	0.5-1.0	0.3-0.5
0.5	2.5	9.8	30.7	30.1	21.2	4.5	0.74	0.33
1.0		4.3	26.9	36.0	26.4	5.1	0.91	0.42
1.5		2.2	23.4	37.8	29.4	5.6	1.04	0.46
2.0		1.4	20.8	38.6	31.5	6.1	1.13	0.50
2.5		0.5	18.0	39.6	33.7	6.4	1.24	0.53
3.0			13.7	41.8	36.0	6.8	1.18	0.63
3.5		0.2	15.9	39.8	35.7	6.7	1.28	0.58
4.0		0.1	14.4	39.7	36.7	7.1	1.40	0.59

These results show that the intensive milling of the components of the batch occurs within the first three hours, when the fraction greater than 20 µm completely disappears. Further reduction in size does almost not happen, but on the contrary, there is agglomeration of particles. Thus, the optimum preparing time of the powder batch in a ball mill is 3 hours. (In the case of using other equipment, e.g., an attritor, the grinding time can be reduced, which requires additional study.)

The microstructure of the powder batch of VNZh 7-3 alloy, ground in a ball mill in the course of 3.0 hours, is presented in figure 1.

Figure 1. SEM image of powder batch after grinding during 3.0 hours.

It is seen that the mixture of powders is homogeneous and has no large agglomerates of particles larger than 10 µm.

The prepared powder batch was transferred into the chamber of the heated blade mixer, where the polymer components were added, and mixing was performed for 35 minutes in the temperature range 160-195°C with increasing temperature at a rate of 1 °C/min. Obtained metal-polymer composition, or feedstock batch, was placed into a preheated die for extruding and forming a cord with subsequent cutting of granules. In the course of the study, five experimental batches of the feedstock of VNZh 7-3 heavy alloy with different content of components were made, the composition of which is presented in table 2.

Table 2. Compositions of experimental batches of the feedstock of VNZh 7-3 alloy.

Number of feedstock batch	Content of components, wt. %			
	Powder batch	Technaset A-110	Low-density polyethylene	Beeswax
No. 1	90	8	1.0	1.0
No. 2	91	8	0.5	0.5
No. 3	92	7	0.5	0.5
No. 4	93	6	0.5	0.5
No. 5	94	5	0.5	0.5

The properties of the produced feedstocks were evaluated by the melt flow index and the deviation from homogeneity of the granules, the values of which are presented in table 3.

Table 3. Properties of experimental batches of the feedstock of VNZh 7-3 alloy.

Feedstock batch number	Average density, g/cm^3	Deviation from homogeneity, %	Melt flow index, g/10min
No. 1	-	-	1150
No. 2	8.42	1.1	430
No. 3	8.87	0.87	365
No. 4	9.76	0.5	230
No. 5	10.11	0.79	60

The melt flow index (MFR) shows the possibility of application of the feedstock to injection molding machines, as well as the correct ratio of metal and polymer parts. The melts of the feedstock batches No. 1 and 2 flowed easily out the die even under its own weight, wherein the cord grew thin and heterogeneous. When reducing the amount of polymer part in the feedstock, the cord coming out under pressure up to 5 MPa, however, he had a uniform cylindrical shape. But the feedstock batch No. 5 contained deficient polymer part and the cord was formed with difficulty. The feedstock batches No. 3 and 4 were the most suitable, their values of MFR correspond to the typical values of MFR of feedstocks on the base of polyformaldehyde (200-500 g/10min), i.e. the optimal ratio between the metal and polymer parts in the feedstock is in the range of 92-93 wt. % metal powders.

Homogeneity shows the degree of uniformity of distribution of metal powders in a polymer binder. As follows from table 3, the feedstock batches No. 3 and 4 have deviation from homogeneity of less than 1%, which is an indicator of high uniformity.

In the study of the microstructure of the feedstocks on electron microscope, pores and voids, large and small inhomogeneous inclusion were detected. Figure 2 shows the microstructure of the feedstock of batch No. 4.The structure of the feedstock is rather uniform, the maximum particle size of the metal phase does not exceed 6 μm.

Four feedstock batches No. 2-5 were used to make control samples by MIM technology to determine mechanical properties and structure of sintered material corresponding to composition of VNZh 7-3 heavy alloy. The results of measuring Brinell hardness of the samples are the following: 285, 310, 333 and 308 HB for batches No. 2, 3, 4 and 5 correspondingly. The hardness of the samples fabricated by MIM technology is 10% higher the hardness of the samples of VNZh 7-3 composition made by technology of traditional powder metallurgy (260-280 HB). The results of measuring the density of the control samples made by MIM technology are the following: 17.0546, 17.0957, 17.0840 and 17.1245 g/cm^3 for batches No. 2, 3, 4 and 5 correspondingly. These values of the density of the

MIM samples are even slightly higher than the density values (16.6-17.0 g/cm^3) specified for VNZh 7-3 heavy alloy, made according to traditional technology of powder metallurgy [1].

Figure 2. SEM magnification of feedstock of batch No. 4.

The microsections were made of sintered control samples, obtained at MIM-technology and traditional technology of powder metallurgy (PM), and their microstructure was analyzed (figure 3).

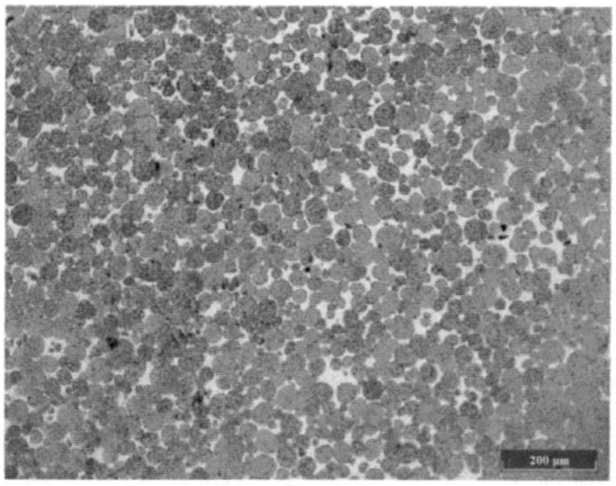

Figure 3. Microstructure of sample sintered from feedstock of batch No. 4.

The microstructure of the MIM samples is fully consistent with the microstructure of the PM samples: rounded tungsten grains surrounded by a Ni-Fe binder. The grain size is 30-40 µm, the individual grains reach 50-70 µm in size.

4. Conclusion

As a result of the study, the technology and composition of the feedstock from domestic metal powders and polymers are developed to manufacture complex-shaped components from VNZh 7-3 heavy alloy by the technology of injection molding and sintering (MIM technology). Microstructure, density and hardness of control samples fabricated by MIM technology from developed feedstock, fully meet, and in some respects superior to the samples of this alloy produced by the technology of traditional powder metallurgy.

Acknowledgements

The authors thank A Yu Illarionov for assistance in SEM investigations.

References

[1] *Metallic powders and powder materials: reference book* 2005 ed Yu V Levinskiy (Moscow: EKOMET)

[2] German RM and Rose A 1997*Injection molding of metals and ceramics*(Princeton: Metal Powder Industries Federation)

[3] Parhomenko A V, Amosov A P, Samboruk AR, Ignatov SV, Kostin DV and Shul'timova AS 2015*Rus. J. of Non-Ferr. Metals* **56**(1) 68–72

[4] Russian researchers work to ensure domestic MIM feedstock supply 2015 *Powder Injection Moulding International* **9**(4) 30

IOP Conf. Series: Materials Science and Engineering **156** (2016) 012034 doi:10.1088/1757-899X/156/1/012034

Sintering oxide ceramics based on Al_2O_3 and ZrO_2, activated by MgO, TiO_2 and SiO_2 additives

S V Matrenin, B S Zenin and R V Tayukin

Tomsk Polytechnic University, 30 Lenin Avenue, 634050, Tomsk, Russia,

E-mail: msv@tpu.ru

Abstract. The positive effect of the addition of MgO and TiO_2 in an amount of no more than 1 wt. % on sintering and physico-mechanical properties of alumina ceramics is established. Addition of 5% of SiO_2 to Al_2O_3 provides the mechanism of liquid phase sintering of ceramics, which leads to increase in its density and strength up to 480 MPa. In ceramic system Al_2O_3 - ZrO_2 - Y_2O_3 highest level of physical and mechanical properties of the composition had a hypereutectic composition 16.6% Al_2O_3 - 76% ZrO_2 – 7.4% Y_2O_3. In this composition two mechanisms of hardening are realized simultaneously, such as transformational hardening by *t*-*m* - ZrO_2 transition and dispersion strengthening with high-modulus particles of α- Al_2O_3.

1. Introduction

Strong ceramics based on Al_2O_3 and ZrO_2 are very promising structural and functional materials. The most common methods of obtaining solid alumina ceramics are powder technology methods, i.e. various kinds of pressing and sintering, modified as applied to the ceramic. The modification consists in choosing the optimal parameters of ceramics consolidation. These parameters are compacting pressure and methods of its application, the temperature of sintering treatment, the environment and the speed of the process. Nevertheless, wide spread practical solid oxide ceramics prevent complexity and low productivity of hot isostatic pressing and hot technology [1, 2], allowing to obtain materials with high tensile strength [3]. A relatively simple technique of uniaxial compression followed by a free sintering usually cannot produce ceramics with high mechanical properties [4].

The purpose of this work was to study the patterns of consolidation of ceramics based on Al_2O_3 and ZrO_2, activated by the addition of the charged powders MgO, TiO_2, SiO_2.

2. Experiment

Industrial nanocrystalline oxide powders (NP) Al_2O_3, Al_2O_3 – ZrO_2 – Y_2O_3, obtained by plasma chemical synthesis (PCS) are used. Oxide powders obtained PHS have a characteristic form of hollow spheres consisting of nanocrystallites and amorphized intergranular phase [5]. The size of the spheres varies between 100…1000 nm, the size of nanocrystals which form the sphere, is between 50…100 nm.

Industrial submicron powders MgO and TiO_2 were used as activating additives. Powder MgO was added to Al_2O_3 with the aim of recrystallization braking during sintering of corundum and preserving fine structure of the sintered ceramic. TiO_2 was used to reduce the sintering temperature of alumina ceramics. Powder SiO_2 was prepared in the laboratory by crushing quartz glass and grind in the planetary mill. Fraction was used with a particle size <40 microns. SiO_2 was added in Al_2O_3 powder in order to implement the mechanism of liquid phase sintering and receive mullite-corundum ceramics in accordance with the phase diagram SiO_2 – Al_2O_3. Powders Al_2O_3, Al_2O_3 – ZrO_2 – Y_2O_3 were annealed in the air atmosphere in a high temperature electric resistance furnace at 1450°C for one hour to convert γ-Al_2O_3 into α-Al_2O_3.

In order to obtain powder compositions, to improve the processing performance and to enhance their activity, annealed powders were treated in the planetary ball mill for 20 minutes at a rotational speed of the grinding vessel 20 Hz. Grinding bodies were zirconia balls. The content of components in the system Al_2O_3 - ZrO_2 - Y_2O_3 varied for obtaining hypoeutectic, eutectic and hypereutectic compositions in accordance with the Al_2O_3 – ZrO_2 phase diagram (Figure 1).

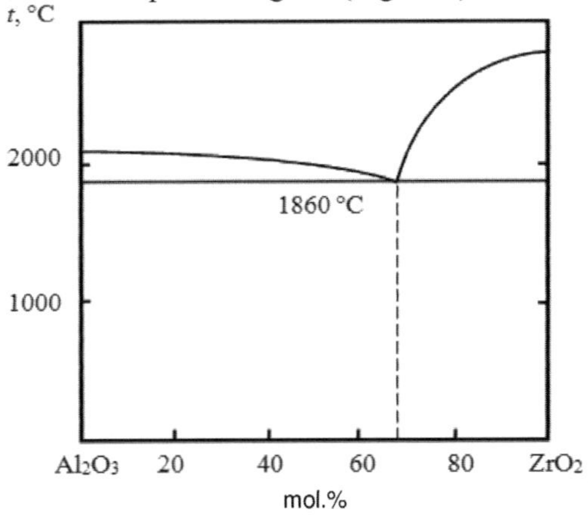

Figure 1. Phase diagram of the Al_2O_3 – ZrO_2 system [5].

The treated powders were sieved for 10 minutes at vibratory drive to obtain a fraction < 40 micrometers and plasticized with a water-solution of carboxymethylcellulose (CMC). Plasticized powders were formed by uniaxial pressing in a rigid mold, compaction pressure was 300 MPa. The obtained compacts were cylinders with diameter 25±0.01 mm and height of 5±0.01 mm.

Sintering the compacts was carried out in a high temperature resistance furnace at a temperature 1700°C, holding time at this temperature was 1 hour.

Density of sintered specimens ρ was determined by hydrostatic weighing in an ethanol with an accuracy ±0.001 g. The relative density Θ of the specimens was also calculated in accordance with equation:

$$\theta = \frac{\rho}{\rho_t} \cdot 100\ \%,$$

where ρ_t is a theoretical density of ceramics.

Indentation was performed using device Nano Indenter G 200. Berkovich pyramid was used as indenter, a load was 500 mN. The device automatically calculates the elastic modulus E_{IT} and microhardness H_{IT} in compliance with ISO 14577.

The strength of the sintered ceramics was determined by the «Scratch Testing» method also using the device Nano Indenter G 200. Essence of the method is to apply a scratch on the surface of the sample by linearly increasing load (as a rule from 0 to 40 mN) and determination of the profile depth and width of the scratch. The strength of the samples was calculated according to the formulas

$$\sigma = \frac{F_n}{A_s \sin\alpha} \quad u \; A_s = \frac{a^2}{2\sqrt{3}\sin\alpha} + \frac{ah}{\cos\alpha},$$

where F_n – normal load, A_s – projection print area, a and h – profile width and depth of the scratch, α=65° (for the Berkovich pyramid).

3. Results of the experiment

Table 1 shows the density values of samples pressed under a pressure of 300 MPa. It should be noted that all samples have a regular cylindrical shape, exfoliate and shedding edges were not observed.

Table 1. The density of compacts.

No.	Composition (wt. %)	ρ_c (g·cm^{-3})	ρ_t (g·cm^{-3})	Θ_c (%)
1	99.6% Al_2O_3 – 0.4% MgO	2.13	3.97	53.65
2	98.6% Al_2O_3 – 0.4% MgO – 1% TiO_2	2.18	3.97	54.91
3	97.6% Al_2O_3 – 0.4% MgO – 2% TiO_2	2.19	3.97	55.16
4	95.6% Al_2O_3 – 0.4% MgO – 4% TiO_2	2.24	3.97	56.42
5	94.6 % Al_2O_3 – 0.4% MgO – 5% SiO_2	2.20	3.87	56.85
6	89.6 % Al_2O_3 – 0.4% MgO – 10% SiO_2	2.16	3.78	57.14
7	76.1% Al_2O_3 – 21.8% ZrO_2 – 2.1% Y_2O_3	2.22	4.29	51.75
8	31.7% Al_2O_3 – 62.2% ZrO_2 – 6.1% Y_2O_3	1.96	5.04	38.89
9	16.6% Al_2O_3 – 76% ZrO_2 – 7.4% Y_2O_3	1.31	5.36	24.44
10	92.9% ZrO_2 – 7.1% Y_2O_3	2.84	5.78	49.13

Table 2 shows the values of density of the sintered samples.

Table 2. Density of sintered samples.

No.	Composition (wt. %)	ρ_s (g·cm^{-3})	ρ_t (g·cm^{-3})	Θ_s (%)
1	99.6% Al_2O_3 – 0.4% MgO	3.60	3.97	90.68
2	98.6% Al_2O_3 – 0.4% MgO – 1% TiO_2	3.66	3.97	92.07
3	97.6% Al_2O_3 – 0.4% MgO – 2% TiO_2	3.57	3.97	89.92
4	95.6% Al_2O_3 – 0.4% MgO – 4% TiO_2	3.67	3.97	92.44
5	94.6 % Al_2O_3 – 0.4% MgO – 5% SiO_2	3.44	3.87	88.89
6	89.6 % Al_2O_3 – 0.4% MgO – 10% SiO_2	3.02	3.78	79.89
7	76.1% Al_2O_3 – 21.8% ZrO_2 – 2.1% Y_2O_3	3.25	4.29	75.76
8	31.7% Al_2O_3 – 62.2% ZrO_2 – 6.1% Y_2O_3	3.39	5.04	67.26
9	16.6% Al_2O_3 – 76% ZrO_2 – 7.4% Y_2O_3	3.07	5.36	57.28
10	92.9% ZrO_2 – 7.1% Y_2O_3	5.16	5.78	89.27

It is seen that the highest density after sintering had a sample of the composition No.4 of 95.6% Al_2O_3 – 0.4% MgO – 4% TiO_2. This is due to the formation of a solid solution of subtraction TiO_2 in α-Al_2O_3, grill of which has a high diffusivity and activates the sintering process. Addition of SiO_2 in alumina ceramics in an amount more than 10% improves the sinterability of the ceramic. Samples of ceramic system Al_2O_3– ZrO_2 – Y_2O_3 investigated compositions also had a low level of relative density after sintering. Table 3 and diagrams (Figure 2) show the physical and mechanical properties of sintered samples determined by nanoindentation technique.

Table 3. Physical and mechanical properties of sintered ceramics.

No.	Composition (wt. %)	E_{IT} (GPa)	H_{IT} (GPa)	σ (MPa)
1	99.6% Al_2O_3 – 0.4% MgO	422.2	19.8	18
2	98.6% Al_2O_3 – 0.4% MgO – 1% TiO_2	400.1	23.2	272
3	97.6% Al_2O_3 – 0.4% MgO – 2% TiO_2	440.4	22.6	78

4	95.6% Al_2O_3 – 0.4% MgO – 4% TiO_2	178.2	99.1	38
5	94.6 % Al_2O_3 – 0.4% MgO – 5% SiO_2	427.2	17.4	480
6	89.6 % Al_2O_3 – 0.4% MgO – 10% SiO_2	292.7	19.9	48
7	76.1% Al_2O_3 – 21.8% ZrO_2 – 2.1% Y_2O_3	356.1	18.5	71
8	31.7% Al_2O_3 – 62.2% ZrO_2 – 6.1% Y_2O_3	329.2	22.1	193
9	16.6% Al_2O_3 – 76% ZrO_2 – 7.4% Y_2O_3	439.2	24.3	274
10	92.9% ZrO_2 – 7.1% Y_2O_3	230.2	14.9	93

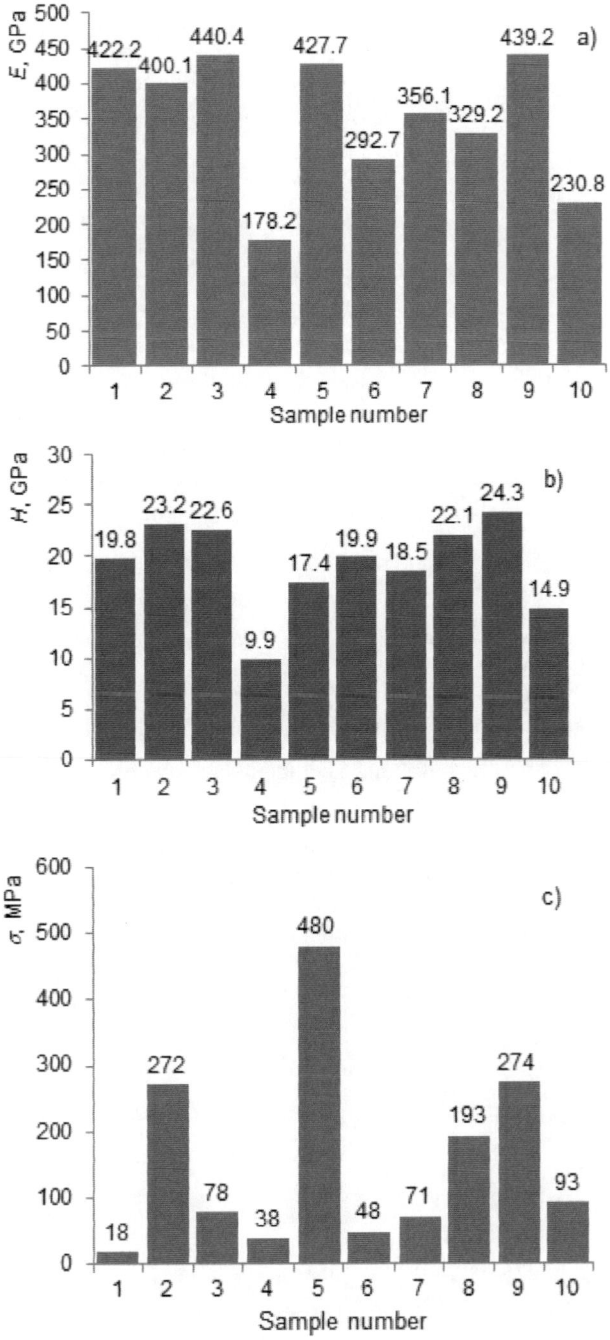

Figure 2. a) Elasticity modulus, b) microhardness, c) strength of sintered ceramics.

It is seen that the elasticity modulus and hardness of ceramics based on the corundum generally correspond tabular (19…21 GPa), while the scatter of strength values determined by the method of scratching is significant. This is due to feature Scratch method: scratching distance of 200 microns, the depth varied between 100…800 nm, the indenter often was stopped on the softening of the grain boundaries (Figure 2). In general, it should be noted that Scratch testing technique requires a very high quality surface preparation of the samples. However, they should have minimal residual porosity.

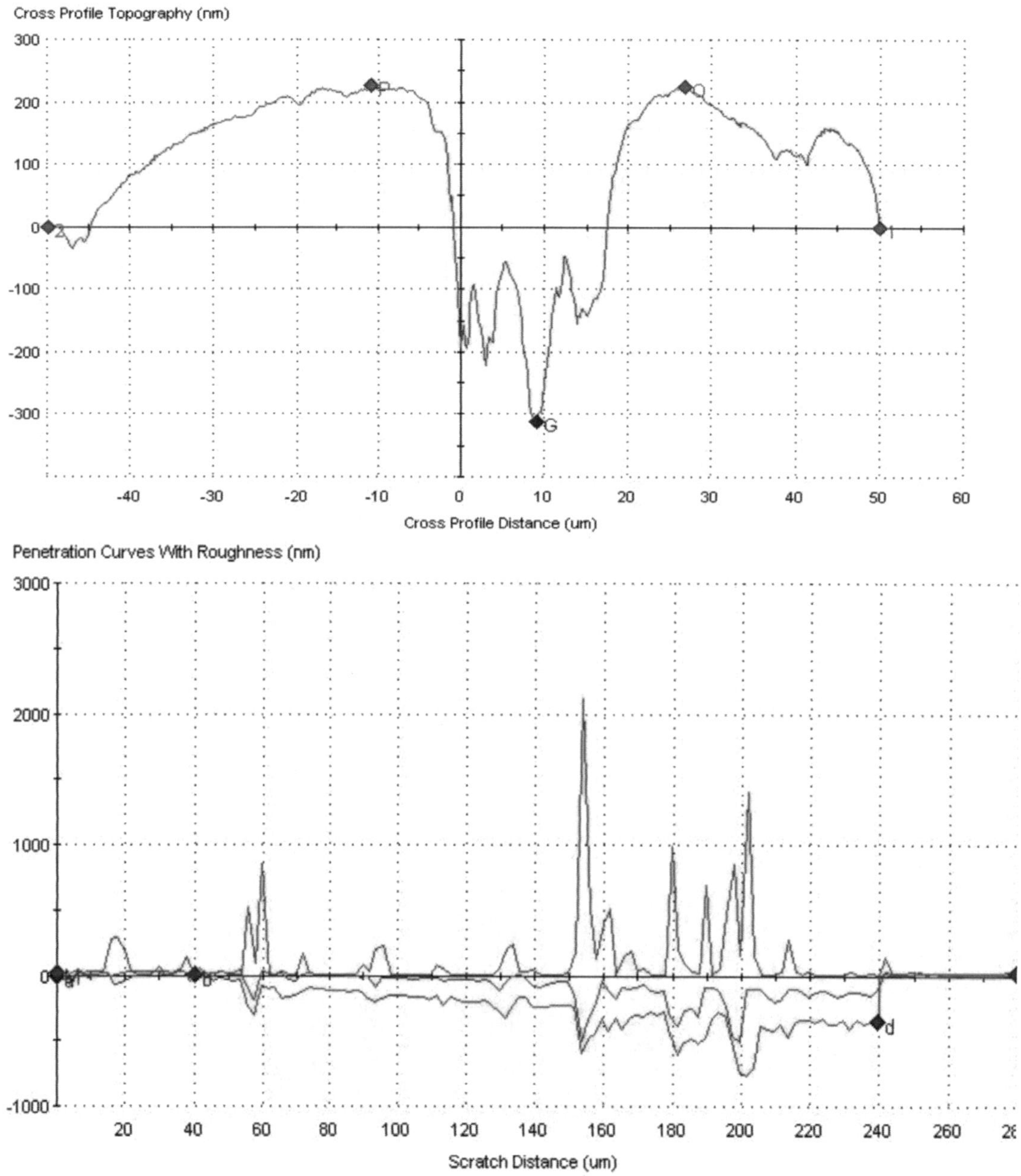

Figure 3. The profile and length of the scratch on the surface of the sample No. 10.

IOP Conf. Series: Materials Science and Engineering **156** (2016) 012034 doi:10.1088/1757-899X/156/1/012034

4. Conclusion

The positive effect of the addition of MgO and TiO_2 in an amount of no more than 1 wt. % on sintering and physico-mechanical properties of alumina is established.

Addition of 5 percent by weight of SiO_2 to Al_2O_3 provides a mechanism of liquid phase sintering of ceramics, which leads to increase of its density and strength up to 480 MPa (table value of flexural strength high density alumina is not more than 400 MPa).

Introduction to the corundum powder additives TiO_2 submicron powder leads to the formation of TiO_2–Al_2O_3 subtraction solid during sintering, which has a high diffusivity and activates the sintering process.

From the investigated compounds ceramic system Al_2O_3 – ZrO_2 – Y_2O_3 highest level of physical and mechanical properties had a hypereutectic composition 16.6% Al_2O_3 – 76% ZrO_2 – 7.4% Y_2O_3. In this composition two mechanisms of hardening are realized simultaneously, such as transformational hardening by *t-m* - ZrO2 transition and dispersion strengthening with high-modulus particles of α-Al_2O_3.

References

[1] Kiparisov S and Padalko O 1988 *Equipment enterprises powder metallurgy* (Moskow: Metallurgy)
[2] Hunold K 1985 Hot isostatic pressing of high temperature ceramics *Interceram.* **39** 2 38, 40-43
[3] Lukin E, Makarov N, et. 2003 Strong and extra strong ceramic based on aluminum oxide and partially stabilized zirconia *Glass and ceramics* **9** 32-34
[4] Smirnov A 1990 *Structural ceramics* (Results of science and technology VINITI. Powder metallurgy) 64-106
[5] Schubert H 1987 *Diss. Dokt. Naturwiss.* (fak. chem. Univ. Stuttgart)

Producing composite materials based on ZrB_2, ZrB_2-SiC

Yu A Mirovoi[1], A G Burlachenko[1,3], S P Buyakova[1,2,3], I N Sevostiyanova[1] and S N Kulkov[1,2,3]

[1]Institute of Strength Physics and Materials Science SB RAS, 2/4, pr. Akademicheskii, Tomsk, 634055, Russia
[2]Tomsk Polytechnic University, 30 Lenin Avenue, Tomsk, Russia, 634050
[3]Tomsk State University, 36 Lenin Avenue, Tomsk, Russia, 634050

E-mail: y.a.mirovoy@gmail.com

Abstract. The effect of mechanical treatment by planetary ball milling on the properties of hot pressed ZrB_2 - SiC ceramics was studied. It has been shown that material densification after mechanical treatment is finished at initial stages of sintering process. Addition of SiC causes a substantial increase in density of the sample to 99% of the theoretical powder containing 20% of silicon carbide, in comparison with samples ZrB2 density not exceeding 76%. It has been shown that all defects which were accumulated during mechanical treatment anneal in hot pressure process and there are no any changes of CDD values in sintered ceramics.

1. Introduction

High-melting compounds are the main part of the materials utilized in high-temperature technologies for heat protection of aerospace machinery and in electronics. Between high melting point materials, ZrB_2-based ceramic composites are the point of special interest. ZrB_2 ceramics with SiC addition demonstrate high thermochemical stability, including outstanding resistance to oxidation under high and ultra-high temperatures [1].

ZrB2-SiC composites are being produced via sintering process under high pressure and temperature of 2000 °C [2,4]. Strong covalent bond in ZrB2 leads to low atoms diffusibility, thus making its processing more complicated. As all the substances with covalent bond, zirconium diboride demonstrates lack of concentration and defect mobility of the lattice cell. Low velocity of the diffusion-viscous flow necessitates activation of consolidation via crystal structure imperfection decrease, diffusion processes acceleration and alleviation of the plastic deformation during baking process. Mechanical treatment of the powders in high-energy mills stays one of the most effective methods for baking temperature reduce and superdense microcrystalline structure production of high-melting compounds without application of activating additives which usually have negative effect on the thermal properties of the material [3].

The aim of this work was to investigate the influence of mechanical treatment of ZrB_2-SiC powder system on the structure of the produced ceramic composites.

2. Experiment

ZrB_2 (d_{50} =2.5μm)-SiC(d_{50} = 4.2 μm) powder blends with silicon carbide content of 10%, 15% and 20% (v/v) were used. The blends were mechanically treated in planetary mill with grinding bodies' centrifugal acceleration of 30 g (where *g* is gravitational acceleration). Treatment time was from 1 to 20 minutes.

Ceramic composites sintering was performed at a temperature of 1800 °C and at a pressure of 75 MPa. Time of isothermal exposure was 30 minutes. Phase composition and crystal lattice parameters were studied by means of XRD analysis with CuK_α irradiation. Scanning electron microscope observations operated at 20-30 kV were used to average grain size using Tescan VEGA-3SBH. Specific surface area was investigated with BET method.

3. Results and Discussion

Dependence of the specific surface area from the duration of mechanical treatment of ZrB_2 and ZrB_2-SiC powders in planetary mill is plotted in Figure 1. Before the mechanical treatment specific surface area of the powders was not differ. Maximum surface area (7.6 m^2/g) was found for ZnB_2 powder. After the mechanical treatment during 20 minutes it was reduced to 5.9 m^2/g. These results don't differ significantly from ones for ZrB_2-SiC powders (7 m^2/g and 5.8 m^2/g before and after the treatment, respectively). Thus, for all the investigated powder systems the reduce of surface area with increase of the mechanical treatment time was found.

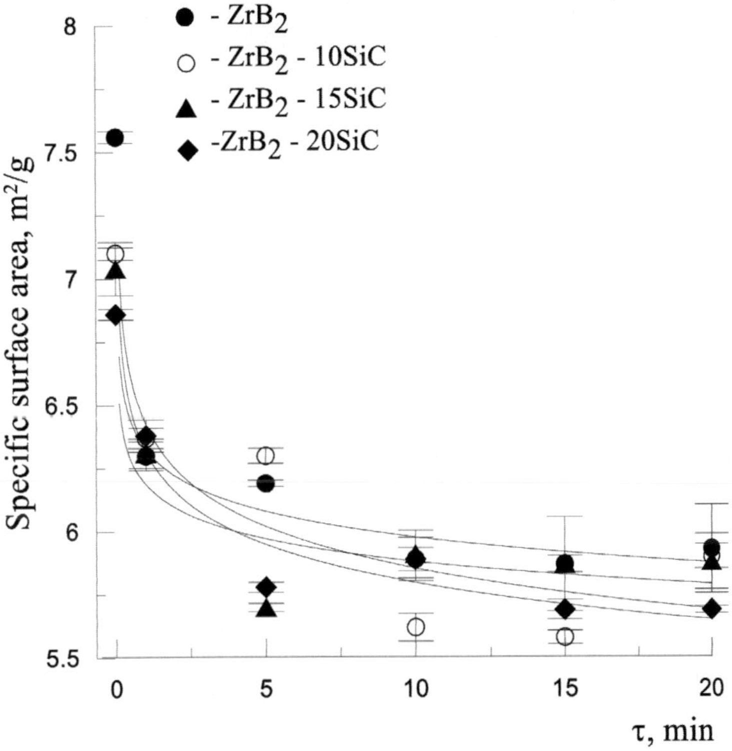

Figure 1. The influence of the mechanical treatment time on BET surface area for ZrB_2 and ZrB_2-SiC powder systems.

Changes of the powder morphology with mechanical treatment duration are shown in Figure 2 and Figure 3 using ZrB_2 - 20%SiC powder system. After 20 minutes of the mechanical treatment, formation of the aggregates with size of 15 μm was observed.

XRD analysis of the ZrB_2 powder and ZrB_2-SiC powder systems indicated no phase changes during the mechanical treatment. Only ZrB_2 reflections were marked in the XRD pattern of the ZnB_2 and ZrB_2-

SiC powder system containing 5% wt. of SiC. On the XRD patterns with higher concentration of SiC its reflections were distinguished.

With increase of the mechanical treatment time ZrB_2 reflection tailing was observed that may be a result of ZrB_2 dispersity rise. Calculations of the X-Ray coherent-scattering regions size demonstrated that with prolongation of the mechanical treatment time it was reduced from 46 to 37 nm (Figure 4). In that time, any changes in X-Ray coherent-scattering regions size with increase of SiC concentration were not observed.

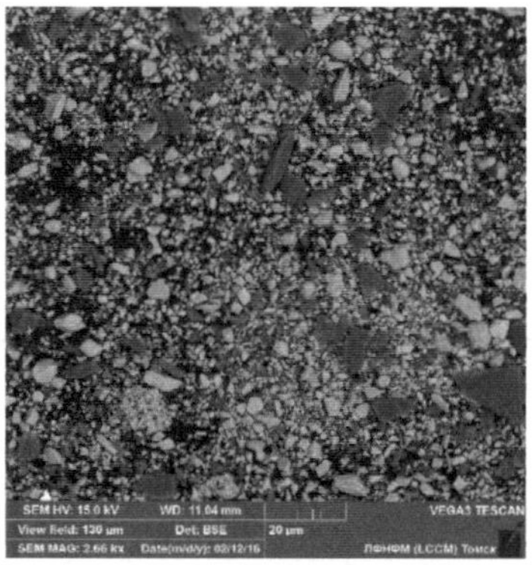

Figure 2. Original blend.

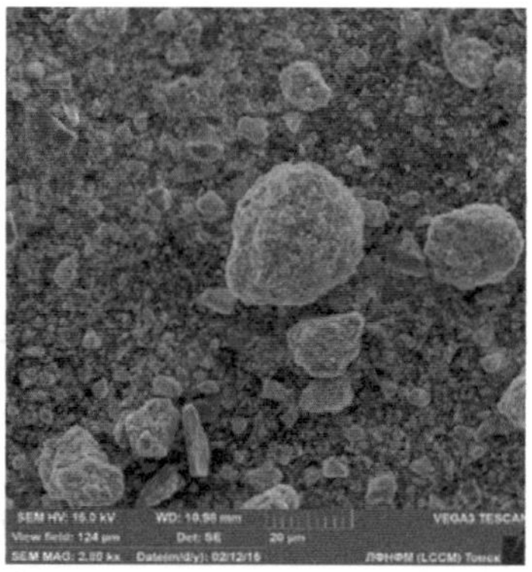

Figure 3. Changing of the powder morphology during the milling process after 20 minutes of the mechanical treatment.

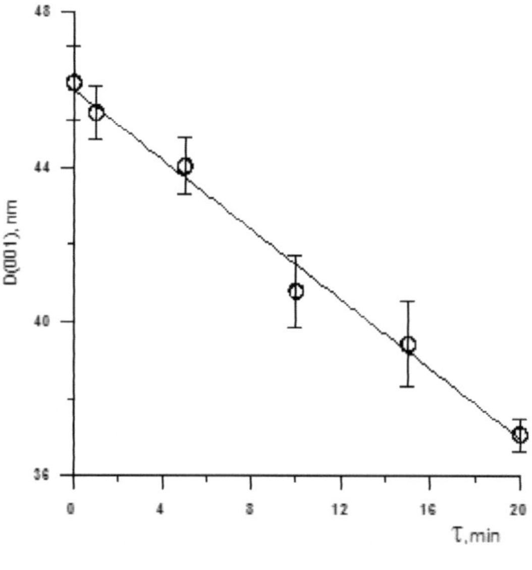

Figure 4. The influence of the mechanical treatment time on the coherent-scattering regions size of the ZrB_2 powders.

Phase structure analysis of the ZrB_2 ceramics and ZrB_2-SiC composite didn't show any differences from powders phase structure. Mechanical treatment influenced ZrB_2 X-Ray coherent-scattering regions size in consolidated materials. In ZrB_2 ceramics and ZrB_2-SiC composites produced from the powders, mechanically treated in planetary mill for 20 minutes, average ZrB_2 X-Ray coherent-scattering regions size was found of 85 nm, which is less than for non-treated powders (98 nm).

Mechanical treatment of the powders influenced density of the produced ZrB_2 ceramics and ZrB_2-SiC composites (Figure 5). For both ceramics and composites increases of the mechanical treatment time lead to the growth of density. However, the largest density was demonstrated by ZrB_2-SiC composites. Thus, relative density of ZrB_2 ceramics was less than 0.76, while for ZrB_2-SiC (80%/20% wt.) it was found at 0.98 (Figure 5).

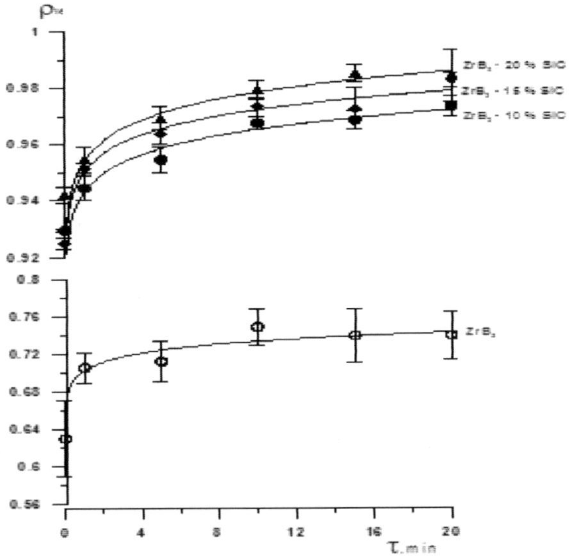

Figure 5. Influence of the mechanical treatment time on relative density of the hot pressured ceramics.

4. Summary

It was demonstrated that mechanical treatment of ZrB_2 and ZrB_2-SiC powders leads to the reduction of their specific area. That fact may be explained by agglomerates formation. Mechanical treatment of the ZrB_2-SiC powders increases density of the produced ceramic materials.

Acknowledgement

This work was partially financially supported by the Ministry of Education and Science of the Russian Federation in the framework of Agreement No. 14.607.21.0056 (RFMEFI60714X0056).

References

[1] Eakins E, Jayaseelan D and Lee W 2011 Toward Oxidation-Resistant ZrB2–SiC Ultra High Temperature Ceramics *Metallurgical and Materials Transactions* A **42** 878-887

[2] Mashhadi M, Khaksari H and Safib S 2015 Pressureless sintering behavior and mechanical properties of ZrB2–SiC composites: effect of SiC content and particle size *Journal of Material Research and Technology* **4** 416-422

[3] Jastin J and Jankowiak A 2011 Ultra high temperature ceramics: densification, properties and thermal stability *Aerospacelab Journal* **3** 1-11

[4] Bongiorno A, Först C and Kalia R 2006 A Perspective on Modeling Materials in Extreme Environments: Oxidation of Ultrahigh-Temperature Ceramics *MRS Bulletin* **31** 410-418

The structure of heat-resisting alloy modified by thermal treatment and alloyed by rhenium and lanthanum

E V Kozlov[1], N A Popova[1], E L Nikonenko[1,2], A A Kondratyuk[2], M E Kuznetsov[1] and A A Klopotov[1,3]

[1]Tomsk State University of Architecture and Building, 2, Solyanaya Sq., 634003, Tomsk, Russia
[2]National Research Tomsk Polytechnic University, 30, Lenin Ave., 634050, Tomsk, Russia
[3]National Research Tomsk State University, 36, Lenin Ave., 634050, Tomsk, Russia

E-mail: vilatomsk@mail.ru

Abstract. The paper presents the scanning and transmission electron microscope investigations of the structure, phase composition, and morphology of heat-resisting alloy modified by thermal treatment and additionally alloyed by rhenium and lanthanum. Rhenium alloy is obtained by the directional crystallization technique. The structural investigations are carried out for three alloy states, i.e. 1) original (after the directional crystallization); 2) 1150°C annealing during 1 h and 1100°C annealing during 480 h; 3) 1150°C annealing during 1 h and 1100°C annealing during 1430 h. It is shown that fcc-based γ- and γ'-phases are primary in all states of the alloy. γ'-phase has L12 structure, while γ-phase is a disordered phase. Rhenium and lanthanum are phase-forming elements. Investigations show that high-temperature annealing modifies the structural and phase conditions of the heat-resisting alloy.

1. Introduction

The properties of heat-resisting nickel alloys are determined for the modern engineering industry by thermal stability of their structure, sizes, shape and amount of strengthening γ'-phase, strength properties of γ-phase solid solution. Usually, such alloys are strengthened by alloying with such phase-forming elements as rhenium and lanthanum. This process increases the operating temperature owing to the melting temperature of forming phases.

2. Materials and methods

This paper mainly focuses on the qualitative and quantitative investigations of the structure, phase composition and morphology of heat-resisting alloy additionally alloyed by rhenium Re and lanthanum La both in the original state and after thermal treatment. Three alloy specimens in different states were investigated: 1) original state (after the directional crystallization); 2) original state after 1150°C annealing during 1 h and 1100°C annealing during 480 h; 3) original state after 1150°C annealing during 1 h and 1100°C annealing during 1430 h. All states of the specimens possessed a single-crystal structure with [001] orientation.

The heat-resisting alloy comprised ~70 at.% Ni; ~17 at.% Al and ~5 at.% Cr. The total amount of the primary alloying elements Mo, W, Ta, Ti, Co, Co was ~7 at.% and 0.4 at.% Re and 0.006 at.% La. The concentration of La in specimen 3 was 0.08 at.%.

The scanning electron microscope (SEM) and transmission electron microscope (TEM) investigations were used in this experiment.

3. Results and discussion

3.1. Phase composition

TEM investigations show the presence of a number of phases in different states of the alloy the qualitative and quantitative composition of which depends on its thermal treatment (see Table 1).

Phases observed in the alloy are classified into primary and secondary. This classification is based on the volume fraction of phases, their contribution to and stationary or single presence in the alloy. According to Table 1, γ'- and γ-phases are primary. Usually, they form the basic structure of the alloy and are present in the form of γ'-phase quasicuboids layered with γ-phase. All other phases are secondary and formed in individual cases. The volume fractures of phases presented in different states of the alloy are also given in Table 1. The error of determination of the volume fractions is ±1%. As this Table shows, the volume fracture of γ'-phase keeps high in all states of the alloy. The γ/γ' ratio in specimens 1 and 2 is practically similar (0.09 and 0.10, respectively). The introduction of La in specimen 3 in a larger amount reduces γ/γ' ratio, because Ni is withdrawn by La more intensively from γ-phase for the formation of Ni_3La_2 phase.

In Table 1, β-phase indicates two phases, namely β-NiAl and AlRe [1,2]. These phases possess CsCl structure and can transform to tetragonal phases of $L1_0$ type. They are mutually soluble and can form the solid solutions. The presence of β-NiAl and AlRe phases on Ni-Al and Al-Re sides of triangular diagram Ni-Al-Re indicates to their ability to form a large area that connects these positions. Since AlRe-phase has no a specific notation herein, except for its composition, β-NiAl and AlRe phases are denoted by β letter.

Table 1. Phase composition and quantitative properties of phases

Phases	Crystal lattice	Space group	Volume fracture, %		
			Specimen 1	Specimen 2	Specimen 3
γ'	Cuboidal	Pm3m	85.60	90.00	87.00
γ	Cuboidal	Fm3m	8.00	9.00	6.00
γ/γ'			0.09	0.10	0.07
β	Cuboidal	$Pm\bar{3}m$	5.00	0	0
χ	Cuboidal	$I\bar{4}3m$	1.40	0	0
σ	Tetragonal	$P4_2/mnm$	0	1.00	2.00
La_2Ni_3	Orthorombic	*Cmca*	0	0	5.00

According to Table 1, σ and χ phases relate to Frank-Kasper phases or topologically close-packed phases [1,2]. These phases are formed due to Re presence in the alloy. Actually, the interaction of Re with Mo and W results in the formation of σ and χ phases. The former one results from high-temperature hardening of the solid solution, the latter one results from its low-temperature breakdown and crystallization. Moreover, σ-phase is formed as a result of Re interaction with Co, Cr, Ni, Al, while χ-phase results from Re interaction with Mo and W as well as with Co and Al.

Finally, Ni_3La_2 phase is formed just in La-alloyed alloy in a larger amount, i.e. in specimen 3 (see Table 1). Except for this phase, other phases are represented by concentrated solid solutions observed in a wide concentration range and involved in all phase transformations.

3.2. Phase morphology

As it is known, the structure of superalloys is represented, first of all, by ($\gamma+\gamma'$)-phase mixture. A presence of such active metals as Re and La in superalloy, complicates its structure and leads to the formation of areas with damaged (γ+γ′)-structure [3-5]. Due to Re, β-phase is formed partially, while σ and χ phases are formed completely. The introduction of La promotes the formation of Ni_3La_2 phase (see Table 1). The formation of these four phases results in a serious damage of the quasi-cuboidal structure of (γ+γ′)-phases. Since the distribution of Re and La in the alloy is not uniform, only a part of (γ+γ′)-phase quasi-cuboids is damaged. The experimental research has assisted in schematic representation of the specimens' structure given in Figures 1-3.

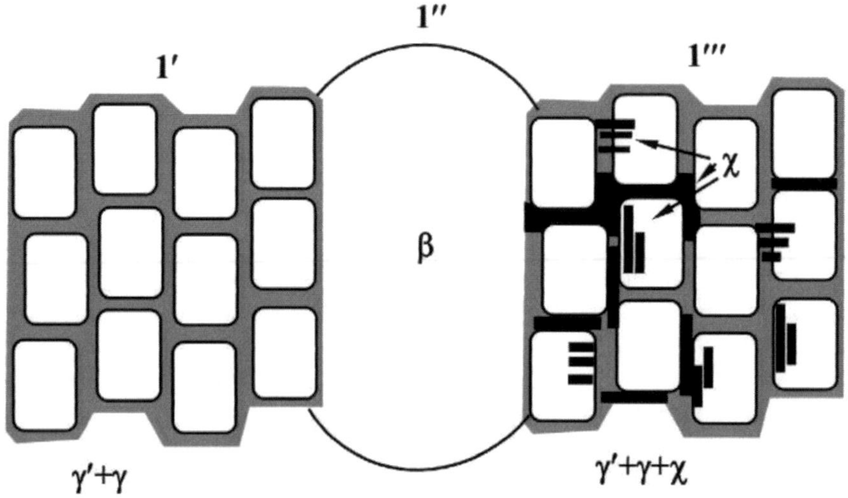

Figure 1. Schematic representation of specimen #1 structure: three states of phase morphology: *1′* – ideal structure of (γ+γ′)-phase quasi-cuboids; *1″* – a part of β-phase; *1‴* – (γ+γ′)-phase quasi-cuboids with χ-phase layers

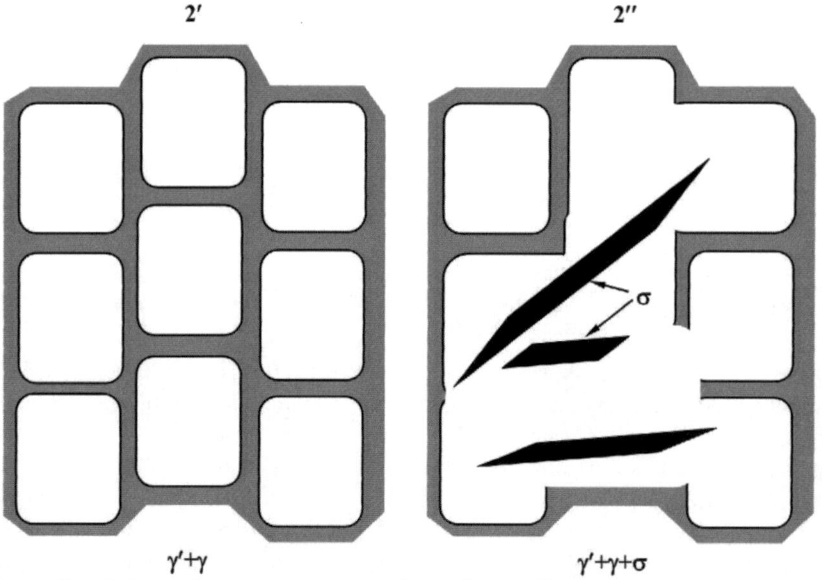

Figure 2. Schematic representation of specimen #2 structure: two states of phase morphology: 2′ – ideal structure of (γ+γ′)-phase quasi-cuboids; 2″ – (γ+γ′)-phase quasi-cuboids with σ-phase separations

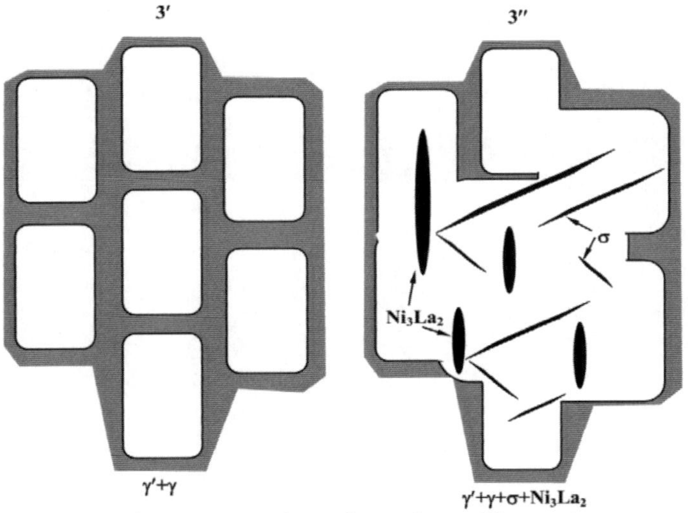

Figure 3. Schematic representation of specimen #3 structure: two states of phase morphology: 3′ – ideal structure of (γ+γ′)-phase quasi-cuboids; 3″ – (γ+γ′)-phase quasi-cuboids with σ-phase and Ni_3La_2 separations.

3.2.1. Specimen 1. Three states 1′, 1″ and 1‴ of phase morphology are schematically represented in Figure 1. State 1′ is the ideal structure of (γ+γ′)-phase quasi-cuboids with a slightly anisotropic distortion. Along with (γ+γ′)-phase quasi-cuboids, state 1‴ includes χ-phase layers. These layers are localized both in γ′- and γ-phase. The orientation of these layers is parallel to the cuboid orientation of γ′- and γ-phases as well as that of γ-phase layers. Quasi-cuboids in state 1‴ are also anisotropic. Figure 4 contains TEM images of states 1′, 1″and 1‴.

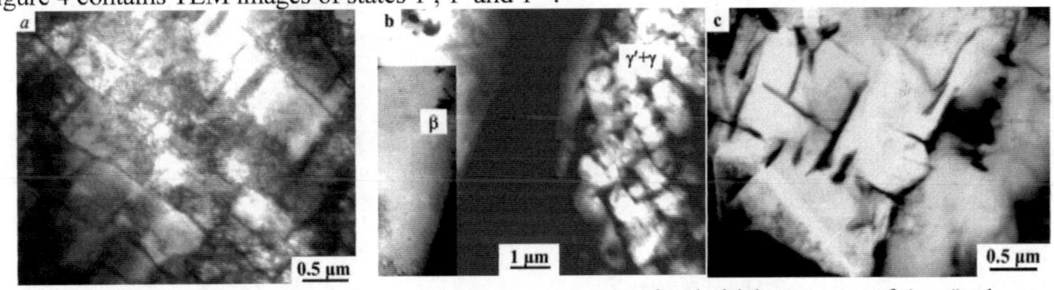

Figure 4. TEM image of specimen #1 structure: *a* - quasi-cuboidal structure of (γ+γ′)-phases (state 1′); *b* – intersection of 1′ (γ′+γ) and 1″ (β) states; *c* – state 1‴ (χ-phase layers are observed)

State 1″ presented in Figure 4 b, is characterized by a completely damaged structure of quasi-cuboids, or rather they are absent. The area of state 1″ is fully occupied, at least, by the three-component solution of NiAl2Re. The latter is a chemical formula of the three-component β-phase. The intersection of this phase and (γ+γ′)-phase mixture shown in Figure 4 b, has [011] direction of γ′-phase. The orientation relation between β-phase and γ/γ′-phases is [110]β II [111]γ′. This relation is typical for the mutual phase transformation of fcc → bcc. In Figure 4 c, the layers of γ′- and χ-phases are parallel. The both phases possess a cubical lattice; their size effects provide a certain crystallographic misorientation.

The SEM images of specimens #1-3 structure are presented in Figure 5. In Figure 5a, the larger volume of the alloy is occupied by quasi-cuboids, both damaged and not damaged. Because of low image magnification, their contrasts do not differ. Rhenium layering is clear due to the separation of β-phase which is predominately stabilized by this element.

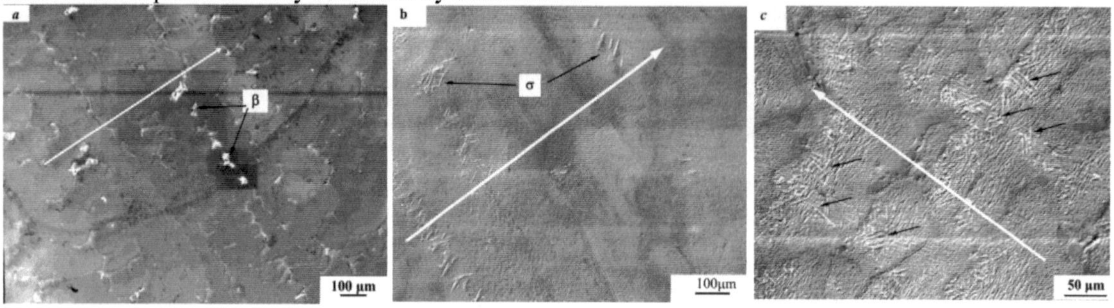

Figure 5. SEM images of alloy structure: *a* – specimen 1. Black arrows emphasize β-phase particles; *b* – specimen 2. Black arrows emphasize σ-phase particles; *c* - specimen 3. Black arrows emphasize La and σ-phase particles. White arrows emphasize the regular structure.

3.2.2. Specimen 2. From the schematic representation of the structure of specimen #2 shown in Figure 2, the alloy comprises (γ+γ′)-phase quasi-cuboids of the ideal structure with the volume fracture of 0.9 and (γ+γ′)-phase quasi-cuboids distorted by σ-phase with 0.1 volume fracture. Let denote these structural and phase conditions as state 2′ and state 2″. TEM images of these states are shown in Figure 6.

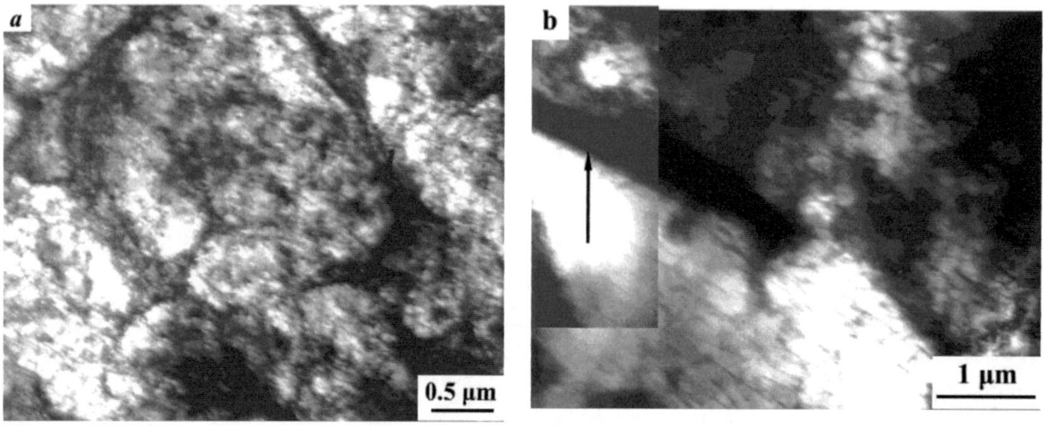

Figure 6. TEM images of (γ+γ′)-phase quasi-cuboids in specimen 2: *a* – state 2′; *b* – state 2″. Black arrows emphasize σ-phase particles

The elongated particles of σ-phase are localized in [011] and [001] directions of γ′-phase. The γ′- and σ-phase intersection is not rather distorted. This fact can be proved by the absence of bend extinction contours in the alloy structure. As shown in Figure 6b, in σ-phase the quasi-cuboid structure is distorted. The effect from the local formation of secondary phases in specimen #2 is more explicit than in specimen 1. Thus, the volume fracture of ideal quasi-cuboids is 0.65 for specimen #1 and 0.9 for specimen #2.

The SEM image of the structure of specimen #2 is presented in Figure 5b. The local distribution of σ-phase particles alternates with the ideal structure of (γ+γ′)-phases. Thus, a good correlation is observed between SEM and TEM investigations.

3.2.3. Specimen #3. According to the schematic representation of specimen #3 given in Figure 3, the ideal quasicuboids occupy just over half (0.55) of the material (state 3′) and 0.45 volume fracture belongs to (γ′ + γ + σ + Ni_3La_2)-phase mixture (state 3″). TEM images of these states are presented in Figure 7. The volume fractures of these both states are comparable. The volume fracture of Ni_3La_2-

phase is larger than that of σ-phase. This is because the additional specimen alloying with La. Moreover, the amount of σ-phase in specimen #3 is larger than in specimen #2. This can be connected both with a longer period of thermal treatment and the indirect influence of La on phase formation. Another interesting fact is that the separation of σ-phase and Ni_3La_2-phase occurs in the same locality. Figure 7b shows that that particles of σ-phase have always an acicular shape, while particles of Ni_3La_2-phase always have the internal structure with a characteristic contrast and finite thickness.

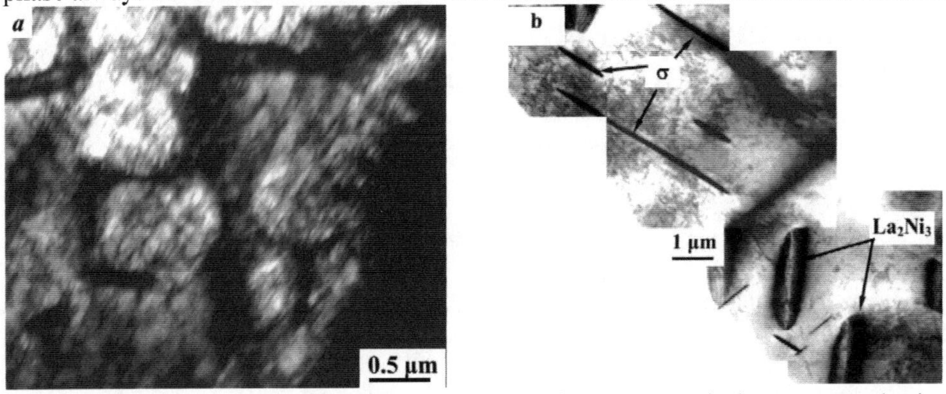

Figure 7. TEM images of specimen #3: *a* – state 3′; *b* – state 3″. Black arrows emphasize secondary phases

The SEM investigations of etched surfaces of specimen #3 shown in Figure 5 c, prove the data obtained as a result of TEM investigations, namely: the presence of two states of the structural and phase conditions (states 3′ and 3″) that have a regular structure; the comparability of volume fractures of states 3′ and 3″.

4. Conclusion

The investigations showed that the introduction of La in the alloy phase composition restrains the formation of γ-phase. Particles of Ni_3La_2-phase have a platelike shape and are localized in certain areas of the alloy. Rhenium and lanthanum alloying resulted in the formation of heat-resisting phases such as β-, σ-, and χ-phases (~1600, ~2600, ~2800°C melting temperature respectively) that promote the strengthening of the alloy.

Acknowledgement

The authors acknowledge the partial financial support from the Russian Foundation for Basic Research Project 16-48-700198 and the partial financial support from project (No. 8.1.42.2015) within the framework of the Program "Scientific Foundation named after D.I. Mendeleev of Tomsk State University" in 2015-2016.

References

[1] Kolobov Yu R, Kablov E N and Kozlov EV 2008 *Structure and properties of intermetallic materials with nano-phase hardening* (Moscow, MISiS) (in Russian)

[2] Kozlov E V, Smirnov A N, Nikonenko E L, Popova N A and Koneva N A 2016 *Phase morphology and transformations due to thermal treatment of Ni-Al-Cri Ni-Al-Co-based superalloys. Scale and concentration effects* (Moscow, Innovatsionnoe mashinostroenie) p 175 (in Russian)

[3] Sugui T, Minggang W, Huichen Y, Xingfu Y, Tang L and Benjiang 2010 *Materials Science and Engineering* **A 527** 4458

[4] Pyczak F, Neumeier S and Goken M 2010 *Materials Science and Engineering* **A 527** 7939

[5] Neumeier S, Pyczak F and Goken M 2011 *Philosophical Magazine* **91 33** 4187

Azide SHS of aluminium nitride nanopowder and its application for obtaining Al-Cu-AlN cast nanocomposite

Y V Titova[1,3], A V Sholomova[1], A A Kuzina[2], D A Maidan[1] and A P Amosov[1]

[1]Samara State Technical University, 244 Molodogvardeyskaya Street, Samara 443100, Russia

[2]Samara National Research University, 34 Moskovskoe Shosse, Samara 443086, Russia

[3]E-mail: titova600@mail.ru

Abstract. Method of azide self-propagating high-temperature synthesis (SHS-Az), using sodium azide (NaN_3) as a nitriding reagent, was used for obtaining the nanopowder of aluminum nitride (AlN) from precursor that was sodium hexafluoroaluminate (Na_3AlF_6). The product of burning the mixture of Na_3AlF_6 + $3NaN_3$ after water rinsing consisted of micro - and nanoparticles of AlN (65%) and the residue of salt Na_3AlF_6 (35%). This product of SHS-Az was mixed with copper powder and pressed into a briquette of nanopowdery master alloy Cu-4%(65%AlN+35%Na_3AlF_6), which was successfully introduced into aluminium melt at a temperature of 850°C. The salt Na_3AlF_6 in the product of combustion played a role of flux during introducing into the aluminum melt and was not included in the final composition of the composite alloy. The microstructure of the obtained cast composite aluminum alloy with the calculated composition of Al-1.2%Cu-0.035%AlN showed that the reinforcing particles of AlN of different sizes, including nanoparticles, were distributed mainly along the grain boundaries of the aluminum alloy.

Keywords: SHS, sodium azide, Na_3AlF_6 precursor, aluminum nitride, nanopowder, Cu-AlN master alloy, Al-Cu-AlN cast nanocomposite.

1. Introduction

Currently, the nanopowder of aluminum nitride is attracting much attention because of the possibility of using it for obtaining new advanced materials: as the basis for the sintering of AlN ceramics with a unique combination of properties (excellent thermal conductivity with low electrical conductivity, high resistance to the action of molten metals and salts, high corrosion resistance and mechanical strength), and also as a reinforcing material in the light-weight aluminum matrix composites Al-AlN having a high heat resistance [1, 2].

Nanoparticles of aluminum nitride are difficult to obtain by mechanical comminution of conventional commercially available AlN powders, so a large number of chemical and physicochemical processes have been developed to produce AlN nanopowders, such as a direct nitriding, plasma-chemical synthesis, carbothermic synthesis, chemical deposition from the gas phase, electrical explosion of aluminum wire and etc. [3-7]. However, these technologies of producing AlN powders are characterized by a large power consumption, complicated equipment, and do not always ensure the nanosize of AlN powder.

The method of self-propagating high-temperature synthesis (SHS) is attractive in this respect because it proceeds at the cost of its own heat of combustion in a simple small-sized equipment and takes a little time [8]. However, the performing the SHS in the classic version, using gaseous nitrogen as a nitriding reagent, does not allow obtaining nanosized powder of aluminum nitride [9]. To solve the problem of obtaining nanopowder AlN by resource-saving technology of SHS, it is promising to use such version as azide SHS technology, which is denoted as SHS-Az and is being developed in Samara State Technical University since 1970 [10]. The SHS-Az technology is based on the use of sodium azide NaN_3 as a solid nitriding reagent and halide salts. The SHS-Az technology is characterized by low combustion temperature and the formation of large amounts of gaseous products, which hinder the coalescence of the initial nanoparticles of the synthesized product and make it possible to retain them in nanosized state. The use in the SHS-Az starting powder mixture of the powder of a precursor, that is aluminum fluoride AlF_3, as a starting reagent instead of the powder of metallic aluminum Al can reduce the temperature of combustion and makes it possible to carry out the reaction in the decomposition of the precursor at the atomic level, which ultimately allows synthesizing AlN powder with a particle size of from 70 to 100 nm [10, 11]. But when burning this mixture, along with the traditional by-product of SHS-Az reactions, that is sodium fluoride NaF, easily removable by water washing, a large quantity of other by-product is generated, that is sodium hexafluoroaluminate Na_3AlF_6, which is difficult to remove and complicates the process of obtaining nanopowder AlN of high purity.

However, these side salts NaF and Na_3AlF_6 represent typical fluxes for refining and modification of melts of aluminium alloys and are used to improve the wetting of the reinforcing ceramic particles during liquid-phase production of aluminum matrix composites (AMCs) [12, 13]. Therefore, when in use of the SHS-Az product for liquid-phase reinforcement of aluminum alloys, it is not necessary to wash this product from the side salts that was shown in the example of modification of aluminum alloys using nanopowdery composition of SiC - Si_3N_4 [13]. Because of the simplicity of technology and equipment, the nanopowders of SHS-Az brand can be 2-3 times less in cost than the similar nanopowders of plasma chemical synthesis.

The aim of this work is to study the possibility of synthesis of AlN nanopowder by SHS-Az method using an alternative precursor, that is sodium hexafluoroaluminate Na_3AlF_6, for subsequent use of the AlN nanopowder as a reinforcing phase in liquid-phase producing the cast aluminum matrix nanocomposite Al-AlN.

2. Materials and methods of investigations

As initial reagents in obtaining aluminum nitride, the powders of aluminum, sodium azide and sodium hexafluoroaluminate were used which are produced by the Russian industry: aluminum of ASD-1 brand (basic substance content not less than 99.7 wt. %; particle size less than 50 μm); NaN_3 (98.71%); Na_3AlF_6 (99.1%). The mixtures of powder reagents were prepared in accordance with the stoichiometric equation of a chemical reaction for producing aluminum nitride in a mode of SHS-Az:

$$Na_3AlF_6 + 3NaN_3 = AlN + 6NaF + 4N_2. \quad (1)$$

All initial powders were sifted on the vibrating screen of LE-203/I type through a set of standard sieves. If necessary, the powders were dried in a vacuum drying box at a temperature of 80°C within 1-2 hours. A mixture of the initial powders, taken in a predetermined ratio, was prepared manually in a porcelain mortar for 5-10 minutes until a uniform distribution of components. The prepared mixture of powders (the charge) was filled in a tracing paper cup with a diameter of 30 mm and a height of 45 mm and was placed in the filter assembly made from carbonic cloth. The assembled structure was mounted on the subject shelf of anazide SHS laboratory reactor of a constant pressure with a volume of 4.5 litres. The pressure of 4 MPa of gaseous nitrogen was created in the reactor. The arrangement of this reactor and methods of work with the measurements of the linear velocity and the maximum combustion temperature by thermocouple method using tungsten-rhenium thermocouples TR 5/20 was described in detail in [10, 11]. Recording the temperature and rate of combustion was carried out using an analog-to-digital converter connected to a computer.

The cooled sample was removed from the reactor and was destroyed by hand in a porcelain mortar to obtain a free-flowing powder. Then water washing the powder product of combustion was carried out to remove by-product residue, which was soluble in water. The powder was diluted with distilled water to ratio of 1:10, the resulting suspension was shaken and filtrated on a vacuum funnel. Rinsing was monitored by measuring the acid-base balance of the wash water (pH).

Shooting X-ray spectra was carried out on an automated diffractometer ARL X trA-138 using Cu-radiation with continuous scanning in the range of angles 2θ from 20 to 80 degrees at a speed of 2 deg/min. A quantitative estimate of the content of different phases was carried out by measuring the ratio of intensities of analytical lines. Morphology and particle size of the powder compositions was carried out using a scanning electron microscope (SEM) Jeol JSM-6390A.

For a preliminary analysis of the burning temperature of the starting components and the composition of the synthesis products, the thermodynamic calculations was performed using the computer program "Thermo", developed at the Institute of Structural Macrokinetics and Materials Science Problems of RAS (Chernogolovka, Moscow region) [8].

Direct introduction of nanopowders in bulk form by mixing into the melt of aluminum alloys is practically impossible [13]. To estimate the potential application of the synthesized ceramic nanopowder AlN of SHS-Az brand for reinforcing and modifying of aluminum alloys, a nanopowdery master alloy was prepared that was a mechanical mixture of AlN nanopowder with plastic powder carrier, which was subjected to mechanical activation and was pressed into briquettes [13]. As a carrier material for light-weight nanopowder AlN, the powder of electrolytic copper of PMS-1 grade was chosen with an average particle size of 15-45 μm having regard to its high density (8900 kg/m^3) relative to the aluminum melt (2400 kg/m^3) and good solubility of copper in the melt. Mechanical mixing of the initial components of the master alloy was carried out in planetary mill "Pulverisette-5" for 90 minutes at a speed of 150 rpm. Cemented carbide balls with a diameter of 10 mm were used as grinding bodies. The weight ratio of balls to powder mixture was 10:1, the fill factor of the drums was 0.5. The homogeneity of mixing was evaluated by randomly selected samples. The charge was considered uniform if at least 95% of randomly selected samples had the specified chemical and granulometric compositions. To obtain the briquettes of nanopowdery master alloy, cold pressing the powder mixture was used. Uniaxial compaction of the powder mixture was carried out with a pressing pressure of 35 MPa in a cylindrical die with an internal diameter of 18.1 mm. The briquettes were introduced into the melt of primary aluminium of technical purity of A7 brand (99.70%).

3. Results and discussion

Table 1 presents the results of thermodynamic calculations showing the dependence of the adiabatic temperature T_{ad} of the formation of aluminum nitride according to reaction (1) from the pressure of nitrogen pumped into the reactor (P_{N2}). It can be seen from the presented data, that T_{ad} is changed only at normal pressure of P_{N2} = 0.1 MPa. T_{ad} does not change at increasing pressures P_{N2} = 1÷6 MPa and is equal to 1269 K. The yield of the products of the reaction corresponds to the stoichiometric one.

Table 1. Results of thermodynamic calculations.

Pressure of nitrogen pumped into reactor, P_{N2}, MPa	Adiabatic temperature of combustion, T_{ad},K	Enthalpy of reaction products, ΔH, kJ	Composition of combustion products, moles				
			$AlN_{s.}$	$NaF_{s.}$	$NaF_{l.}$	$NaF_{g.}$	N_2
0.1	969	–3259	1	5.31	0	0.69	4
1	1269	–3259	1	5.45	0.55	0	4
2	1269	–3259	1	5.45	0.55	0	4
3	1269	–3259	1	5.45	0.55	0	4
4	1269	–3259	1	5.45	0.55	0	4
5	1269	–3259	1	5.45	0.55	0	4
6	1269	–3259	1	5.45	0.55	0	4

IOP Conf. Series: Materials Science and Engineering **156** (2016) 012037 doi:10.1088/1757-899X/156/1/012037

The results of an experimental study of the dependence of temperature and velocity of combustion of the mixture "Na_3AlF_6 + $3NaN_3$" from the pressure of nitrogen in the reactor are shown in table 2.

Table 2. Results of experimental investigation of combustion parameters.

Pressure of nitrogen in reactor, P_{N2}, MPa	Temperature of combustion, T_c, °C	Velocity of combustion, U_c, cm/s
0.1	720	0.40
1	850	0.50
2	910	0.56
3	930	0.60
4	950	0.62
5	950	0.62
6	950	0.62

Five experiments were carried out for each point of these dependencies. From the presented data, it is visible, that the temperature and velocity of combustion increase as the pressure increases up to 4 MPa, and then their values no longerdepend on pressure. The maximum values of T_c and U_c are 950°C and 0.62 cm/s, respectively.

Figure 1 shows the results of X-ray phase analysis of the products of combustion of mixture "Na_3AlF_6 + $3NaN_3$" at a pressure P_{N2} = 4 MPa.

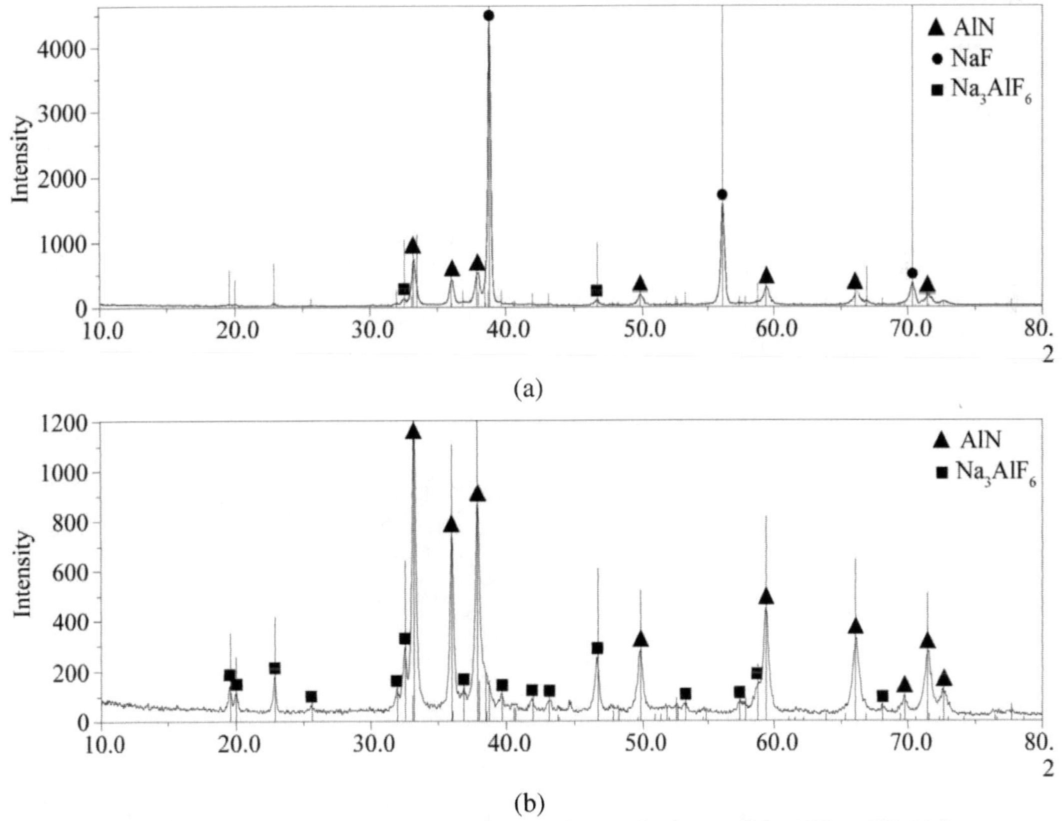

Figure 1. XRD patterns of combustion products of mixture "Na_3AlF_6 + $3NaN_3$": (a) until washing; (b) after washing.

The phase composition of unwashed combustion products is a mixture of aluminum nitride, sodium hexafluoroaluminate and sodium fluoride (figure 1 (a)). Due to its good solubility in water, the sodium fluoride is completely removed from the combustion products by rinsing in distilled water (figure 1 (b)). On the contrary, sodium hexafluoroaluminate can also be observed in the washed reaction products due to poor solubility in water, and its content is 35%.

Figure 2 shows the micrograph of the powder synthesized from the mixture of "Na_3AlF_6 + $3NaN_3$".

(a) (b)

Figure 2.SEM magnification of combustion products of mixture "Na_3AlF_6 + $3NaN_3$": (a) until washing; (b) after washing.

Comparing the pictures presented in figures 2 (a) and 2 (b), and having regard to the results of the X-ray phase analysis, it can be concluded that during combustion of the mixture "Na_3AlF_6 + $3NaN_3$", the equiaxed particles of irregular shape of aluminum nitride and sodium hexafluoroaluminate are formed, as well as fibers of sodium fluoride, which are present until washing in figure 2 (a) and are not present after washingin figure 2 (b). The aluminum nitride is represented by a mixture of micron-sized particles (several microns) and a large number of nanoparticles with size from 50 to 100 nm.

The washed product was used for mixing and pressing nanopowder master alloy of Cu-4%(65%AlN+35%Na_3AlF_6) composition in the form of disk-shaped briquette of 2.5 g mass and 18.1 mm diameter. At a compression pressure of 35 MPa, the briquette had a height of 1.5 mm, which corresponded to the density of 8290 kg/m^3 and a porosity of 28%. The microstructure of compacted nanopowder master alloy is presented in figure 3.

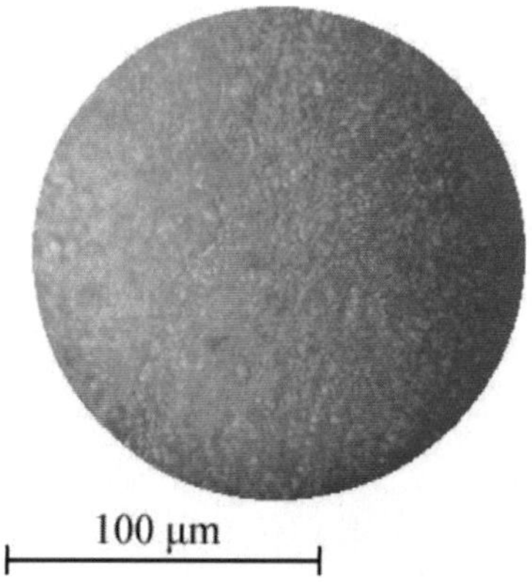

Figure 3. Microstructure of master alloy of Cu-4%(65%AlN+35%Na_3AlF_6).

The average size of powder particles formed after intense grinding-mixing in the master alloy is 6-15 μm. Pores are also observed in the structure of the briquette, the presence of which may facilitate its dissolution during subsequent introducing into the melt.

Such briquette of nanopowder master alloy of Cu-4%(65%AlN+35%Na_3AlF_6) composition dissolved in the melt of A7 aluminum of 197.5 g mass at a temperature of 850°C and allowed introducing the reinforcing particles of AlN into the aluminum alloy. This can be confirmed by X-ray diffraction of the sample of cast composite after crystallization, represented in figure 4.

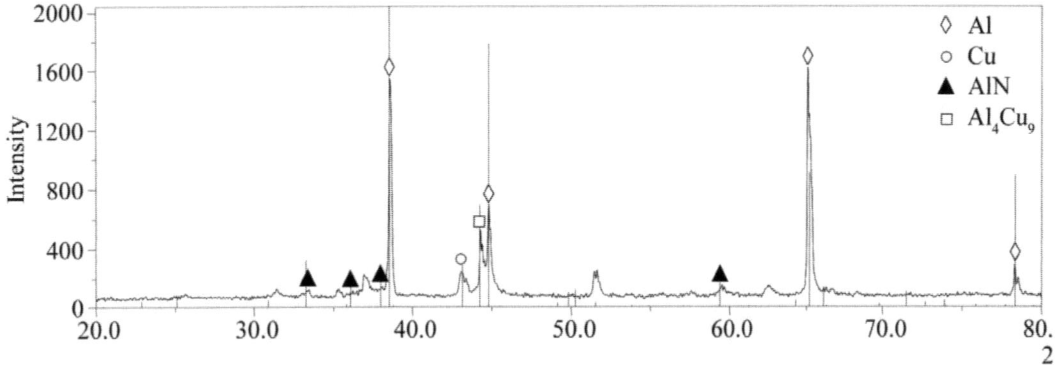

Figure 4. XRD pattern of cast aluminum composite.

XRD analysis of the cast sample shows the presence of aluminum (base alloy), copper, copper aluminide Al_4Cu_9 and aluminum nitride AlN, which indicates an absorption of the nanopowder of aluminum nitride of SHS-Az brand in composition of nanopowder master alloy of Cu-4%(65%AlN+35%Na_3AlF_6) by aluminum melt. Figure 4 also shows the lack of lines of Na_3AlF_6 halide salt, i.e. the salt played the role of a flux in the aluminum melt, but was not included into the composition of the solidified alloy. The produced cast composite alloy has a calculated composition of Al-1.2%Cu-0.035%AlN. The microstructure of the sample of solidified alloy is presented in figure 5.

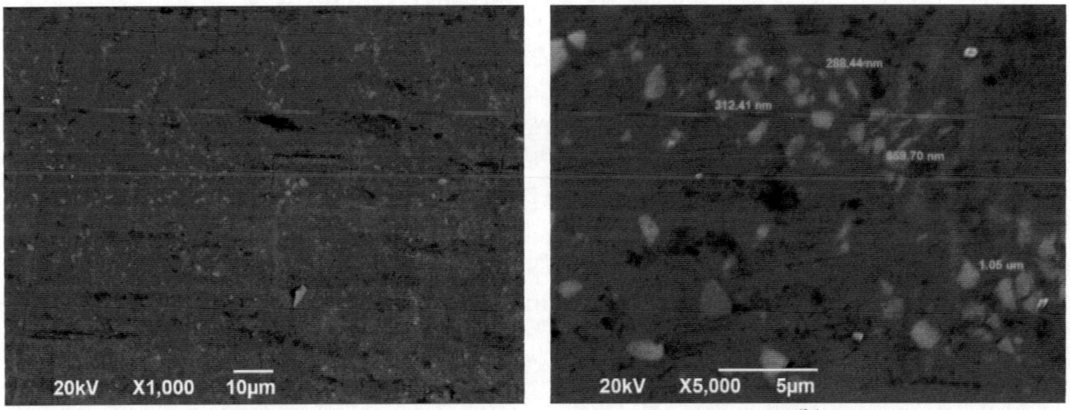

(a) (b)

Figure 5. Microstructure of cast aluminum composite alloy:
(a) magnification ×1000; (b) magnification ×5000.

The SEM images demonstrate AlN particles of different size (of micron size, ultrafine and nanoscale), which are distributed over the body of composite alloy not uniformly, butmainly along the grain boundaries of aluminum alloy. This distribution is typical for highly dispersed reinforcing ceramic particles in cast aluminum composite alloys [14].

4. Conclusion

The use of halide salt Na_3AlF_6 as a precursor in the method of azide SHS allowed us to obtain a product of combustion, consisting after water rinsing of micro - and nanopowder of aluminum nitride (65%) and residues of salt Na_3AlF_6 (35%). This product of combustion in the composition of compacted powder master alloy of Cu-4%(65%AlN+35%Na_3AlF_6) was introduced into the aluminum melt. This made it possible to produce the cast composite aluminum alloy with calculated composition of Al-1.2%Cu-0.035%AlN, in which the reinforcing particles of AlN of different sizes, including nanoparticles, are distributed mainly along the grain boundaries of the composite alloy. The salt Na_3AlF_6 in the product of combustion played a role of flux during introducing into the aluminum melt and was not included in the final composition of the aluminum matrix nanocomposite.

Acknowledgements
Work is executed at financial support of RFBR, grant No. 16-08-00826.

References

[1] Trinkler L, Berzina B and Palcevskis E 2015 *Nitride Ceramics: Combustion Synthesis, Properties, and Applications*, ed AA Gromov and LN Chukhlomina (Weinheim: Wiley-VCH) p 265-93
[2] Borgonovo C, Apelian D and Makhlouf MM 2011*JOM* **63** (2) 57-64
[3] Jia L, Kondoh K, Imai H, Onishi M, Chen B and Li S 2015 *J. Alloys Comp.* **629** 184-7
[4] Kim K 2005 *J. Crystal Growth.* **283** 540-6
[5] Shinji H, Tetsuya M and Tsutomu I 1989 *J. Jap. Inst. Metals* **30** 1035-1040
[6] Wu N, Tsai M, Wang M and Liu H 2000 *J. Crystal Growth.* **208** 189-96
[7] Beketov I V, Kotov Yu A and Azarkevich E I*Sec. All-Rus. Conf. on Nanomaterials "NANO 2007"* (Novosibirsk, Russia) p 109
[8] Amosov A P, Borovinskaya I P and Merzhanov A G 2007 *Powder Technology of Self-propagating High-temperature Synthesis of Materials* (Moscow: Mashinostroenie-1)
[9] Zakorzhevsky V V, Borovinskaya I P and Sachkov H B 2002 *Inorg. Mater.* **38** 1340-50
[10] Bichurov GV 2015 *Nitride Ceramics: Combustion Synthesis, Properties, and Applications*, ed AA Gromov and LN Chukhlomina (Weinheim: Wiley-VCH) p 229-63
[11] Titova Yu V, Shiganova L A, Maidan D A and Bichurov G V 2012 *Izvestiya vuzov. Poroshkovaya metallurgiya i funktsionalnye pokrytiya* **3** 25-9
[12] Kurdyumov AV, Pikunov M V, Chursin V M and Bibikov V L 2011 *Production of Castings from Alloys of Non-Ferrous Metals*ed 3(Moscow: MISIS)
[13] Amosov A P, Titova YuV, Maydan DA, Ermoshkin AA and Timoshkin IYu 2015 *Rus. J. Non-Ferr. Met.* **56**(2) 222-8
[14] Vinod Kumar GS, Murty BS and Charaborty M2005*J. Alloys Comp.* **396** 143-50

Influence of synthesis conditions on the crystallinity of hydroxyapatite obtained by chemical deposition

N E Toropkov[1], V I Vereshchagin[1], T S Petrovskaya[1, 2] and N S Antonkin [1]
[1]Tomsk Polytechnic University, 30 Lenin ave., Tomsk, 634050, Russia
[2]Skolkovo Institute of Science and Technology, Moscow, Russia

E-mail: zerogooff@gmail.com

Abstract. The hydroxyapatite synthesis on a variety of substrates under various conditions was studied. It was shown that the increase in the temperature of the reaction medium increases the amount of nanocrystalline phase with an average crystallite size of 25 nm. Studies revealed that in addition to the pure hydroxyapatite, β-Ca3(PO4)2 along with calcium carbonates and carbonate-substituted hydroxyapatites were formed. A significant increase in phase crystallinity during the heating of reactants up to the reaction temperature was shown.

1. Introduction

Calcium phosphates are widely used as an artificial material to treat or replacement the bone since they reduce reactions of inflammation and rejection of the implant, and are involved in the processes of calcium metabolism and osteogenesis [1,2,3]. Besides, hydroxyapatite ($Ca_{10}(PO_4)_6(OH)_2$, HA) and tricalcium phosphate ($Ca_3(PO_4)_2$) are successfully used as the components of surface coatings of metal implants [2,3,4].

As a bone component, HA generates a crystalline acicular structure and forms a scaffold [5,6]. Therefore, depending on the functionality the implants must include HA in a designated form to provide bone restoration or integration [3, 5].

The crystalline HA is usually obtained by the methods of liquid or solid-phase synthesis. The latter is carried out at high a temperature that leads to a reduction in defect structure of crystal lattice, particle growth and a decrease of the specific surface of a powder. This makes the powder not suitable for bulk implants obtaining [7,8]. In addition, the products of solid-phase synthesis usually have chemical and mineralogical inhomogeneity, along with chemical inertness. The liquid-phase synthesis on the contrary, provides product with an inherent chemical homogeneity. The dispersity of a product is determined by the concentration of precursors; fine-crystalline product is produced from precursors with lower concentration [11]. Furthermore, there is a possibility to control the phase composition and the crystallinity of the product. [9, 10].

The aim of this study is to determine the liquid phase synthesis conditions to obtain fine-crystalline calcium phosphates.

2. Materials and research techniques

HA synthesis was performed according to the following reactions:

$$6(NH_4)_2HPO_4 + 10Ca(NO_3)_2 + 8NH_4OH = Ca_{10}(PO_4)_6(OH)_2 + 20NH_4NO_3 + 6H_2O \quad (1);$$

$$6(NH_4)_2HPO_4 + 10Ca(OH)_2 = Ca_{10}(PO_4)_6(OH)_2 + 12NH_4OH + 6H_2O \quad (2);$$

The precursors with a high purity were used. Solutions were prepared by degassed boiling of distillate with a concentration ratio $[Ca^{2+}]/[PO_4^{3-}]$=0.5M/0.3M and pH=8.73. The value of pH was controlled by electronic pH-analyzer Mettler Toledo S700 with a gel electrolyte.

Details of synthesis conditions are shown in Table 1. Samples 1-1, 1-2, 1-3, 1-4, 1-5 and 1-6 were obtained by reaction (1), samples 2-1, 2-2, 2-3 2-4, 2-5, 2-6 were obtained by the reaction (2). A part of the reaction products were soaked in the mother liquor (1-3, 2-3, 2-5, 2-6) for fixed time. All products were separated of the mother liquor, then washed in the distillate water at least 4 times until neutral pH was reached, then were centrifuged, dried and dehydrated at 105 °C, except the sample 1-6.

The synthesis temperature has been defined by calculation of Gibbs free energy and provided shift the equilibrium toward the reaction product. After drying, the samples were subjected to XRD studies performed on a DRON-3M diffractometer ($Cu_{K\alpha}$- radiation, wavelength 1.54056 Å) with the following scanning parameters: accelerating voltage 90 kV, beam current of 30 mA, scan step of 0.03 ° in the 10-70 ° 2θ range, preset time of 1s. The degree of the crystallinity of samples was calculated using the software packages POWDERCELL 2.4 and Crystallographica Search-Match. As a reference, HAP 2007 ICDD №01-074-0565 was used, it had X-ray reflexes obtained under similar scanning conditions [12]. Morphology of the powders was studied by scanning electron microscopy (SEM) JEOL JSM 6000 with its attachment that was used for elemental analysis. Infrared (IR) spectra of the powders were recorded on a Nicolet 5700 device.

Table 1. Synthesis conditions for HA obtained.

Sample	Type of mixing	Heating	Reaction temperature, °C	ΔG, kJ	Curing in the mother liquor		The calcination temperature, °C
					soaking, days	Temperature, °C	
1-1	dripping	resistance	60	-1134	-	-	-
1-2	draining	resistance	60	-1134	-	-	-
1-3	dripping	microwave	60	-1134	-	-	-
1-4	draining	resistance	100	-3381	-	-	-
1-5	draining	resistance	100	-3381	1	25	-
1-6	draining	resistance	60	-1134	-	-	400
2-1	draining	resistance	100	-824	-	-	-
2-2	draining	resistance	100	-824	-	-	800
2-3	draining	resistance	100	-824	1	37	-
2-4	draining	resistance	100	-824	1	37	800
2-5	draining	resistance	100	-824	14	37	-
2-6	draining	resistance	100	-824	14	37	800

3. Results and discussion

3.1. XRD analysis

The X-ray diffraction patterns of the reaction (1) products and reference sample are shown in Figure1. All of the samples are presented of HA.

The average size of crystallites is determined using by characteristic HAP reflex (002) according to the Scherrer formula (data are presented in Table 2):

$$L = k\lambda / \beta \cos\theta$$

where L means the average crystallite size, k – constant close to unity, λ – X-ray wavelength ($Cu_{K\alpha}$= 1.54056 Å), β - broadening of the X-ray reflex $\beta = (B^2 - b^2)^{1/2}$, where B – full-width at half-maximum (FWHM) of the (002) diffraction line for the sample, b – FWHW of the diffraction line of the reference sample, θ – Bragg's diffraction angle for (002) line [11]. As one can see from the data above, the highest concentration of HA in a product, namely 95%, was reached by drip mixing along with synthesis at 60° C (sample 1-1); the least concentration of 52% (sample 1-2) was obtained by pouring together precursors and maintaining the temperature of 60° C; respectively sizes crystals were 40 and 35 nm.

Table 2. The average size of the crystallites of synthesis products

Sample	1-1	1-2	1-3	1-4	1-5	1-6	2-1	2-2	2-3	2-4	2-5	2-6	ICDD 2007 01-074-0565
Crystallite size, nm	~40	~35	~37	~35	~30	~24	~37	~40	~35	~40	~20	~27	~14
HA content wt. %	95	52	58	68	70	70	68	90	70	90	70	90	99

Complete pouring together of heated precursor and reacting at 100° C resulted in a formation of fine-grained product with a crystallite size of 24 nm (1-6); wherein the HA content of 70% was achieved when the additional heat treatment of the precipitate at 400° C was performed. Maturing precipitate in the mother liquor for one day at room temperature does not give the same effect as heating for the sample 1-5 [13]. The use of microwave heating was not conducive to the increase of the yield of HA or the decrease of the crystallite size. Obtained data are explained by the Onsager law, namely by tensor direction of the reaction toward the higher temperatures, and by the fact that product (precipitation) is fast extracted from the reaction zone.

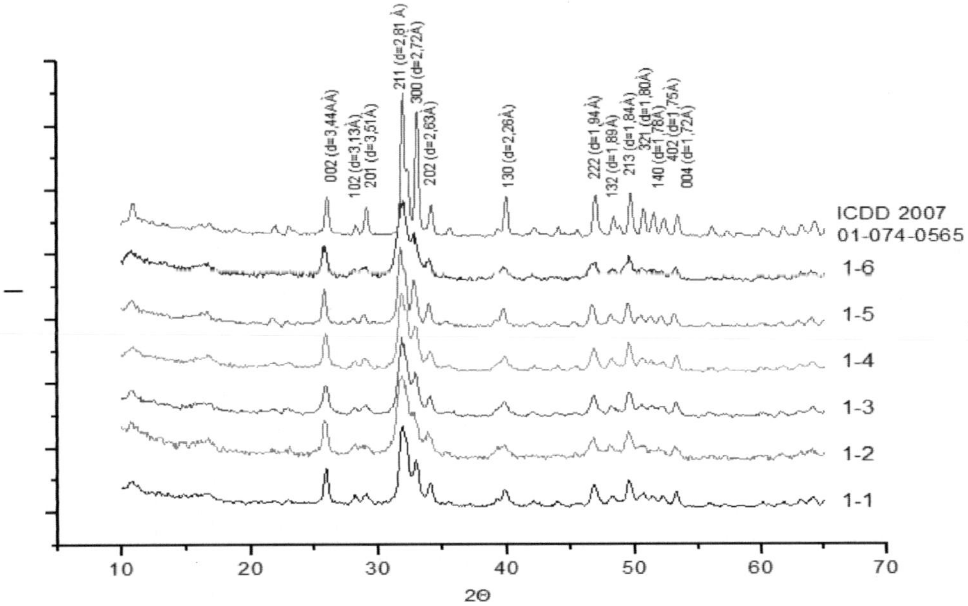

Figure 1. X-ray diffraction patterns of the reaction (1) products and of reference sample.

As shown in Figure 2, the powders obtained by the reaction (2) consist of HA mainly. The average crystallite size in the samples 2-1, 2-3, 2-5, obtained at 37° C is respectively 37 nm, 35 nm, and 20 nm. At that 2-3, 2-5 were kept for 1 day and two weeks respectively in the mother liquor. All these X-ray diffraction patterns indicated comparatively low intensity of the reflexes. However, after the calcination at 800 ° C, the crystalline phase content of the HA increased up to 90%.

Thus, the increase of the degree of product crystallinity is observed when preheated precursors are mixed at 100° C and precipitation is kept in the mother liquor for more than a day. Rapid removal of the product from the solution entails to an increase of the portion of the fine-crystalline phase.

Comparison of the thermodynamic conditions of the (1) and (2) reactions demonstrates that the decrease in Gibbs free energy effects positively on increase of the portion in the product of HA with the smallest crystallite size (1-4, 1-5, 1-6, 2-5).

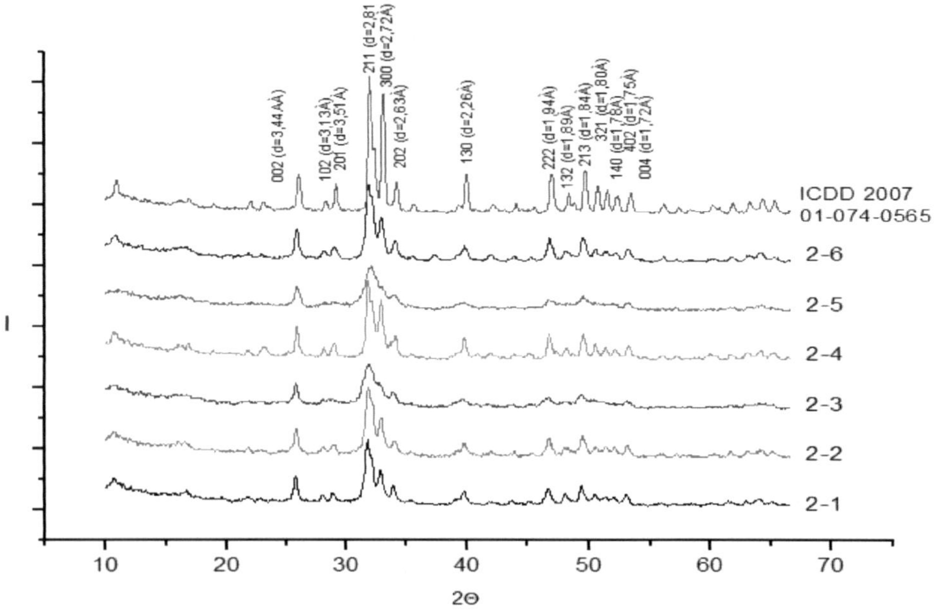

Figure 2. X-ray diffraction patterns of the reaction (2) products and of reference sample.

3.2. Analysis of IR spectra

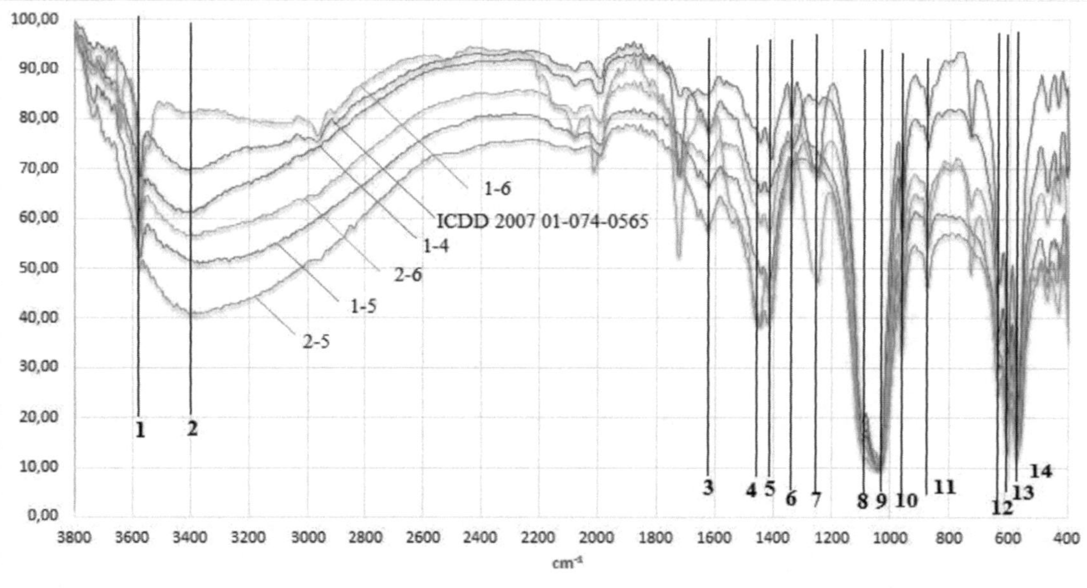

Figure 3. IR spectrum of the sample.

The IR spectra of the samples (Figure 3, Table 3) include the bands corresponding to the vibrations of the phosphate groups in the range of wave numbers 955-962, 520-660 and 1020-1090 cm^{-1}[15]. A broad band of deformation vibrations OH^{-}-groups at 3440-3570 cm^{-1} corresponds to the water adsorbed. The low intensity of the band at 3570 cm^{-1} and 630 cm^{-1} responsible for the stretching and deformation vibrations of OH^--groups in HA obtained without maturing the precipitate in the mother liquor (sample 1-6) indicates a low content of HA phase in a product. Vibrations corresponding to the bands in the ranges 870-875 cm^{-1} and 1300-1500 cm^{-1} are associated to the C=O groups in the carbonates. The presence of carbonate groups in all samples was caused by adsorption of carbon dioxide from the air during the synthesis. Besides, in the spectra of the calcined samples (1-6, 2-6), the intensity of the respective peaks in the ranges of 1300-1500 cm^{-1}, 3440-3570 cm^{-1} and 1630-1650 cm^{-1} was significantly reduced due to the desorption of molecules of water and carbon gas from the powder surface and decomposition of the carbonate-HAP [14]. The obtained data demonstrates the presence of the carbonate groups in the products of reactions.

Table 3. Infrared vibrational modes of synthesized hydroxyapatite, standard and published data.

№	Characteristic frequencies	Vibrational frequency, ν cm^{-1}						
		1–4	1–5	1–6	2–5	2–6	reference	literature [15,16]
1	OH structural	3570	3571	3573	3570	3572	3572	3572
2	OH deformation	3400–3570	3400–3570	–	3400–3570	3400–3570	3400–3570	3420–3570
3	CO_3 structural	1632	1634	1635	1631	1634	1632	1630–1650
4	CO_3 group v^3	1453	1456	1453	1454	1454	1459	1450–1460
5	CO_3 group v^3	1414	1416	1413	1420	1420	1420	1429
6–7	groups of adsorbed H_2O	1100–1300	1100–1300	–	1100–1300	–	–	1100–1300
8	PO_4 bending v^3	1022	1040	1032	1030	1044	1040	1050
9	PO_4 bending v^3	1090	1089	1087	1086	1088	1089	1090
10	PO_4 stretching v^1	963	962	960	963	961	965	961
11	CO_3 group v^3	879	873	870	880	881	879	879
12	OH structural	627	630	630	650	631	631	635
13	PO_4 bending v^4	602	602	598	610	601	601	607
14	PO_4 bending v^4	573	573	573	564	574	574	571

As per IR spectra, maturing products in the mother liquor entail the increase the degree of its crystallinity as well as the increase of the content of carbonate groups, presumably as a part of the calcium carbonate and carbonate-containing HA. It should be noted that the inclusion of the carbonate anion in the structure of the HA in a small amount did not change the phase composition and was not observed in the X-ray diffraction patterns [17]. No differences in IR spectra of samples 2-5 and 2-6 were observed, thus, the calcination does not result in the increase of the content of crystal phase in a product.

3.3. Scanning electron microscopy

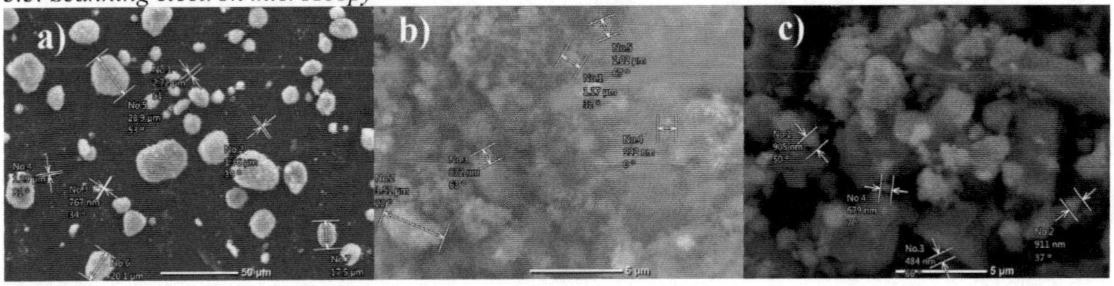

Figure 4. SEM images of the samples a) – 1-4; b) – 1-5; c) – 2-5.

All products represent as powder systems which include the aggregates of 10-100μ sized that were formed due to the surface activity (Figure 4a). The aggregates and precipitates themselves have a loose structure (Figure 4b) in which the crystalline phase is distributed statistically. The only exception was the sample 2-5 (Figure 4c), which had been kept in the mother liquor for 2 weeks. The sample 2-5 structure includes predominantly amorphous particles of irregular shape ranging in the size from 100 up to 500 nm, and larger particles (greater than 500 nm) have denser structure. Apparently, the evolution of the precipitate went the same way as for the other samples. During the first day, the precipitate had a loose structure, wherein the partial crystallization occurred. While keeping precipitation in the solution for 2 weeks, the accumulation of a crystalline product and formation of a lamellar crystal with a regular faceting took place.

The synthesis process has no significant effect on the Ca/P ratio, as shown by elemental analysis (Table 4). Maturing precipitation in the mother liquor bring the Ca/P ratio close to the stoichiometric value.

Table 4. Elemental Composition (atomic %) of the synthesized samples.

Index	Ca	P	O	C	Ca/P
1-4	16.02	10.16	39.93	10.81	1.61
1-5	19.98	12.09	35.52	17.74	1.65
2-5	21.01	12.55	30.11	15.04	1.67
Standard	14.01	8.01	40.8	20.87	1.75

4. Conclusion

It was found that the products synthesized in reactions (1) and (2) primarily consist of hydroxyapatite. The conditions for obtaining a nanocrystalline product (20-27 nm) with a HA fraction of 70 to 90 wt.% were determined. Draining precursors at a temperature close to the boiling point provides active convection, uniform distribution of the reagents in the bulk that accelerates the synthesis. It is preferable to keep the precipitation in the mother liquor more than a day and up to two weeks to increase the content of nanocrystalline phase. Further calcination the product is not reasonable because of the low energy efficiency of the process.

It is shown that the initial state and the method of the heating precursor influence insignificantly on the crystallinity, Ca/P ratio and the content of the nanocrystalline phase.

References

[1] Matsumoto T, Tamine K and Kagawa R 2006 *Ceramic Society of Japan* **114** 760-764
[2] Supova M 2015 *Ceramics international* **41** **8** 9203
[3] Lilley K J, Gbureck U, Wright A J, et al. 2005 *Materials science: materials in medicine* **16** 1185.
[4] Barinov S M 2005 *Bioceramics of calcium phosphate* (Moscow: Nauka).
[5] Bulanov E N 2012 *Obtaining and study on nanostructured biocompatible materials based on hydroxyapatite* (Nizhny Novgorod: Nizhny Novgorod State University).
[6] Tevlin R, McArdle A, Atashroo D, et al. 2014 Biomaterials for craniofacial bone engineering *J. Dent. Res* **93** 1187
[7] Tretyakov Yu D and Putlyaev V I 2006 *Introduction to the chemistry of solid-state materials* 400
[8] Cao J M, Feng J, Deng S G, et al. 2005 *Materials science* **40** 6311
[9] Kolk A, Handschel J, Drescher W, et al. 2012 *Craniomaxillofac Surg* **40** 706
[10] Safronova T V, Shekhirev M A, et al. 2007 *Inorganic Materials* **43** **8** 901
[11] Vanetsev A S and Tret'yakov Yu D 2007 *Russian Chemical Reviews* **76** **5** 397
[12] JCPDS-ICDD, 1995, PDF-2 Sets 1—45 database, Newton Square, PA 19073, USA, card # 9—432
[13] Dorozhkin S V 2015 *Materials Science & Engineering* **55** 272

[14] Filipov Y Y, Klimashina E S, Putlyaev V I, et al 2011 *Journal of physics: Conference series* **291** **1** 12
[15] Kim D W, Cho I S, Kim J Y, et al. 2009 *Langmuir* **26** **1** 384
[16] Prabakaran K, Balamurugan A 2005 *Bulletin of Materials Science* **28** **2** 115
[17] Jonas K, Vassanyi I, Ungvari I 1980 *Phys. Chem. Minerals* **6** 55

IOP Conf. Series: Materials Science and Engineering **156** (2016) 012039 doi:10.1088/1757-899X/156/1/012039

Developing porous ceramics on the base of zirconia oxide with thin and permeable pores by crystallization of organic additive method

K S Kamyshnaya and T A Khabas

National Research Tomsk Polytechnic University, Tomsk, 634050, Russia

E-mail: ksenia@tpu.ru

Abstract. In this paper porous ceramics on the base of ZrO_2 nanopowders and micropowders has been developed by freeze-casting method. A zirconia/carbamide slurry was frozen in mold and dehydrated in $CaCl_2$ at room temperature. This simple process enabled the formation of porous ceramics with highly aligned pores as a replica of the carbamide crystals. The samples showed higher porosity of 47.9%. In addition, these materials could be used as membrane for air cleaning.

1. Introduction

Porous ceramics is widely used since 1950's and has a wide range of application in different fields now [1, 2]. Ceramics on the base of refractory oxides in particular ceramics on the base of ZrO_2 has a greatest demand. Zirconium dioxide ceramics has good physical and chemical properties (strength, chemical resistance, crack resistance and etc.). Zirconium dioxide ceramics can be used as membranes for air cleaning from dangerous for organism particles (less 5μm).

There are a lot of methods for obtaining porous ceramics but significant disadvantage of these methods is impossibility to control pores morphology. Application of organic additives crystallization in suspension because of change of temperature method can be used to solve this problem [3, 4]. This method allows to regulate pores morphology in the sample and does not pollute sample by impurities from burnable additives.

The aim of this research is to develop method for obtaining porous ceramics based on ZrO_2 powders of various dimensions with application of organic pore agent.

2. Experimental Part

Zirconium dioxide powder of micron size (0.23 μm average particle size, produced by the Chepetsk Mechanical Factory, Russia) and nanopowder (30 nm average particle size, produced by the Seversk Chemical Factory, Russia) were used to obtain porous ceramics. Specific surface area of powders was determined by Surface Area and Pore Size Analyzer Quantachrome Nova-2200-e. Powders were mixed with mass ratio 50:50 in ball mill with zirconia mill and balls. Samples from 100% micronpowder and nanopowder were produced for comparison. Morphology and structure of powders were studied using analytical Field Emission SEM JEOL JSM-7500FA.

Suspension of carbamide prepared by dissolved of carbamide granules in hot water (t=80 ^{0}C) with mass ratio 2:1 (carbamide: water) was used to obtain porous ceramics. The prospect of the use of carbamide is the ability to produce aligned crystals (in the form of needles) of carbamide during crystallization. Behavior of the carbamide crystals during heating was analyzed using thermal analyzer NETZSCH STA 449 F3 Jupiter®. Samples were formed by casting method in conditions dramatic change temperature. Figure 1 demonstrates the process scheme. Complex mold has been developed for crystallization process and instantaneous temperature change at the bottom of the form (to form elongated in one direction, crystals). The bottom of the mold was made from metal (thermal conductivity is 197 W/(m^{0}C), walls of the mold were made from polyurethane (thermal conductivity is 0.315 W/(m^{0}C) In this mold instantaneous change of slurry temperature was achieved only on its bottom by combination of materials with different thermal conductivity. Walls and the top part of the mold were isolated from the external environment by a thin film to achieve gradual cooling of the slurry from bottom to walls of the mold. Drying of the samples was made in desiccator over calcium chloride granules to remove excess water. Samples were heat treated at 1580°C in air, rate of temperature rise ~ 2°C per a minute, curing at finishing temperature for 2 hours.

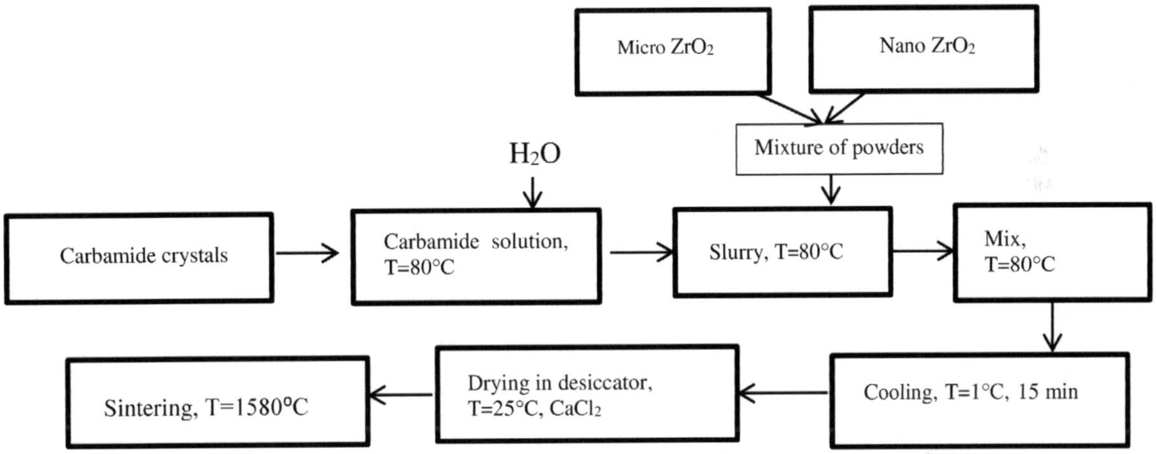

Figure 1. Scheme of porous ceramics obtaining by crystallization method of burning addition

3. Results and Discussion

Differential scanning calorimetry (DSC) and thermogravimetry of needle carbamid crystals were carried out to research the process proceed in samples at carbamide crystals heating (figure 2). Type of DSC graph demonstrates that complex endothermal reaction is carried out at heating. Two steps fixe on mass loss curve. The first stage is characterized by greater weight loss (49.06%) than the second stage (44.58%). The first weight loss occurs at the temperatures ranging from 130 to 256^{0}C due to the fact that melt of carbamide is carried out at the heating from 130 ^{0}C to 150^{0}C, biuret ($C_2H_5N_3O_2$ or $(H_2NCO)_2NH$) forms from 232^{0}C to 256^{0}C under simultaneous production of ammonia.

$$2H_2NCONH_2 \longrightarrow (H_2NCO)_2NH+NH_3$$

The second loss of weight at the full decomposition of carbamide is due to rejection of isocyanic acid (HNCO) in the temperature range from 365 to 405 ^{0}C. The cuts of samples before firing were made to investigate carbamide crystals obtaining in samples.

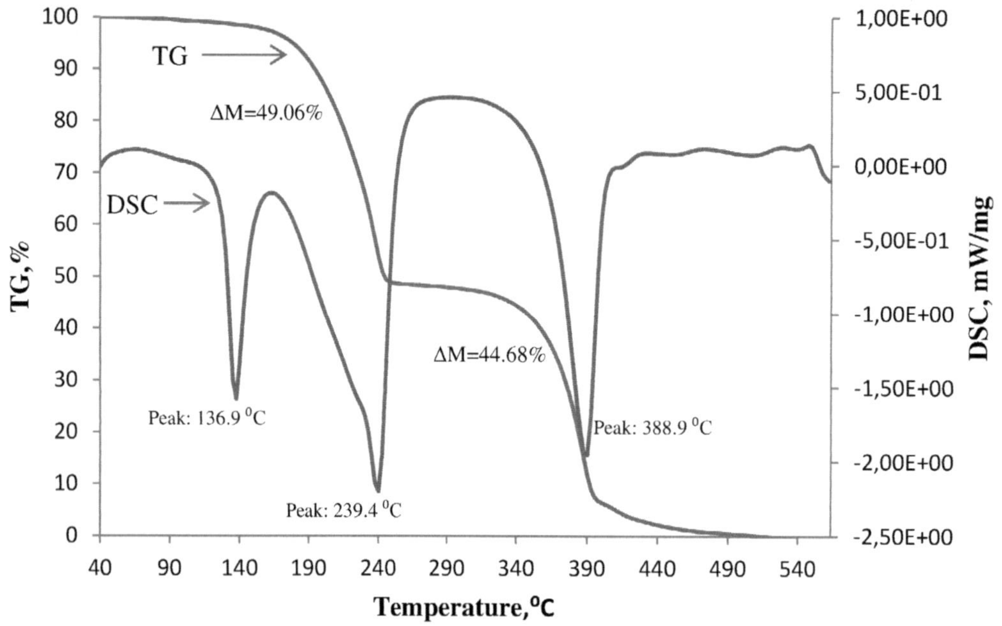

Figure 2. Differential scanning calorimetry and thermogravimetry of needle carbamide crystals

Figure 3 demonstrates carbamide crystallization in form of thin needles in sample volume. In this case particles of oxide powders surround each crystal. Carbamide crystals are burned in firing process with pores forming. Sintered samples have permeable pores in form of needles. Type of porosity is clearly visible on side section and top surface of sample (figure 3).

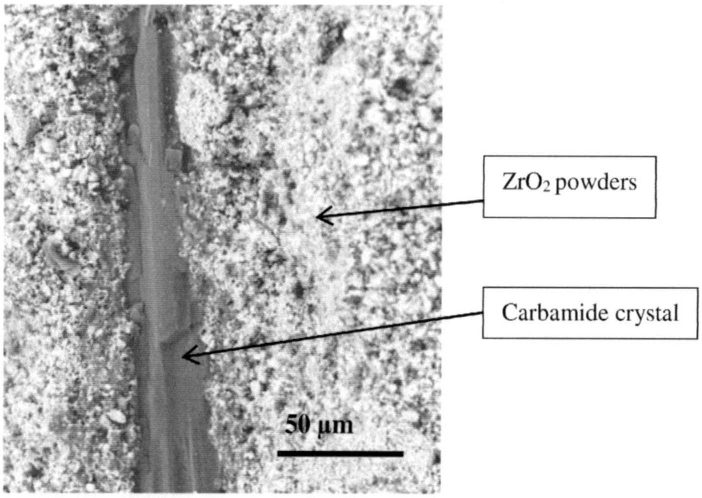

Figure 3. Photo of raw samples consist of 50 mas.% nanopowder ZrO_2 and 50 mas.% micron powder ZrO_2

Pores formed after removing carbamide crystals have an aligned form (figure 4a). Length of channels achieves 5 mm, pores width changes from 0.2 to 200 µm (figure 4b).

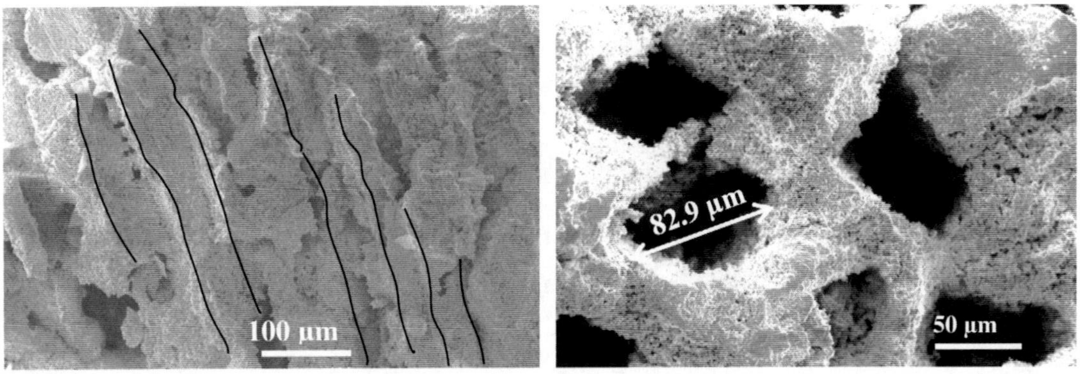

a b

Figure 4. Photo of sintering samples consist of 50 mas. % nanopowder ZrO_2 and 50 mas. % micron powder ZrO_2 (a-side elevation, b- plan view)

4. Conclusion

Application of organic additives crystallization in slurry method allows to obtain ceramics with thin permeable pores and total porosity 48%. Ceramics can be prepared with a predetermined level and morphology of pores by varying the powder composition, the slurry temperature, the amount of carbamide in solution and the mold. Samples obtained in the experiment have pores from 0.2 to 200μm size, that allows to use this type of ceramics for gas purification from large and small particles (until 0.5 μm).

References

[1] Nettleship I 1996 Applications of porous ceramics *Key Eng Mat* **122–1** 305–324
[2] Zhu X L and Su X J. 2000 Porous ceramics materials *China Ceram* **36(4)** 36–39
[3] Sujith Vijayan, Narasimman R and Prabhakaran K 2013 A urea crystal templating method for the preparation of porous alumina ceramics with the aligned pores *Journal of the European Ceramic Society* **33(10)** 1929-1934
[4] Lei Qian and Haifei Zhang 2011 Controlled freezing and freeze drying: a versatile route for porous and micro-/nano-structured materials *Journal of Chemical Technology and Biotechnology* **86(2)** 172-184

Influence of dispersion state of initial AlN powder on the hydrolysis process in air environment

A A Ditts, I B Revva, N Y Grishko and R V Tarnovskiy
Department of Silicate and Nanomaterials Technology, Tomsk Polytechnic University, Tomsk, 634050, Russia

E-mail: ditts@tpu.ru

Abstract. The research results of the hydrolysis processes of aluminum nitride powders received by the SVS method in dependence on humidity of the storage environment, and grain size distribution are presented in this work. Oxidation kinetics was estimated by means of X-ray Diffraction (XRD) and scanning electron microscopy (SEM). The induction period of the hydrolysis process for various powders, its dependence on powder dispersion and thickness of the oxide layer on surface of particles have been defined.

1. Introduction

Aluminum nitride (AlN) is advanced material for substrates of electron devices due to combination of electrical, physical, and dielectric properties [1,2] and high thermal conduction (320 W / (m · K)$^{-1}$) [3]. Production of ceramics on the basis of aluminum nitride increases every year. Thermal conduction of sintered AlN depends on impurities in AlN grains, the largest action is performed by oxygen impurity because oxygen, replacing nitrogen, makes vacancies on aluminum which scatter phonon fluctuation of the lattice [3, 4-8]. Therefore, AlN powders with the low oxygen content are preferable for production of high-heat-conducting products. Change of oxygen content can happen in the course of production of powder, technological processing and at its storage.

The aim of this work is the study of the kinetics of interaction reaction of wet air with powdery aluminum nitride at storage in the conditions of real production.

2. Materials and methods of researches

Aluminum nitride powders produced by method of the self-propagating high-temperature synthesis (SHS) (Chernogolovka, Russia) of SHS-S, SHS-I and SHS-K grades were chosen for this research. The phase composition of the initial powders and products was determined with help of the Shimadzu XRD-7000 (Japan) x-ray diffractometer, we used the ICDD PDF-2 database (2013) for interpretation. According to the obtained data all initial powders are presented by the only phase – aluminum nitride (PDF 000-08-0262).

Definition of particle size distribution was performed with the help of the SALD 7101 laser diffraction particle size analyzer (Japan). Results are presented in (table 1).

The obtained data are in accord with the results of calculation of average diameter of particles (d_{av}) according to BET data obtained with the Nova 2200-e Quantachrome installation (USA). Oxygen content was determined with the LECO installation of the ONH836 grade, table 1. If to assume that all oxygen is in form of Al_2O_3 which is evenly distributed on the surface of particles, thickness of the oxide layer – l_{cal}, can be calculated according to method [9]. The thickest oxide layer (table 1) coated aluminum nitride powder particles of SHS -K grade, the thinnest oxide layer was on the surface of SHS -S particles.

Table 1. Properties of initial powders of aluminum nitride

Grade of AlN powder	Particle size, μm			S_{spec}, m^2/g	d_{av}, μm	Content of O_2, wt. %	thickness of the oxide layer $l_{cal.}$, nm
	D_{10}	D_{50}	D_{90}				
SHS-S	0.23	2.11	5.05	7.04-8.01	2.20	1.60	0.12
SHS-I	1.64	3.10	5.12	1.55-2.50	7.40	1.14	0.32
SHS-K	1.92	12.29	29.71	0.50-0.70	26.30	0.92	0.88

Data of electron microscopy of aluminum nitride powders performed by the JEOL JSM 6000 scanning electron microscope are presented in figure 1. Powder of SHS-S grade (figure 1, a) is presented by particles of irregular shape with size from 1 to 5 μm. Aluminum nitride of SHS-S grade (figure 1, b) is formed by volume particles of rounded (spherical) form with size not more than 5 μm. One can see large volume particles of clastic form with shell-like fracture with sizes to 20 μm on microphotographs of aluminum nitride of SHS-K brand (figure 1, c). Fine particles with sizes from 2 to 5 μm and less are located between large particles. The received microphotographs confirm these definitions of particle size distribution of initial powders.

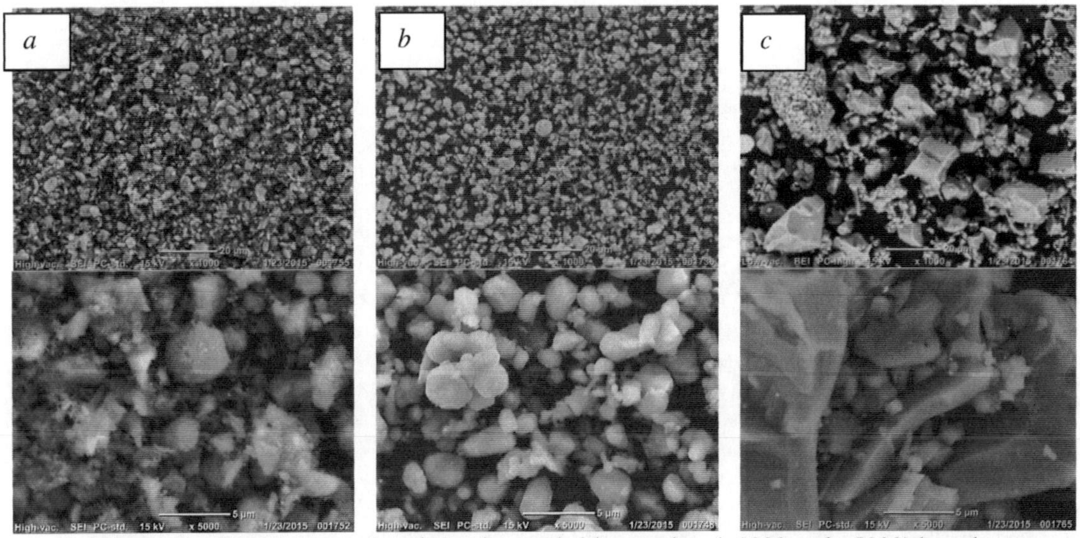

Figure 1. Microphotographs of aluminum nitride powders (×1000 and ×5000) brands: a) SHS-S; b) SHS-I; and c) SHS-K

Difference in the granulometry of powders is defined by features of their behavior at heating. Data of synchronous thermal analysis, which has been performed using F3 Jupiter Netzsch STA 449 device (Germany), are given in table 2 and in figure 2. Conditions of analysis were: environment – air, T_{max} – 1500 °C, the rate of heating 15 ° / min., standard α-Al_2O_3.

Table 2. Thermal analysis data

Aluminum nitride powder grade	Characteristic temperatures				ΔH, J/°
	$T_{b.o.}$	T_{p1}	T_{p2}	$T_{e.o.}$	
SHS-S	1045	1149	-	1349	15568
SHS-I	1116	1171	-	1300	21208
SHS-K	1120	1186	1361	1500	25946

Similarity of nature of oxidation of aluminum nitride powders of SHS-S and SHS-I grades is explained by similar particle size distribution. Powders are completely oxidized in one stage up to the temperature of 1500 °C in air. Two stages oxidation of SHS-K grade powder is connected with bimodal particle size distribution. Differences in temperature of the beginning of oxidation ($T_{b.o.}$) are connected with protective action of oxide film on particle surface, which thickness correlates with temperature of the beginning of oxidation.

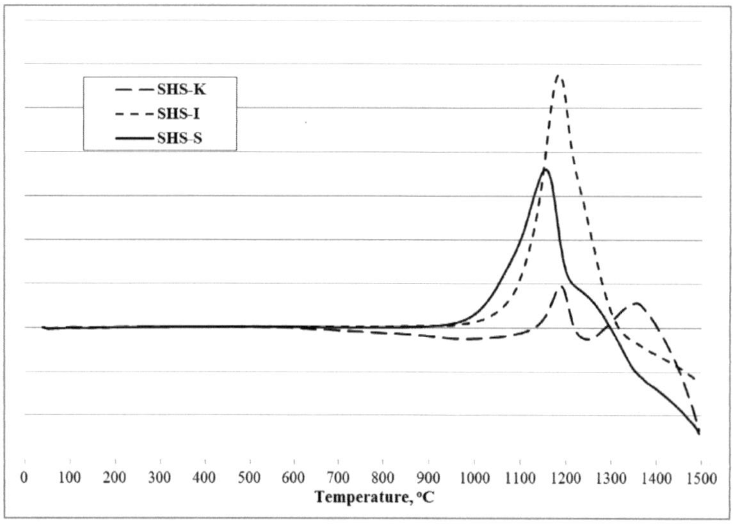

Figure 2. Thermograms of aluminum nitride powders

Method of carrying out experiment. Initial aluminum nitride powder was poured in the Petri dish with layer with thickness not more than 2 mm to increase area of interaction and was located in two exsiccators. One exsiccator contained air environment with high humidity (96-98%) created and maintained with the help of distilled water. Humidity of air environment in other exsiccator was not more than 15% (conditionally "dry" environment).

Aluminum nitride at interaction with water begins to hydrolyze according to following reactions [10]:

$$AlN + 2\,H_2O \rightarrow AlOOH_{(\mathrm{amorph})} + NH_3$$

$$AlOOH_{(\mathrm{amorph})} + H_2O \xrightarrow{OH^-} Al(OH)_{3(crystal)}$$

Assays were sampled with certain periodicity, then they were analyzed on content of oxygen, nitrogen, hydrogen, as well as X-ray phase analysis was performed.

3. Results and discussion

According to data of definition of phase structure of the assays stored in dry atmosphere, all powders irrespective of grade and duration are presented by the only phase of aluminum nitride (PDF 000-08-0262). Change of content of oxygen, nitrogen, hydrogen in them is in measurement error limit. Data of RFA of assays of aluminum nitride of SHS-S, SHS-I, and SHS-K grade after storage during 63 days in dry environment are presented in figure 3.

At storage of assays in the wet atmosphere change of phase composition for different grades of aluminum nitride begins at various terms of interaction. The powder of SHS -S grade with the minimum particle size and the thinnest layer of protective oxide film from all presented powders has the greatest reactivity. Emergence of $Al(OH)_3$ peaks on XRD patterns is fixed for 9th day of experiment. With increase of storage periods intensity of aluminum hydroxide peaks increases that testifies intensification of interaction processes. At the same time intensity of aluminum nitride peaks naturally decreases.

SHS-K grade powder has the smallest activity that is connected with larger thickness of protective oxide layer on surface of AlN particles. Emergence of aluminum hydroxide peaks (figure 4) is fixed at the 32nd

day of storage in wet medium. Powder of the SHS-I grade has average activity, $Al(OH)_3$ is revealed on XRD patterns on the 14th day of experiment.

Analysis of powders surface after 63-day interaction with moisture of air (figure 5) showed that powders of all grades have similar surface structure. There are loose formations on the powder particles, which have less clear sides in comparison with initial structure. Assays kept initial size distribution. Powders of the SHS-S (figure 5, a) and SHS-I (figure 5, b) grades are formed by particles with sizes from 2 to 5 μm. At the same time fine powders with more uniform sizes agglomerated in strength granula-like formations. Aluminum nitride of SHS-K grade (figure 5, c) is presented both by fine (2-5 μm), and by large particles (to 20 μm).

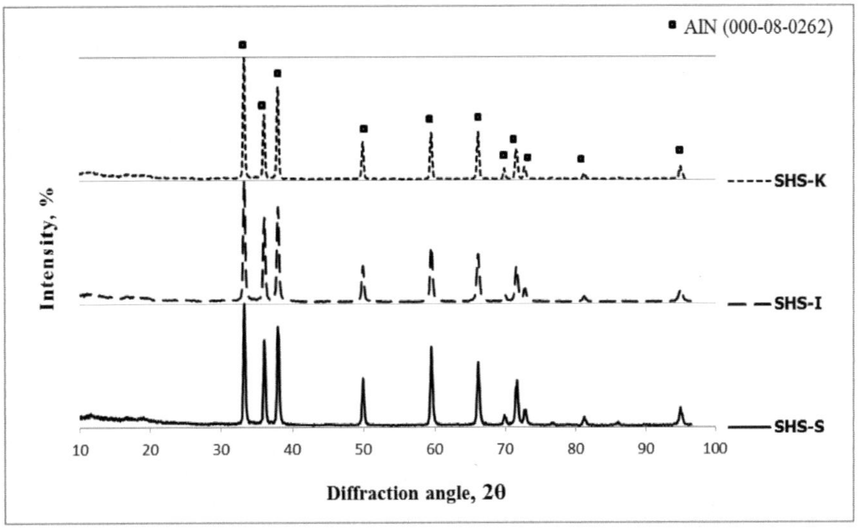

Figure 3. XRD pattern of aluminum nitride assays stored in the dry atmosphere during 63 days.

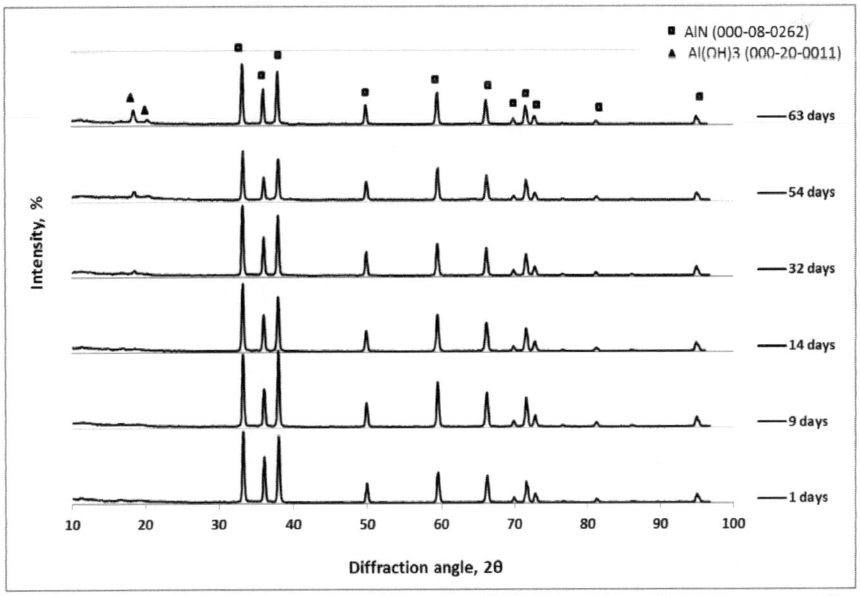

Figure 4. XRD patterns of aluminum nitride assays of SHS-K grade at different terms of interaction, stored in wet conditions

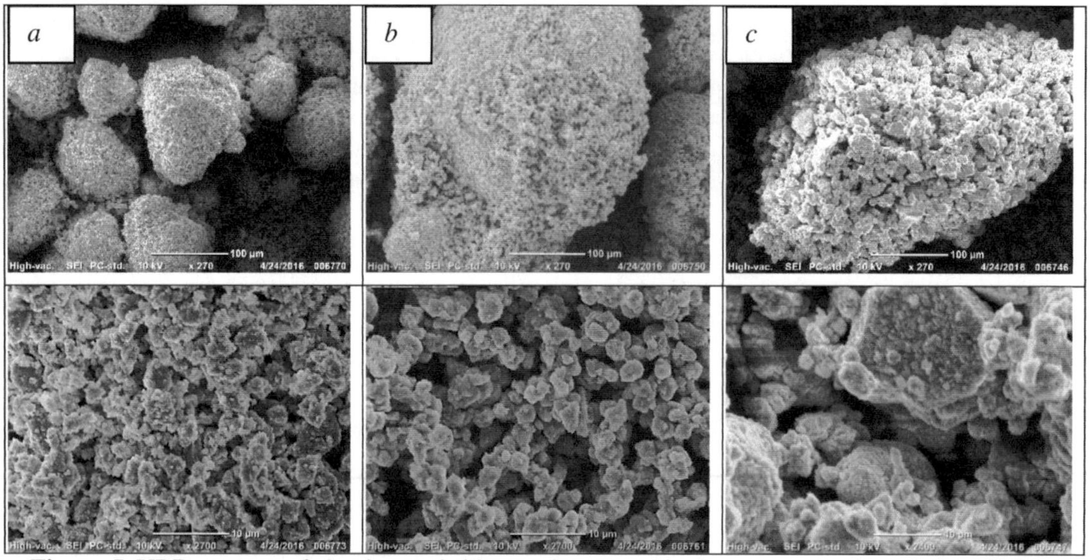

Figure 5. XRD patterns of aluminum nitride powders stored in the wet atmosphere during 65 days (×1000 and ×5000): a) SHS-S; b) SHS-I; and c) SHS-K.

To describe mechanism of interaction of initial powders with moisture of air environment we calculated degree of transformation. According to obtained data (figure 6) one can see that all powders have the induction period which duration depends on thickness of surface oxide layer of particles. Large particle powder of aluminum nitride of SHS-K grade (20 days) has the longest induction period, since its surface aluminum oxide layer is larger by 2.5 times than one of other powders. With decrease of thickness of surface layer of Al_2O_3 on particles of initial powder, duration of induction period is reduced.

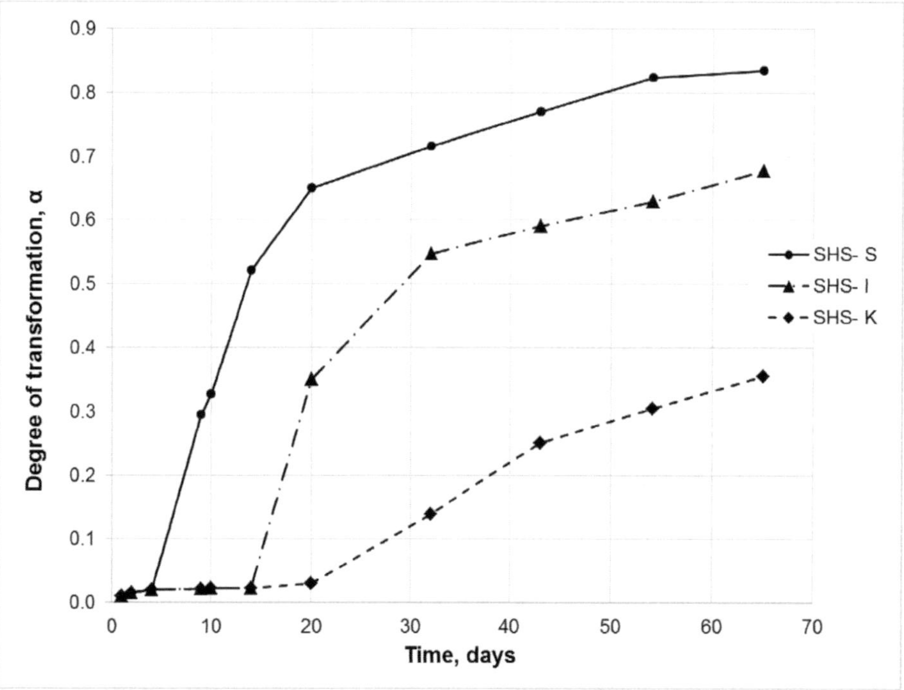

Figure 6. Degree of transformation of aluminum nitride of various grades at storage in wet conditions

4. Conclusion

Disturbance of storage procedure for aluminum nitride powders can lead to considerable changes of phase structure. Quantitatively these changes will depend on external environment, storage periods, dispersion state of initial assay, and thickness of the surface oxide film.

References

[1] Weimer A W 1997 Carbide, Nitride, and Boride Materials Synthesis and Processing. (London: Chapman & Hall) 671

[2] Sheppard L M 1990 Aluminum Nitride: A Versatile But Challenging Material. *American Ceramic Society Bulletin.* **69** 1801–1812

[3] Slack G A, Tanzill R A, Pohl R O, and Vandersande J W 1987 The Intrinsic Thermal Conductivity of AlN *Journal of Physics and Chemistry of Solids* **48** 641–647

[4] Jinwang Li *at al* 2006 Mechanism and Kinetics of Aluminum Nitride Powder Degradation *Journal of the American Ceramic Society* **89** 937–943

[5] Watari K, Kawamoto M and Ishizaki K 1991 Sintering Chemical Reactions to Increase Thermal Conductivity of Aluminium Nitride *Journal of Materials Science* **26** 4727–4732

[6] Watari K, Ishizaki K and Fujikawa T 1992 Thermal Conduction Mechanism of Aluminium Nitride Ceramics *Journal of Materials Science* **27** 2627–2630

[7] Watari K, Ishizaki K and Tsuchiya F 1993 Phonon Scattering and Thermal Conduction Mechanisms of Sintered Aluminium Nitride Ceramics *Journal of Materials Science* **28** 3709–3714

[8] Ditts A A. *at al* 2015 Obtaining high thermally conductive materials by pressing from the granulate *IOP Conference Series: Materials Science and Engineering* **71** 1–7

[9] Borovinskaya I P *et al* 1987 Influence of grinding operation on composition of surface layer of aluminum nitride powders *Powder Metallurgy and Metal Ceramics (Poroshkovaya metallurgia in Russian)* **5** 1-6

[10] Tolbanova L O and Ilyin A P 2010 Water durability of nitride containing ceramic materials synthesized by combustion in air Bulletin of the Tomsk Polytechnic University **316** 12-17

IOP Conf. Series: Materials Science and Engineering **156** (2016) 012041 doi:10.1088/1757-899X/156/1/012041

Influence of temperature and glass composition on aluminum nitride contact angle

R Tarnovskiy and A Ditts
National Research Tomsk Polytechnic University, Tomsk, Russian Federation

E-mail: tarnovskiy@tpu.ru

Abstract. Results of research of different glass compositions for possibility of their application in metallization pastes intended for ceramics based on aluminum nitride are presented in this article. It includes research of contact angle of aluminum nitride with glasses of different compositions at different temperatures and different roughness of ceramics.

1. Introduction

Ceramics based on aluminum nitride is a perspective material used in electronics as housings and substrates for electronic chips due to its unique combination of high thermal conductivity and high dielectric properties. More than 90% of aluminum nitride ceramics market are presented by metallized ceramic products, therefore it is necessary to develop metallization pastes intended for ceramics based on aluminum nitride. At the moment the process of aluminum nitride metallization is poorly understood. One of the most effective ways of ceramics metallization is using glass as binding additive in metallization pastes.

In accordance with world trends of development, one of demanded market segments is electronics and electrical engineering (ceramic elements for electronic industry, including thermal loaded elements of semiconductor devices, high-power LEDs; insulators for various applications for electrical engineering and energetics). One of promising directions is producing high thermal conductive materials and products based on aluminum nitride. Aluminum nitride possesses a row of unique properties compared with other materials used in microelectronics [1]. Aluminum nitride application as base in high-power LEDs manufacturing requires applying a conductive paste on it. Conductive low-temperature silver-containing pastes developed in a large amount are intended for oxide materials, so they have low adhesion to aluminum nitride. It is possible to increase adhesion of metallization paste to aluminum nitride by introduction glass binder to their composition [2], which must wet aluminum nitride at low temperatures.

The aim of this work is development of glass additive which wets aluminum nitride at temperatures up to 1000 °C.

2. Experiment

Samples in the form of aluminum nitride discs sintered at relative density 98% were used in the work. The discs were subjected to grinding and cleaning. Surfaces were cleaned from organic pollutants by isopropyl alcohol and dehydrated by annealing at temperature 600 °C. To

IOP Conf. Series: Materials Science and Engineering **156** (2016) 012041 doi:10.1088/1757-899X/156/1/012041

determine contact angle an installation has been constructed, a schematic diagram is shown in Figure 1.

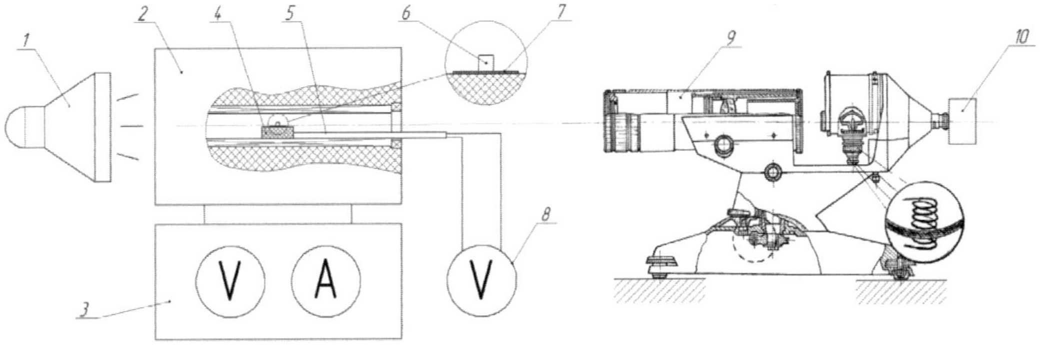

Figure 1. The installation schematic diagram

The installation consists of: 1 –lighting system; 2 –tubular furnace at a controlled heating rate; 3 – heating control system; 4 – sample stand; 5, 8 – thermocouples for the temperature control of the sample; 6 – glass sample; 7 – substrate (disc) of AlN; 9 – optical system; 10 – digital camera with an image displayed on the computer display.

The investigated glass was pressed in the form of cylinder with 3 mm diameter and 5-6 mm tall. In the pressing process does not use organic binders. After setting sample the furnace was heated with rate from 7 to 10 °C/min. When the temperature reached 600 °C the sample began to be photographed every 10 °C. Image processing was carried out using the software «Micro-Analiz Pro» supplied with the digital camera.

For the work were selected glass compositions in the system B_2O_3-SiO_2-Na_2O, which provide low cooking and softening temperatures. The compositions are shown in the Table 1.

Table 1. Ratio of components in the selected glasses.

Letter	Content of components, mol. %							Total
	Na_2O	CaO	B_2O_3	SiO_2	V_2O_5	K_2O	PbO	
a	20	5	10	65	–	–	–	100
b	20	5	10	55	10	–	–	100
c	10	5	10	65	–	10	–	100
d	20	5	10	55	–	–	10	100
e	10	5	10	75	–	–	–	100

Calculation of properties of the selected glasses was carried by specialized software SciGlass version 7.0. The one especially important of them is dependence of viscosity on temperature. Calculation results are shown in Figure 2.

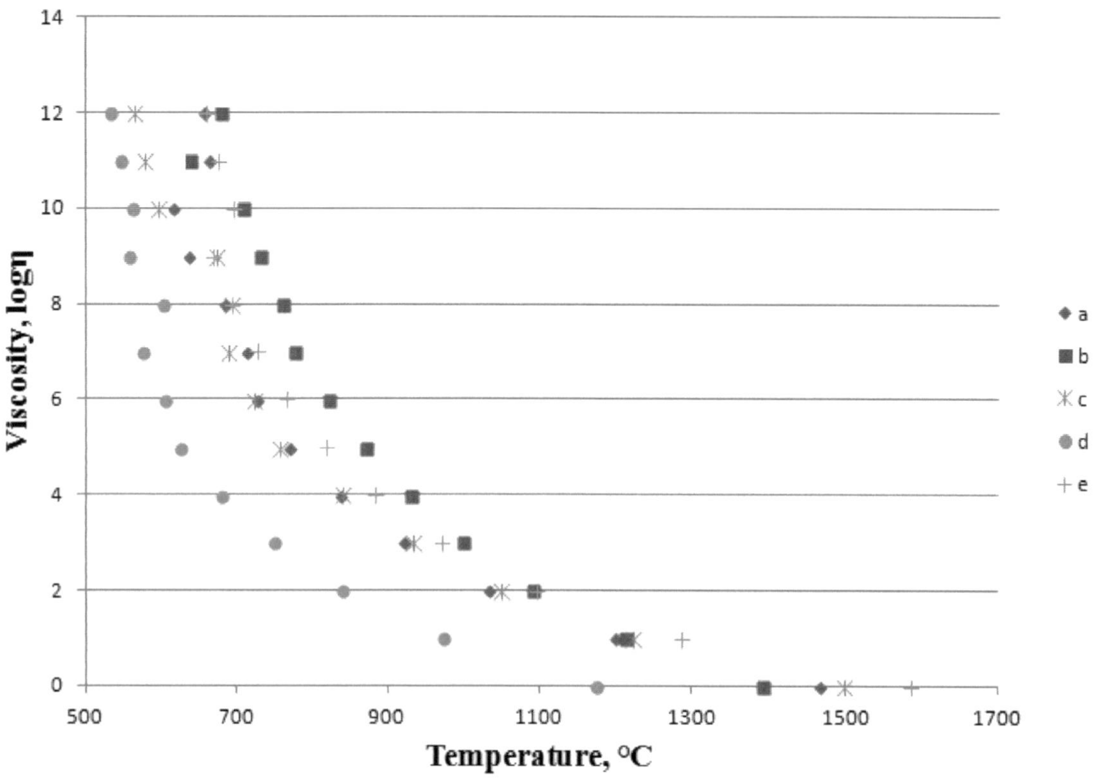

Figure 2. Dependence of viscosity on temperature

Due to calculated data one can see that the least refractory composition is composition "d" containing PbO, and the most refractory is composition "b" containing V_2O_5.
The glasses of proposed compositions were cooked whereupon they were grinded till full passing through the sieve with cell size 63 mkm. The glass powder was pressed into samples in the form of cylinder with diameter-to-height ratio equal 1/2. In Fig 3, 4 are shown sample photo at the different temperatures and data of contact angle definition.

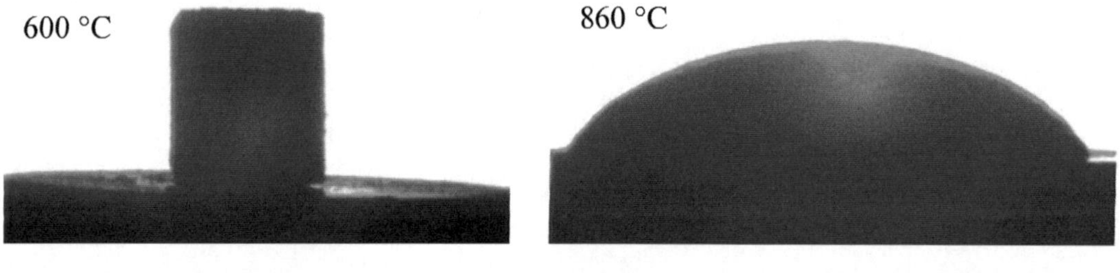

Figure 3. The glass of composition «d» samples photo at the temperatures 600 and 860 °C

Table 2. Glass samples images wetting polished and unpolished aluminum nitride samples at different temperatures.

		600	700	800	900	1000	1100	1200
a	polished							-
	unpolished							
b	polished						-	-
	unpolished							
c	polished							
	unpolished							
d	polished				-	-	-	-
	unpolished						-	-
e	polished							
	unpolished							

Images of glass samples at different temperatures on polished and unpolished aluminum nitride discs are shown in the Table 2. In this table one can trace process of glass spreading on the aluminum nitride substrate, as well as clearly see that in practically all cases the polished aluminum nitride samples are wetted by the glass better than unpolished. Apparently it's connected with the formation of air pockets reducing wettability on the surface of unpolished samples.

It is also noteworthy that the glass of composition "e" turned up the most refractory one contrary to calculated data.

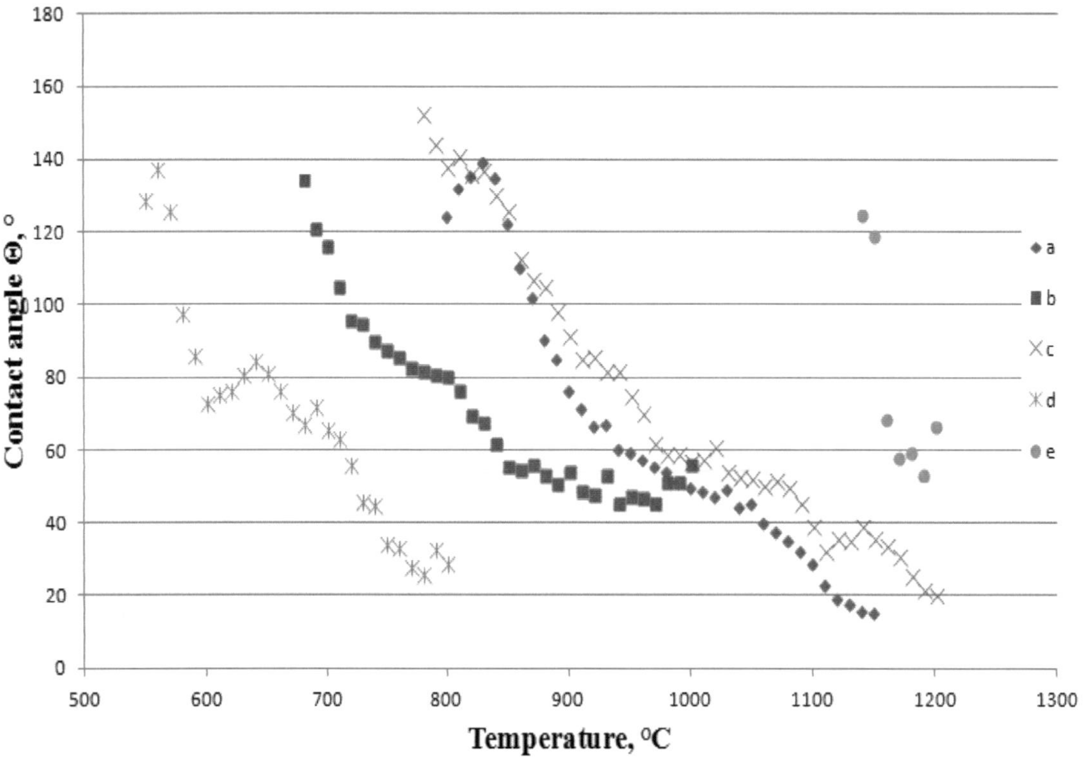

Figure 4. Dependence of aluminum nitride contact angle on temperature

Figure 4 shows diagrams of dependencies of contact angle of polished aluminum nitride samples with glass on temperature. One can see that these dependencies are similar in nature. Firstly, there is a maximum when the drop becomes spherical shape, then the drop begins to spread, and contact angle is decreasing. Secondly, there is an inflection associated with foaming glass: initially drop volume increases, then glass begins to be clarified and the foam is settling. Further, the contact angle goes to zero.

3. Conclusions

As one can see in experimental data shown in Table 2 and Figure 4, the glasses of compositions "d" and "b" respectively containing lead and vanadium begin to wet aluminum nitride at the lowest temperatures. These glasses can be used for metallization pastes based on silver which are burned at temperatures 800÷920 °C. The glass of composition "e" begins to wet aluminum nitride only at temperatures of about 1150 °C and in the future, possibly, it may be used in the molybdenum-manganese metallization pastes which are burned at temperature 1200÷1300 °C.

References

[1] Kosolapov A, Ditts A, Revva I and Pogrebenkov V 2015 *High-temperature materials obtained by method of pressing* (Tomsk: Tomsk Polytechnic University Press)
[2] Makarov N 2004 *Ceramics Metallization: Tutorial* (Moscow: D.Mendeleev University of Chemical Technology of Russia)
[3] Summ B and Goryunov Yu 1976 *Physical and chemical bases of wetting and spreading* (Moscow: Chemistry)

[4] Yuan Y and Lee R 2010 *Contact Angle and Wetting Properties* (Springer - International Publisher Science, Technology, Medicine)
[5] Kwok D and Neumann A 1999 *Contact angle measurement and contact angle interpretation* (Advances in Colloid and Interface Science)

Increase in composite binder activity

R Fediuk, A Smoliakov, N Stoyushko
Far Eastern Federal University, 20, Ajax St.,Vladivostok, Russky Island, 690000, Russia

E-mail: roman44@yandex.ru

Abstract. The binder of portland cement (51-59 wt.%), fly ash of thermal power stations (36-44 wt.%), limestone crushing waste (4-9 wt.%) and dry hyper plasticizer (0.2 wt.%) has been created. It can be used in the building materials industry for production of high-strength concrete. The composite binder is obtained by co-milling of the components in vario-planetary mill to a specific surface area of 550-600 m^2/kg. The technical result is the possibility of obtaining a composite binder with significant replacement of cement with industrial waste, cost-effective and superior to portland cement for construction and technical properties, increased activity. This allows producing concrete for walling with a compressive strength of 100 MPa, while using more than 50% of industrial waste.

1. Introduction.

One of the most important tasks of our time is the reduction of energy consuming in manufacturing efficient building composites; environmental protection; the optimization of the system "man-material-environment". These problems are typical for the Far Eastern region of Russia, the priority development of which is the most important task of the state.

Building materials industry widely uses concrete on cement binder and natural aggregates as a structural material. At the same time large-capacity ash waste and rubble crushing of the rocks of different composition are formed as a result of mining industry companies and energy economy activities in the Far East.

Reengineering of concrete mixtures structure formation through the use of industrial waste seems to be necessary, which will increase strength characteristics. It will help to improve the environmental situation in the region through the use of industrial waste.

A large number of works by both Russian and foreign scientists [1-3] is devoted to the problem of increase in composite binder activity. It is known that one of the ways to improve the performance of concrete is the use of highly active additives of various composition and genesis both micro- and nanosized level, which contribute to the optimization of structure formation processes by initiating the formation of hydrated compounds. So the works carried out previously, prove effectiveness of using nanostructured silica-modifier composition as active mineral additives. Besides, capabilities of increase in composite binder activity by mechanical grinding of its components have been studied. However, with replacement of a large part of cement with production waste, there is a low strength of cement stone. It is inherent for such materials as fly ash and portland cement comprising wt. %: acidic fly ash of TPP - 30-40, bauxite and nepheline sludge - 20-30, the complex additive - 3-5, portland cement jointly milled to a surface area of 450-550 m^2/kg [4]. Thus, development of a dense structure of cement stone matrix, providing high physical and mechanical characteristics is rational. The goal of this is increasing strength of the cement stone, as well as obtaining high quality products on the basis of composite binding material with significant replacement of cement with industrial waste.

2. Composite binding material development

To achieve this goal, the composite binders obtained by co-grinding of portland cement, hyper plasticizer, ash and limestone in the vario-planetary mill have been used.

Fly ash of the Primorye largest thermal power plants such as Vladivostok TPP (figure 1) and Artem TPP, Primorye SRPS and Partizansk SRPS was used as a component of the composite binder. An important factor is the capacity of dry separate ash extraction that is currently implemented at these power plants.

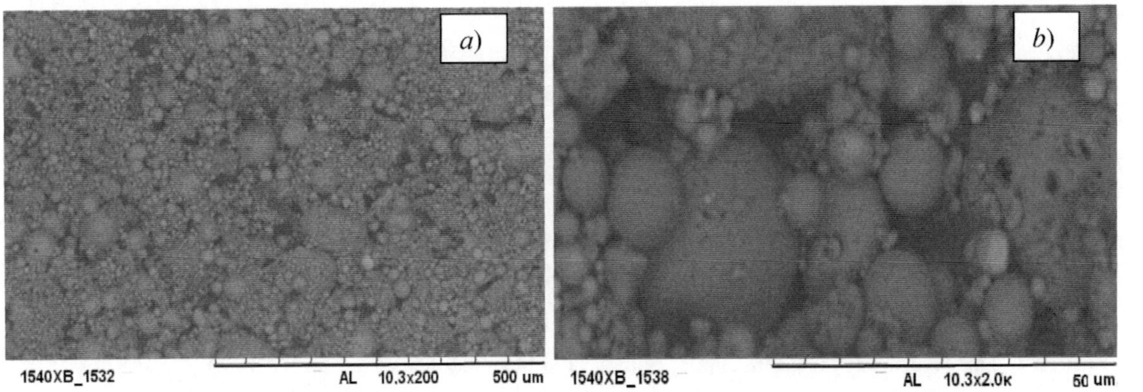

Figure 1. The general view of the Vladivostok TPP fly ash; 200x magnification *(a)* and 2000x magnification *(b)*

The use of technogenic raw materials in the manufacture of building materials contributes to the solution of the following main objectives: saving energy and raw resources; waste management; improvement of the environmental situation in the regions. Thermal power plants fly ash is an effective material for the production of active mineral and fine admixtures [5-6].

Considering the focus of research for the development and use of environmentally friendly materials, we have evaluated ash radioactivity (table 1).

Table 1. Specific effective activity of fly ash depending on the composition

Indicator	The measurement result (A), Bq / kg			
	Primorye SPRS	Vladivostok TPP	Artem TPP	Partizansk SRPS
Activity ^{40}K	496.9±101	362±89	342±68	516.9±101
Activity ^{232}Th	153.6±20.3	31.5±19.7	29.5±15.7	193.2±22.3
Activity ^{226}Ra	163.1±9.36	37.63±6.32	27.23±5.93	113.1±6.37
$A_{eff} = A_{Ra} + 1.31A_{Th} + 0.085A_K$	>398	80±30	93±20	>410

It was found that the ash of Vladivostok TPP and Artem TPP belongs to the first class of materials ($A<370$ Bq/kg) in accordance with GOST 30108-94 "Building materials and products. Determination of the specific effective activity of natural radionuclides" and can be used for all kinds of construction work [7-9].

Thermal studies showed that in the range of low temperatures physically bound water is removed from the ash sample. Exotherm with maximum at about 400°C indicates burnout of organic substances and endotherm at 712°C shows dissociation of calcite to CaO and CO_2, which is confirmed by X-ray diffraction (figure 2).

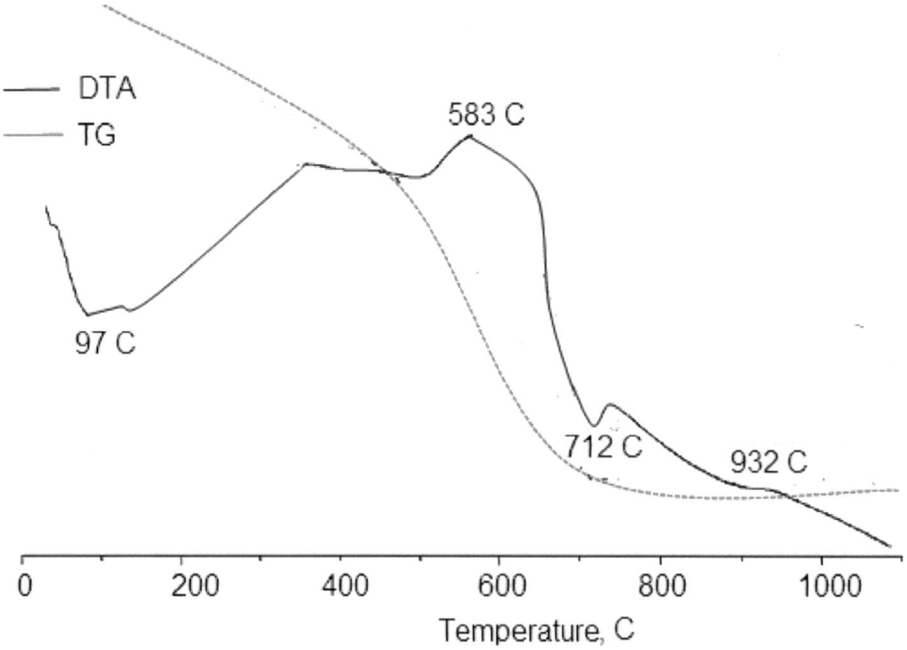

Figure 2. The results of DTA and TG of fly ash the Vladivostok TPP

Process optimization of structure formation by hydration of composite binder components creates a dense structure of the matrix, what is necessary to solve the problem of the creation composite for the production of high-strength concrete. This can be realized by co-grinding of portland cement, polyfunctional mineral admixtures and reducing the water-cement ratio of the concrete mix by using hyper plasticizer [10-12].

To reduce the water demand of concrete mix powder hyper plasticizer was selected of the six most common in the Far East market construction materials. Cement paste broke was measured with Hagermann cone. Spassky cement CEM I 42,5N was used for grout. Water-cement ratio is 0.3. Dosage plasticizer is 0.3%. Time of measurement start of Water Polution Class was recorded after the end of mixing the cement paste.

High values of WPC are achieved on binder raw mixture with hyper plasticizer PANTARHIT PC160 Plv (FM) (table 2).

Table 2. Broke of the cement paste with different hyper plasticizers

Time measurements, min	Melflux 1641 F, Germany	Melflux 5581 F, Germany	PANTARHIT PC160 Plv (FM), Russian Federation	FOX™-8H (Pwd), Russian Federation	PC-1030, China	JK-04 PPM, China
			broke, mm			
0	290	350	370	250	240	130
5	380	390	400	260	280	120
30	390	350	390	240	190	98

For further research 7 composite binder compositions have been developed. Hyper plasticizer PANTARHIT PC160 Plv (FM) was added in each of them an amount of 0.3%, the ratio of binder:

sand - 1: 3. To determine the optimal number of components in the "cement-lime-ash" system they were ground to a specific surface of 600 m^2/kg at various ratios (table 3).

Table 3. The compositions and properties of composite binders

No	Cement content, wt. %	Fly ash, wt. %		Limestone, wt. %	Compressive strength, MPa		
		Vladivostok TPP	Artem TPP		3 d	7 d.	28 d.
1	100 (without grinding)	–	–	–	17	32,5	47,5
2	30	–	50	20	30.2	40.1	50.4
3	35	45	–	20	34.2	43.1	53.2
4	40	–	45	15	36.6	48.2	56.6
5	45	45	–	10	39.2	50.1	59.2
6	50	–	40	10	45.1	54.9	65.8
7	55	40	–	5	47.2	54.1	70.2
8	100	–	–	–	60.3	81	103.2

Note: The control formulation number 1 (without final grinding); number 2-8 compositions milled to Ssp = 600 m^2/kg

3. Results

The positive dynamics of growth the composite binder strength under the joint influence of the fine constituents of ash, limestone crushing screenings and hyper plasticizers with the maximum increase in the activity of the binder at 62% has been achieved.

Active mineral components of the composite binder contribute to the binding of $Ca(OH)_2$ produced during cement hydration in additional amount hydrosilicate tumors. At the same time optimization of the process of structure formation is achieved by polydispersity of composite components. Highly spherical ash particles act as nucleation sites and act as a nano- and micro- filler. In conjunction with larger particles of the mineral component denser filling of intergranular spaces in cement concrete structure occurs with the reduction in the number of pores and microcracks.

This is confirmed by micrographs of cement paste on the composite binder, produced by joint grinding of clinker and industrial wastes of the Far Eastern region. Cement stone structure is a very dense packing of fine grains in the total weight of the crystalline tumors (figure 3).

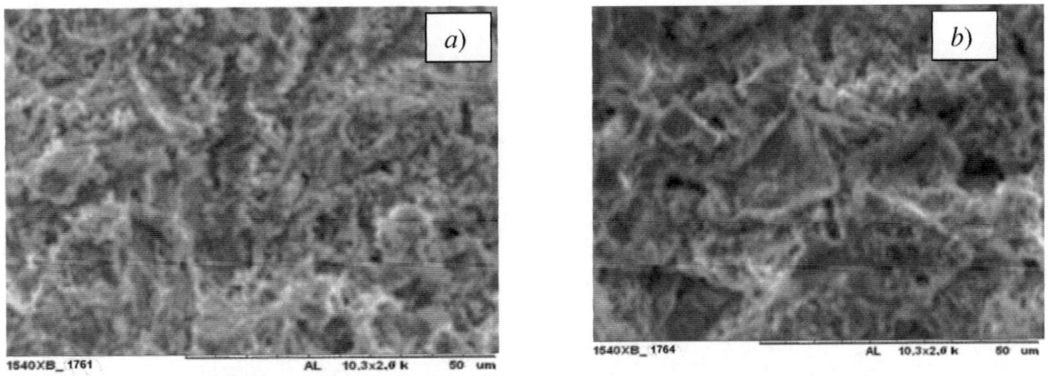

Figure 3. The cement stone microstructure: on CEM I 42.5 N *(a)* on the composite binder (composition number 7 in Table. 3) *(b)*

An additional amount of hydrated crystalline phases contributes to filling of the voids at the micro level in the crystalline matrix of calcium hydrosilicates on the boundary of the contact area, increasing the degree of adhesion of the binder with filler.

4. Conclusion

The composite binder of portland cement (51-59 wt.%), fly ash of thermal power stations (36-44 wt.%), limestone crushing waste (9.4 wt.%) and dry hyper plasticizer (0.2 wt.%) may be used in the building materials industry for production of high-strength concrete. Binder composite is obtained by co-milling of the components in vario-planetary mill to a specific surface of 550-600 m^2/kg. The technical result is the possibility of obtaining of a composite binder with significant replacement of cement with industrial waste, cost-effective and superior to portland cement for construction and technical properties, increased activity. This allows producing concrete for walling with a compressive strength of 100 MPa, using more than 50% of industrial waste.

References

[1] M B Ramli and O R Alonge 2016 *Construction and Building Materials* **121** 599-611
[2] S U Khan and T Ayub 2016 *Construction and Building Materials* **120** 540-557
[3] V S Lesovik 2013 *Construction of housing* **1** 9-13
[4] V T Erofeev, V T Fomichev and D V Emelianov 2015 *Fundamental researches.* **2-6** 1175-1181.
[5] Camila Barreneche, Aran Solé, Laia Miró, Ingrid Martorell, A Inés Fernández and Luisa F Cabeza 2012 *Meas. Sci. Technol.* **23** 085606
[6] X Fan and M Zhang 2016 *Composites Part B: Engineering* **93** 174-183
[7] Hideyuki Nakano, Akira Kouchi, Shogo Tachibana and Akira Tsuchiyama 2003 0004-637X **592** 1252
[8] P Billberg 2002 *Proceedings of the First North American Conference on the design and use of SCC* **177**
[9] M Herranz, S Rozas, R Idoeta and N Alegría 2014 *J. Radiol. Prot.* **34** 133
[10] R S Fediuk and D A Khramov 2016 *International Research Journal* **1 (43)** 77-79
[11] M Eichmann, T Krause, D Flühs and B Spaan 2012 *Phys. Med. Biol.* **57** 421
[12] R S Fediuk and D A Khramov 2016 *Modern Construction and Architecture* **1** 57-60

Fly ash particles spheroidization using low temperature plasma energy

V V Shekhovtsov[1], O G Volokitin[1], A A Kondratyuk[2] and R E Vitske[2]

[1]Tomsk State University of Architecture and Building, Applied Mechanics and Material Sciences Department, Tomsk, 634003, Russia
[2]National Research Tomsk Polytechnic University, Tomsk, 634050, Russia

E-mail: shehovcov2010@yandex.ru

Abstract. The paper presents the investigations on producing spherical particles 65–110 μm in size using the energy of low temperature plasma (LTP). These particles are based on flow ash produced by the thermal power plant in Seversk, Tomsk region, Russia. The obtained spherical particles have no defects and are characterized by a smooth exterior surface. The test bench is designed to produce these particles. With due regard for plasma temperature field distribution, it is shown that the transition of fly ash particles to a state of viscous flow occurs at 20 mm distance from the plasma jet. The X-ray phase analysis is carried out for the both original state of fly ash powders and the particles obtained. This analysis shows that fly ash contains 56.23 wt.% SiO_2; 20.61 wt.% Al_2O_3 and 17.55 wt.% Fe_2O_3 phases that mostly contribute to the integral (experimental) intensity of the diffraction maximum. The LTP treatment results in a complex redistribution of the amorphous phase amount in the obtained spherical particles, including the reduction of O_2Si, phase, increase of $O_{22}Al_{20}$ and Fe_2O_3 phases and change in Al, O density of $O_{22}Al_{20}$ chemical unit cell.

1. Introduction

Cenospheres are produced by thermal power plants (TPP) as a byproduct of coal combustion. A cenosphere is a lightweight, inert, hollow sphere that ranges in size from 50 to 500 μm. The continuous and nonporous cenosphere walls have the thickness ranging from 2 to 10 μm. The content of spherical particles is represented mostly by nitrogen and carbon dioxide [1-3].

The unique combination of the low density, small size, spherical shape, high hardness, melting temperature, and chemical inertness of cenospheres stipulates for their wide use in construction and other industries. Cenospheres have a mixed vitreous transparent structure comprising 76% glass, 22% mineral substance, and 2% impurities and are represented by aluminosilicate glass, silica, mullite, calcite, iron oxides, calcium silicates, and sulfates.

The aim of this work is to investigate the production of fly ash-based spherical particles with the size range between 20-100 μm using the low temperature plasma (LTP) treatment. The particle morphology is determined by the properties of fly ash (dispersiveness, chemical composition) and electrophysical properties of plasma generator [4]. Due to the heterogeneous chemical [5, 6] composition of fly ash, a number of preliminary transformations should be carried out to obtain particles with the predetermined properties.

The synthesis process includes three stages as presented in Figure 1. The first two stages relate to the preparation of fly ash for the subsequent production of spherical particles.

At Stage 1, the particles are grinded down in a planetary ball mill to submicrometer dimensions. At Stage 2, the spray drying technique is used to obtain agglomerated powders ranging in size from 30 to 120 μm. At Stage 3, agglomerated powders are subjected to a high-temperature plasma flow to produce spherical particles.

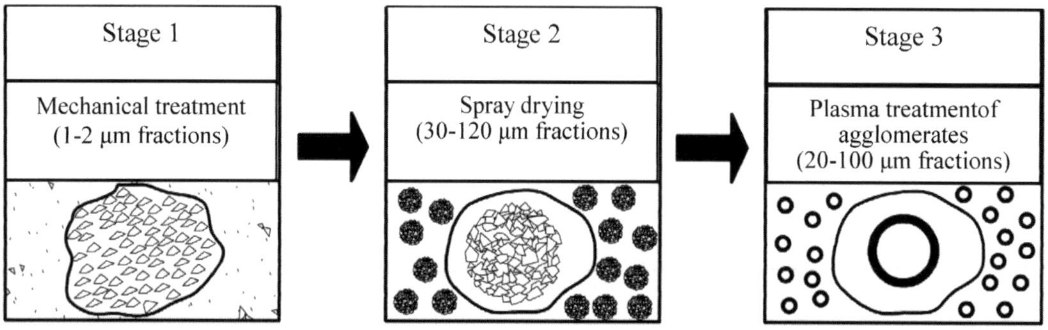

Figure 1. Flow-chart of producing fly ash-based spherical particles using plasma energy

2. Materials and Methods

Fly ash, produced from the combustion of coals was collected in dust collectors at Seversk TPP and used as the material to be treated. The results of the chemical composition analysis of fly ash are given in Table 1.

Table 1. Averaged chemical composition of fly ash collected at Seversk TPP

Oxide content	Composition, (wt.%)
SiO_2	51.16
Al_2O_3	35.07
Fe_2O_3	3.62
CaO	8.33
MgO	0.91
R_2O	0.23
Ignition losses	0.68
Viscosity modulus	5.23

Phase composition of the original specimens and obtained spherical particles was investigated on the XRD-6000 X-ray Diffractometer from Shimadzu which was modified for digitalsignalprocessing. The Rietveld refinement was used to quantify phases [7, 8] by modeling the amorphous phases dominating in the integral intensity of the diffraction maximum. The determination of amorphous phases was based on investigations of the original state of fly ash crystal lattices. Energy dispersive X-ray spectroscopy on the scanning electron microscope JSM-7500F (JEOL, Japan) was carried out to investigate the morphology, shape, and size of obtained spherical particles.

3. Experimental

The test bench was designed to carry out spheroidization of silicate particles. The structural scheme of producing spherical particles is presented in Figure 2.

The powder feeder is used to perform the fly ash delivery to the area of plasma jet. The delivery is provided with nitrogen as a carrier gas, which reduces the enthalpy and the rate of plasma jet when enters it together with the prepared mixture. As a result, the mixture discharges into the plasma jet mainly due to the inertial forces. While the particles are present in the high-temperature plasma flow, they are heated and then transit to a viscous flow state. Owing to their surface tension force, the particles take on a spherical shape. These spheres travel to the collection unit represented by a box filled with water.

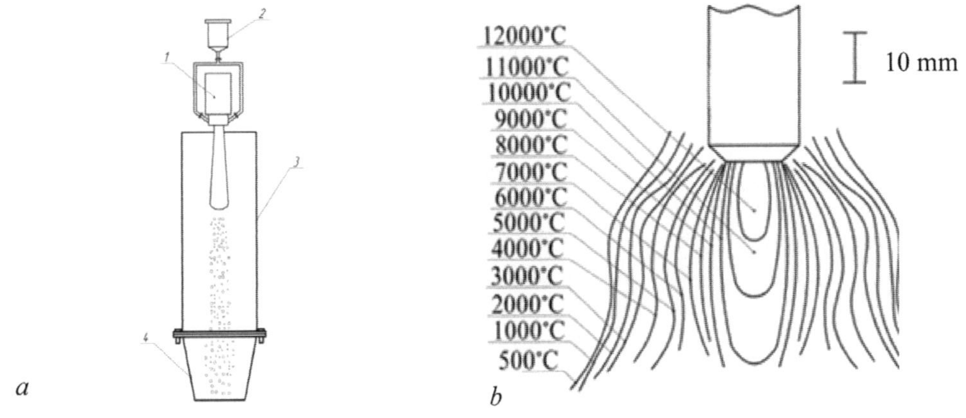

Figure 2. Structural scheme of producing spherical particles (*a*): *1* – plasma torch; *2* – powder feeder, *3* – processing chamber, *4* – particle collection unit. Distribution of plasma temperature field (*b*)

In Figure 2b, the distribution of plasma temperature field is shown. According to the data found in the literature, the melting temperature for fly ash is 1700 °C. Thus, the conditions favorable for the particle transition to a viscous flow state are formed within the temperature range observed at 20 mm distance from the jet center. This fact proves the efficiency of the LTP treatment for the particle spheroidization.

4. Results and Discussion

4.1. X-ray phase analysis

After the LTP treatment, fly ash and spherical particles produced thereof were subjected to the X-ray phase analysis (XRP). The XRP analysis shows that such stable phases as SiO_2, Al_2O_3, Fe_2O_3, TiO_2, CaO, MgO are present in the original state of fly ash that is also supported by its chemical composition (see Table 1). Fly ash composition is represented by a complex mixture of SiO_2, Al_2O_3, Fe_2O_3 components. The amount of SiO_2 phase is ~56.2 wt.%, while that of Al_2O_3 and Fe_2O_3 phases is 14.41 and 17.55 wt.%, respectively. The XRD pattern of fly ash in its original state is plotted in Figure 3.

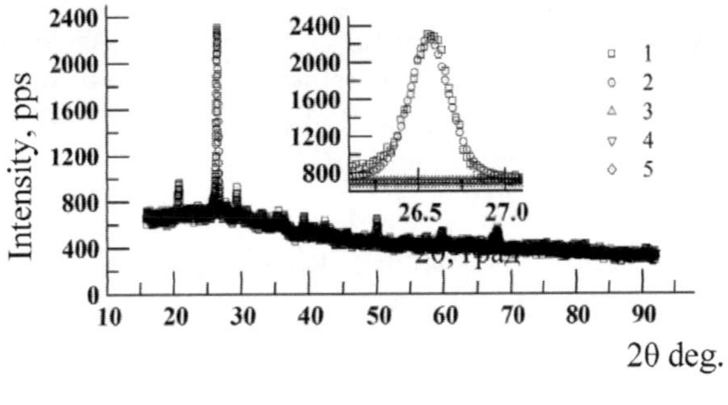

Figure 3. XRD pattern of fly ash in its original state:
1 – experimental; 2 - SiO_2 phases; 3 - Fe_2O_3 phases; 4 – TiO_2 phases.

The XRD pattern of fly ash is complex and characterized by both intensive and weak overlapping reflections. The Rietveld refinement allows determining the fly ash composition of 56.23 wt.% SiO_2, 20.61 wt.% Al_2O_3 and 17.55 wt.% Fe_2O_3 phases that make a dominating contribution to the integral (experimental) intensity of the diffraction maximums. The quantitative phase analysis (QPA) using the Rietveld method shows that the total phase content is 94.39 wt.%. This proportion is adequate to the results obtained for phase identification in fly ash.

The spherical particles produced as a result of LTP treatment, have an amorphous structure identified by the QPA by the Rietveld method. For this purpose, the model of amorphous SiO_2, Al_2O_3, Fe_2O_3 phases is created for their original state. The obtained atom density of these phases determines the size of the chemical unit cell with atoms of SiO_2, Al_2O_3, Fe_2O_3 phases. The XRD patterns of plasma-treated fly ash are presented in Figure 4.

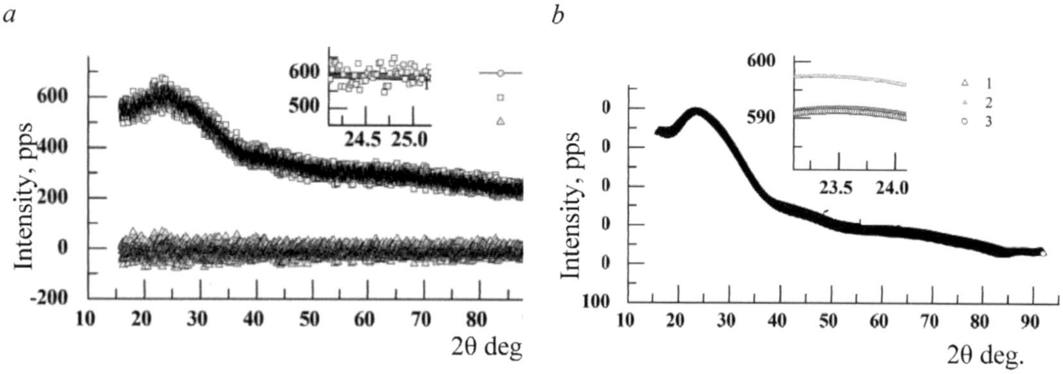

Figure 4.XRD patterns of fly ash particles after LTP treatment:
a – experimental (1); theoretical (2); b – amorphous phases: 1 – Fe_2O_3; 2 – $O_{22}Al_{20}$; 3 – SiO_2.

The optimum sizes of chemical unit cells and the equilibrium atom arrangement are detected at the following stage. The achieved amorphous state of $O_{22}Al_{20}$, Fe_2O_3 and O_2Si phases is characterized by the known space distribution of atoms in the chemical unit cell of the optimized size. In case of the primitive lattice of the space group, these phases can be identified by the chemical unit cells of $O_{22}Al_{20}$, Fe_2O_3 and O_2Si phases. The structural factors and, hence, the integral intensities of diffraction patterns can be obtained for optimized chemical unit cells of $O_{22}Al_{20}$ and O_2Si phases. The amorphous $O_{22}Al_{20}$, Fe_2O_3 and O_2Si model phases are used to determine the phase composition of fly ash. Figure 4 b contains the XRD patterns of fly ash particles analyzed by QPA by the Rietveld method. A superposition of the amorphous phase intensity is 93.13 wt.% of the XRD pattern shown in Figure 4a. The investigation results show that the spherical particles produced by the LTP treatment possess the amorphous structure. Additionally, the LTP treatment results in a complex redistribution of the amorphous phase amount in the obtained particles, i.e. the considerable decrease of O_2Si phase; increase of $O_{22}Al_{20}$ and Fe_2O_3 phases and change in Al, O density of $O_{22}Al_{20}$ chemical unit cell.

4.2. SEM Investigations

Figure 5a presents the scanning electron microscopy (SEM) image of the exterior view of the fly ash-based spherical particles obtained after the LTP treatment.

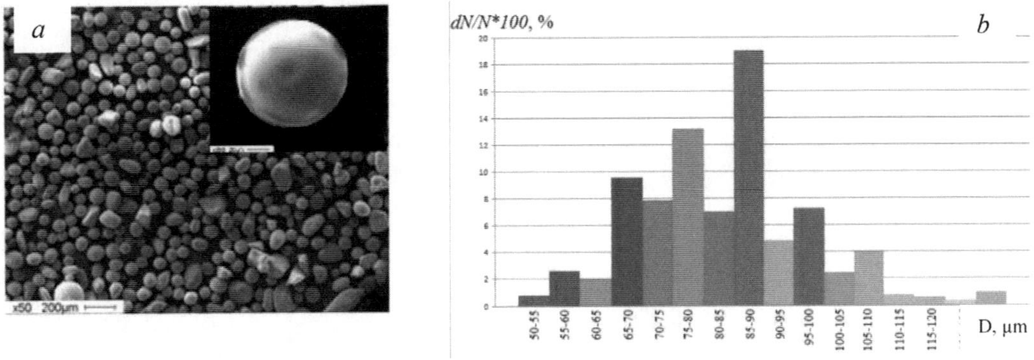

Figure 5. SEM image of spherical particles obtained after fly ash LTP treatment:
a – exterior view, *b* – size distribution

According to Figure 5b, the greater number of spherical particles ranges in size from 65 to 110 μm. These particles have no defects and are characterized by a smooth exterior surface. The presence of coarse particles (130-300 μm) of irregular shape indicates the formation of big drops during the plasma treatment, which are then sintered but not able to melt.

5. Conclusions

These investigations allowed producing spherical particles having the size range of 65–110 μm based on fly ash generated from the coals combustion collected at Seversk TPP, Tomsk region, Russia. The spherical particles were obtained as a result of fly ash treatment with low temperature plasma energy. The exterior surface of the obtained spherical particles is smooth and without defects. The test bench was designed to carry out the experiments on spheroidization of fly ash-based particles. Thus, the conditions favorable for the particle transition to a viscous flow state were achieved within the temperature range observed at 20 mm distance from the jet center. The fly ash particles underwent the XRP analysis both grinded (original state) and plasma-treated. Using the Rietveld method, it was determined that fly ash contains 56.23 wt.% SiO_2, 20.61 wt.% Al_2O_3 and 17.55 wt.% Fe_2O_3 phases that make the dominating contribution to the integral (experimental) intensity of the diffraction maximums. The LTP treatment resulted in a complex redistribution of the amorphous phase amount in the obtained spherical particles, i.e. the considerable decrease of O_2Si phase; increase of $O_{22}Al_{20}$ and Fe_2O_3 phases and change in Al, O density of $O_{22}Al_{20}$ chemical unit cell.

References

[1] Iordanidis A, Buckmanb J, Triantafyllou A and Asvesta A 2008 Fly ash–airborne particles from Ptolemais–Kozani area, northern Greece, as determined by ESEM-EDX *International Journal of Coal Geology* **73** 63–73

[2] Bajukov O, Anshits N, Petrov M, Balaev A and Anshits A 2009 Composition of ferrospinel phase and magnetic properties of microspheres and cenospheres from fly ashes *Materials Chemistry and Physics* **114** 495–503

[3] Wang D, Zhang Y, Dong A, Tang Y, Wang Y, Xia J and Ren N 2003 Conversion of fly ash cenospheres to hollowmicrospheres with zeolite/mullite composite shells *Adv. Funct. Mater.* **13** 563–567

[4] Zhukov A, Arkhipov V, Bondarchuk S and Gol'din V 2013 Evaluation of the morphology of particles produced by plasma-chemical synthesis of ceramic powders *Russian Journal of Physical Chemistry* **7** 777-782

[5] Volokitin O,Volokitin G and Skripnikova N 2014 Mineral fiber production based on ash from the republic of Kazakhstan using low-temperature plasma apparatus *Glass and Ceramics* **70** 340-343

[6] Volokitin G, Skripnikova N, Volokitin O and Volland S 2011 Technology for producing mineral

fibers by recycling ash-sludge and oil-shale wastes *Glass and Ceramics* **68** 239-241
[7] Klopotov A A, Abzaev Yu A, Potekaev A I and Volokitin O G 2012 Principles of X-Ray Analysis in Materials Studies (Tomsk: TGASU)
[8] Abzaev Yu A, Sarkisov Yu S, Klopotov A A et al. 2012 Fullprofile x-ray analysis of clinker mineral C4AF *Vestn. TGASU* **12** 200-209

IOP Conf. Series: Materials Science and Engineering **156** (2016) 012044 doi:10.1088/1757-899X/156/1/012044

Energy-efficient LED irradiator for greenhouse cropping

T V Grechkina, V I Korepanov, A Sh Omarkhan and S B Turanov
Departments of laser and lighting engineering, Tomsk Polytechnic University, 30 Lenin Avenue, Tomsk, 634050, Russia

Abstract. The paper presents the results of modeling and calculation of highly directional optimal light intensity curve (LIC). The authors proved the effectiveness of using the calculated LIC for lighting of greenhouse plants. A method for placing fixtures with LIC in a typical greenhouse was recommended

1. Introduction

Ideal conditions for plant development at any time of the year in any region of the earth can be created in the greenhouse. Constantly increasing scale of this mode of production of agricultural products, the need to improve the productivity of greenhouses requires introduction of new technologies, including those for supplementary lighting of plants. This became possible after the development of high-power LEDs that make possible to design irradiation facilities (IF) of different radiation spectra and to provide the control of exposure parameters.

Many researchers note the advantages of LED irradiation, and manufacturers are working intensively to create special LEDs and irradiators for greenhouse IF [1–4]. The basic principles of the energy-efficient IF for greenhouses are reported in [5]. In this paper, we present the results that allow more specific requirements for the parameters, design and radiation spectrum of a universal adaptive irradiator for supplementary lighting of various kinds of plants cultivated in greenhouses. The optimal requirements to radiation sources for plant growing were as follows:

- proper plant development requires the irradiation well balanced in spectrum and flux in the range of 380–750 nm.
- plants are adapted to the quality and flux of the light under which they formed.
- the requirements for radiation quantity and quality may vary during plant development.

Consequently, the irradiator for supplementary lighting of plants should provide the required reference radiation spectrum and the possibility of its variation. The irradiator design should be convenient for its installation and operation, and provide admissible fluxes and their uniform distribution across the area (volume) of the irradiated plants.

IF is to be adaptive to seasonal, daily, hourly, and continual changes in solar radiation parameters and changes in the quantity and quality of the radiation required for plant development.

2. Experimental methods

The main objective when designing the adaptive LED IF is the choice of the reference radiation spectrum through selection of LEDs emitting in different spectral regions. Selection of the emission spectrum is typically associated with the action spectrum of photosynthesis in plants [6-9]. N.N. Protasova, the author of the most frequently referred research [10], defined the "optimal" ratio of the radiance (W/m2) for plants for the wavelength range (nm): 400–500/500–600/600–700 equal to 30/20/50, respectively. The reference spectrum can be changed by variation of the LED current.

As it was shown in [5], the optimal and variable ratio in the LED module radiation spectrum can be created from various combinations of colored and white LEDs. With respect to the action spectra of photosynthesis and photomorphogenesis, the optimal and universal irradiator is a three-color version: white (warm color), blue (λp=420–465 nm) and red (λp=660 nm). Three types of LEDs allow change in the total flux, and simultaneously they can be used to create any combination of the spectrum.

White LED is the basis for creating a reference spectrum. It covers a wide spectral range and can provide excitation of all pigments which cause the photochemical processes in plants, including the photocontrolling one. The latter requires significantly less light energy compared to photosynthesis. The LED Nichia NS3W183TS, which has the widest spectral range (420–750 nm), is most suitable for this purpose.

For energy-intensive photosynthesis and control of photochemical processes in the plant, white LED should be mixed with blue and red LEDs. The blue LED Nichia NCSB119T is suitable for this purpose. The PHILIPS LED LXM3-PD01 can be used for the red spectral region.

The ratio derived by N.N. Protasova was taken as a criterion to select the number of white, blue and red LEDs in the irradiator. The SPECTRA software was used for LED selection [11]. We found that two white LEDs (1W), one blue LED (1W) and one red LED (3W) provide the flux ratio in the blue/green/red regions of the total spectrum equal to 26/15/59 that is close to the target flux ratio equal to 30/20/50. The radiation spectrum of the irradiator with this flux ratio is shown in figure 1.

A more economical version is a two-color irradiator which consists of the white Nichia LED and red Philips LED. However, this combination cannot provide the Protasova's value. A special LED with an enhanced blue component in the radiation spectrum is required to implement this version. However, the spectrum of the three-color irradiator is easier to change if compared to that of the two-color one.

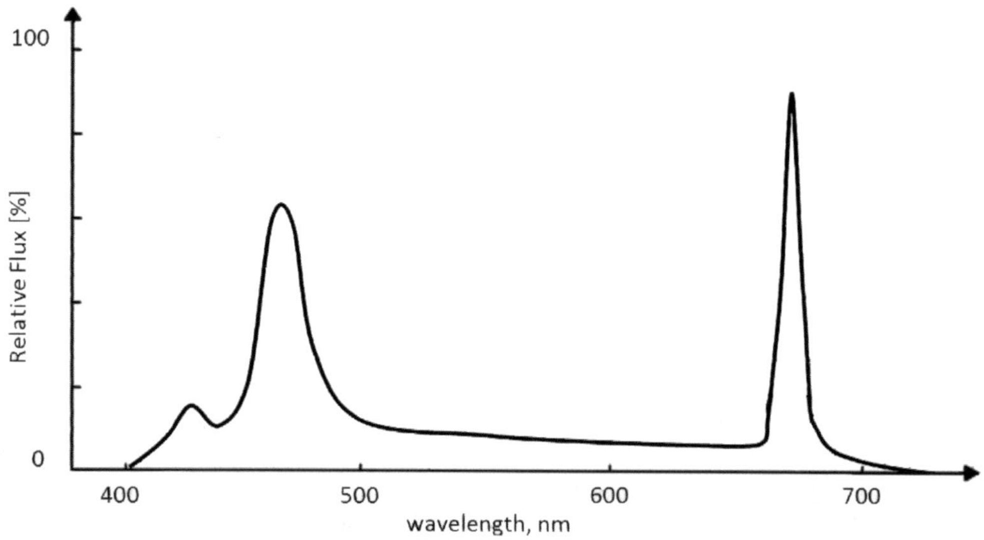

Figure 1. Irradiation spectrum of the combination of the warm white and blue Nichia LEDs and red Philips LED.

Another important task is to develop an optimal design of the emitter and optimal light intensity curve (LIC). The main purpose of the proposed irradiator is supplementary lighting of tall plants in industrial greenhouses. The irradiator should be of the sizes to provide the lowest shading coefficient. LED sources make possible to develop irradiators of virtually any shape and power. Since tall plants (cucumbers, tomatoes) in a greenhouse are planted in a long row one meter apart, it is advisable to make a fixture with a length of 1–1.5 meter with an approximate cross-section of $0.05x0.1m^2$ and to place it along the rows of plants. PHILIPS manufactures the irradiators of this size [1].

First of all, we find the location of the irradiator in the greenhouse and the predominant direction of the flux incident on tall plants. The analysis showed the optimal location of fixtures at a height of 4–6 meters (typical greenhouse) above the rows of plants. In this case, the side lighting of the plant can be arranged to provide sufficiently uniform irradiation of all the leaves along the stem. The selected scheme of lighting fixtures exposure and location is shown in figure 2 (a).

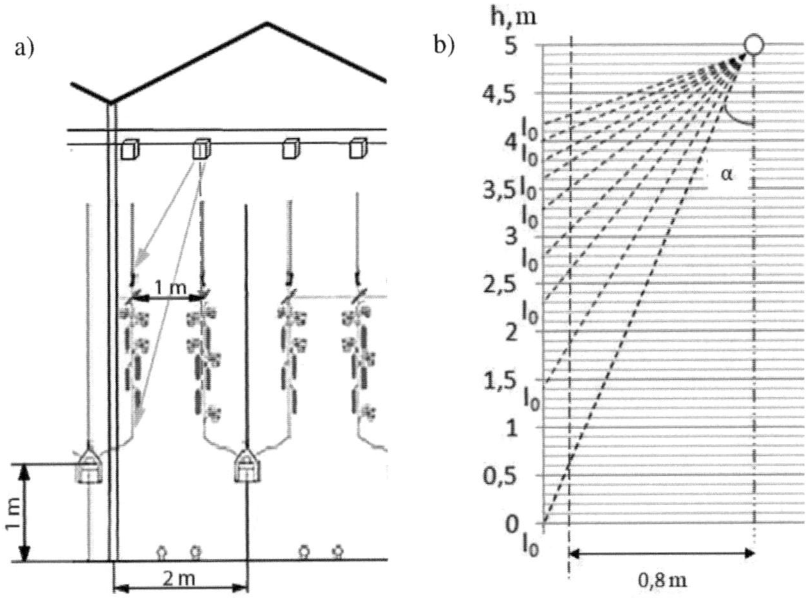

Figure 2. Arrangement of plants, fixtures, areas, and direction of incident beams (a), and the schematic diagram of the LIC model for LED irradiator (b).

To calculate IF, it is required to determine the model of the plant. For closely planted low-growing plants, IF can be modeled as a flat horizontal plate [12, 13]. Modeling of IF for tall plants several meters high is more complicated. In our opinion, a good approximation is a distribution of irradiance when it is uniform in the areas located perpendicular to the direction of propagation of the beams incident on the plant at different angles along its axis (figure 2a, b). This modeling is similar to the arrangement of the leaves of plants.

We have considered two versions of light intensity curve formation. The first version considers a point radiation source. These properties can be inherent in irradiators formed by groups of LEDs. In this case, the irradiance in the area perpendicular to the direction of propagation of the beam I0 (figure 2, b.) is proportional to l_2, where l is the distance from the source to the point of the beam incidence on the vertical plane along the plant. Using this law, we found the target values of the initial intensities I01 for beams emitted by the point source at different angles with I_0 constant at any point of the calculated vertical plane. To perform the calculation, we placed the vertical plane at a distance of 0.2 m from the plant stem, i.e. at a distance of 0.8 m from the light source.

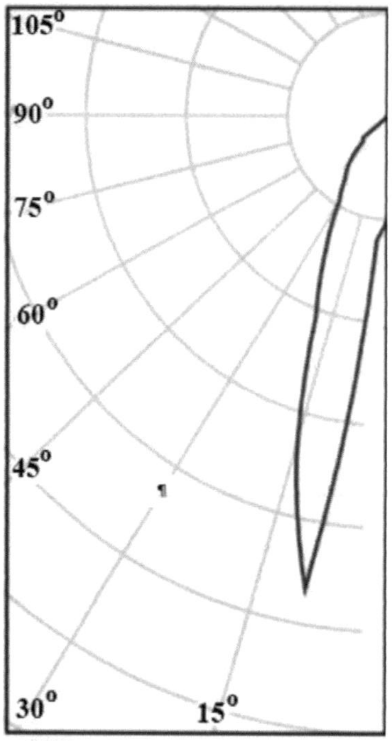

Figure 3. LIC of the irradiator in the IES format for point light source

The calculation data was used to plot LIC in the form of the dependence $I_{01}=f(\alpha)$, where α is the angle between the vertical plane and the direction of beams. To calculate and analyze the irradiance, the IES file (figure 3) was formed using the DIALux software.

After that, we simulated the radiance distribution for different planes of the model greenhouse facility with lamps of the calculated LIC. The analysis of the results suggests the following findings and conclusions.

3. Conclusions

An optimal version of the reference spectrum for adaptive irradiator is a three-color irradiator with white LED emitting in a broad spectral range (400–750 nm). Energy-efficient two-color irradiator can be designed based on red and white LEDs with increased blue component in the radiation spectrum. An optimal irradiator for tall plants should have a spot light distribution diagram. To calculate IF (irradiance standardization) for greenhouses with tall plants, it is necessary to develop a plant model taking into account the geometry of leave arrangement along the stem. Irradiators with the calculated LIC can uniformly irradiate both tall and low growing plants, so they can be used for growing all kinds of plants in greenhouses of various types.

Acknowledgments
This study was carried out under the state assignment of the Ministry of Education and Science of Russian Federation for 2014-2016 (research work № 8.2500.2014/K).

References

[1] Catalog Philips: www.philips.com/horti (accessed date: 01.07.2016)
[2] Catalog Osram: http://www.osram.ru/osram_ru/ (accessed date: 27.06.2016)

[3] Catalog Atom Svet: http://www.atomsvet.ru/ (accessed date: 27.06.2016)
[4] Catalog ArtLeds: http://www.artleds.ru/ (accessed date: 27.06.2016)
[5] Korepanov V I and Kozyreva I N 2014 Methods for creating adaptive energy conserving irradiation units for greenhouses *Russian Physics Journal (in Russian)* **57 9/3** 89-93
[6] Voskresenskaya N P 1965 *Photosynthesis and spectral content of light* (Moscow: Nauka) p 311
[7] Sventitckiy I I 1972 Evaluation of photosynthetical efficacy of optical radiation *Svetotekhnika (in Russian)* **4** 23–25.
[8] Ritchie R J 2010 Modelling photosynthetic photon flux density and maximum potential gross photosynthesis *Photosynthetica* **48 4** 596–609.
[9] Ichiro Terashima, Takashi Fujita, Takeshi Inoue, Wah Soon Chow, Riichi Oguchi 2009 Green light drives leaf photosynthesis more efficiently than red light in strong white light: revisiting the enigmatic question of why leaves are green *Plant and cell physiology* **50(4)** 684–697.
[10] Protasova N N 1987 Photoculture as a way to identify the potential productivity of plants *Plant Physiology* **34 – 4** 812–822
[11] Spectrum simulator: www.spectra.1023world.net (accessed date: 27.06.2016)
[12] Kositsyn O A 1978 Mathematical modeling of spatial characteristics of biological radiation detectors *Lighting Engineering* **6** 15–16.
[13] Kositsyn O A, Suetinov G S and Ovsyannikov E A 2013 Simulation of the irradiance indicatrix for closely spaced bodies with spherical surfaces *Collection of research papers MSAU (Agroengineering)* **3 18** 20–21.

Evaluation of the effect of led irradiator spectral content on the development of greenhouse plants

S B Turanov, A N Yakovlev, V I Korepanov, D A Buzmakova and T V Grechkina

Departments of laser and lighting engineering, Tomsk Polytechnic University, Tomsk, 30 Lenin Avenue, 634050, Russia

E-mail: tyrsb@yandex.ru

Abstract. The work presents the results of an experiment studying the effect of radiation spectral content (considering its equal intensity in terms of photosynthetically active radiation) on the growth and development of greenhouse model plants. The results demonstrate that the effective development of model subjects requires the adaptation of radiation spectral content according to the growth period and plant type, unlike the illumination level. The obtained results demonstrate the necessity of the creation of adaptive irradiation unit.

1. Introduction

The effective development of plants requires a certain complex of favorable external conditions. Nature not always provides a complete set of such conditions. Hence, the humanity strives to artificially achieve them, for instance, by growing plant in greenhouses. Considering the increasing scale of greenhouse business, one of the determinative factors of its development is the implementation of energy conserving technologies, which is required by the Federal Law On energy consumption and increase of energy efficiency [1].

One of the primary subjects of energy efficiency improvement is the system of artificial illumination in greenhouses. For this reason, Russian agricultural complex demonstrates increased interest in application of LED irradiators that have been actively developing in recent years exhibiting decreasing price and augmenting luminous efficacy. Over time, initial and operational costs for LED illumination will become lower than those for conventional light sources (metal halide, sodium-vapor and fluorescent lamps). In the present time, the understanding of the whole complex of actions towards increased energy efficacy of LED-based irradiation units (IUs) for greenhouses remains unclear. In this connection, the majority of LED lamps for plants (phyto lamps) currently presented on the market [2, 3] feature the radiation spectrum primarily formed by blue (440–470 nm) and red (630–660 nm) monochromatic LEDs. Such phyto lamps like conventional light sources lack the ability of optimizing radiation spectral content, which is known to determine the energetics of photosynthesis and control this process [4-10].

The work [11] attempts to formulate the principles for increasing energy efficacy of IUs for greenhouses that are based on the following ideas:

- formation of a spectral content (by combining color LEDs) and radiation intensity necessary and sufficient for plant development with enhanced (if possible) fraction of low-energy red radiation;

- development of energy efficient, low-cost custom greenhouse plant growing LEDs and IUs on their basis (for instance, GaN-based with different combination of luminophores);
- control of spectral content, radiation intensity according to the requirements of a specific type of plant at given development stage, i.e. creation of adaptive IUs.

Practical implementation of such principles requires the understanding of the character and extent of radiation spectral content effect on the growth process at all development stages of each plant type. Such information is widely published in monographs and periodicals. Nevertheless, each of such publications has its own goals and only indirectly can be used for practical realization of energy consumption principles for IUs, especially for those based on LEDs. Real implementation requires a data bank containing the information about LED-illumination effect (for certain type of color industrial LEDs) on the plant growth processes. Such data will enable the creation of an IU with optimal basic radiation spectral content featuring adjustable color range. Such light will be more affordable. Evidently, it should contain various color LEDs and should have adjustable light flow within the boundaries of photosynthetically active radiation (PAR).

This work is aimed at substantiating the choice of industrial LEDs and their combination for creating a greenhouse adaptive lamp with optimal basic radiation spectral content. The substantiation will be based on experimental studies of the effect of LED illumination with different colors on the growth and development of model plants in phytotrons.

2. Experimental methods

Model plants were grown in three prototypes of greenhouses (figure 1), phytotrons (PhT). The analysis of the processes of growth and development of model plants involved the determination of arithmetic mean values of the following parameters: shoot height (from the ground to the first truss), width (for the widest part of a leaf) and length of a leaf (along the length of a lamina). The increment of the parameters was determined at each plant development stage by comparing the parameter value in the beginning and in the end of a stage. To decrease the error and eliminate the influence of initial seed quality, the measurements were made using two (for tomato and cucumber) and six (for lettuce) shoots.

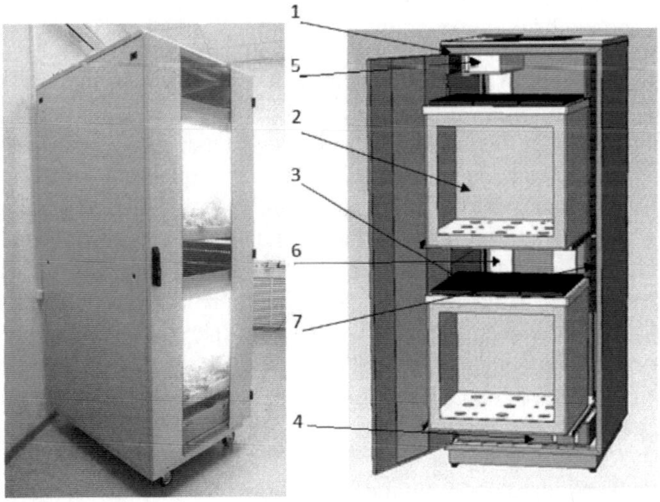

Figure 1. Prototype of "smart" greenhouse, phytotron, where 1 is the case of the research installation; 2 are phytotron cells; 3 are lamps with independent adjustment of plant irradiation regime; 4 and 5 are elements of watering system (lower and upper tanks); 6 is microclimate support system; 7 is electronic control and power supply system.

Plants in each of phytotrons were irradiated by special LED irradiators that allow forming in the process of the experiment any color gamma of the irradiation unit with the wavelength from 400 to 700 nm. Working modes of each LED comprising LED modules were constant and selected in such a way that IUs in all phytotron cells had equal total flux in the range of photosynthetically active radiation (PAR), but had different irradiation spectrum. Selection of modes was performed by altering LED current. This is the easiest and aptly controllable method for changing the intensity of LED radiation, i.e. selecting irradiation spectrum of a whole irradiation unit.

3. Results and Discussion

Were studied the growth of three types of model plants under LED irradiation: «Battler greenhouse tomato», «April greenhouse cucumber» and «Moscow greenhouse lettuce». In the course of the experiment, the duration of continuous plant irradiation during the day (phytoperiod) was 12 hours. Regarding the photoperiodic response, the chosen specimens are neutral-day plants. So, the chosen regime lies within the region of optimal duration of daylight irradiation exposure for the plants under study.

Table 1 contains the selected modes of LED module operation, illumination intensities and corresponding fluxes in the range of PAR. We should note that the spectral characteristic of plant leaf sensitivity to photosynthesis differs from that of human eye [12, 13]. Hence, the evaluation of irradiation unit efficacy using term "illumination" is not adequate and is used only for comparison purposes, since in many practical cases, greenhouse businesses specifically use illumination intensity to evaluate light flux incident at plant formation.

Table 1. LED module operation mode

No.	Operation mode	Current (mA)			E	PF_{PAR},
		Red	Blue	White	(klx)	(μmoles/m^2*s^{-1})
PhT1	continuous spectrum, prevailing red component	350	0	53	3.57	
PhT2	continuous spectrum, prevailing white component	197	0	350	4.04	120±10
PhT2	discrete spectrum, red and blue component	176	350	0	1.51	

To eliminate the influence of illumination intensity on the processes of plant growth and development, the operation modes of LEDs were selected in a such manner that photosynthetic photon flux densities (PPFD) in phytotrons with IU having different spectra were equal.

To accurately determine the photosynthesis efficiency under the influence of radiation fluxes with different spectrum, it is necessary to evaluate the PPFDs in quantum units, since elementary act of photochemical reaction is determined only by the quantity of light quanta absorbed by a chlorophyll molecule. Nevertheless, the understanding of the amount of quanta falling on the plant cover is not sufficient, for the efficacy of radiation absorption by a plant depends on the energy of incident quantum. The absorption spectrum of leaves can be represented by the absorption spectrum of chlorophyll molecule and basic pigments. Sometimes, the spectral characteristic of leaf susceptibility to incident radiation can be depicted by Sventiskiy curve [12, 13]. However, this curve is known to be non-universal, especially for different plant growth stages [5, 6, 14]. Besides, in the spectrum region where the chlorophyll molecule demonstrates no absorption, the effective absorption of incident radiation is specific for molecules of other pigments, which can transfer excitation energy to chlorophyll. Thus, at a first approximation, the estimation of the amount of energy absorbed by plants in quantum units is possible, assuming that the absorption coefficient of leaves is constant within PAR. The estimation of PAR was based on the illumination spectra of modules measured in given operation modes of LEDs by AvaSpec-2048 USB2 spectrometer (figure 2). The spectra were generated with a due consideration of spectral response characteristic of spectrometer's measuring track.

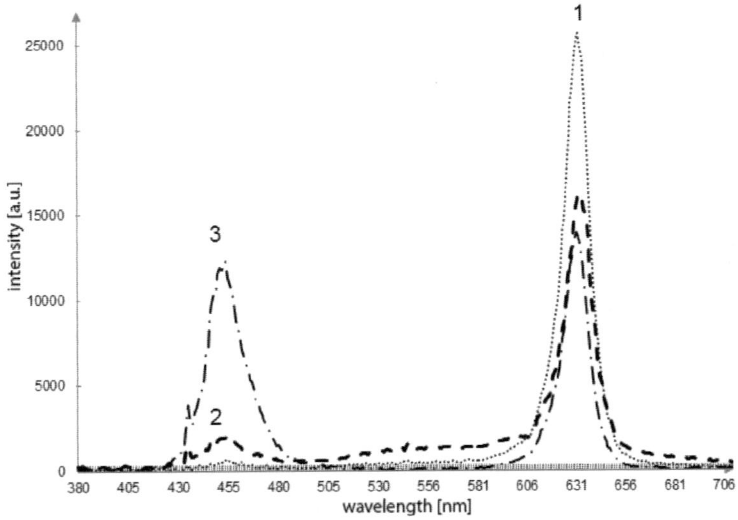

Figure 2. Radiation spectra of experimental LED modules.
1) is PhT No. 1, 2) is PhT No. 3 and 3) is PhT No. 3.

Obviously, in illumination spectra of LED modules, radiation bands in blue, green and red regions prevail. Spectrum 1 contains primarily red component of radiation and small part in blue and green regions. Spectrum 2 contains blue, green and red components of radiation spectrum. Spectrum 3 contains practically equal parts of blue and red radiation. Selection of such spectra was stipulated by the participation of these spectrum components in the process of photosynthesis. The goal of increasing the energy-efficiency is confined in selecting those compositions of blue, green and red components, which would provide for necessary conditions for photosynthesis and would be more beneficial from the point of view of LED module energy efficiency. Obviously, the long wave radiation is more effective; but application of solely long-wave radiation is impossible [4–8].

The first development stage lasts from the moment of embryo formation (i.e. up-coming plant) to the reveal of a shoot on a ground surface.

The duration of the first development stage from seeding to the appearance of a sprout on the ground surface amounted to 5, 7 and 8 days for model plants of lettuce, cucumber and tomato, respectively. The earliest shoots were found in the phytotron with PhT3 mode. The latest shoots were found in the phytotron with PhT2 mode.

The second development stage starts from the appearance of a sprout on the ground surface and ends with the formation of leaf primordia on the main stem. The results of model plant development at this stage are presented in table 2.

Table 2. Measurements at the second stage of model plant development

Phytotron	Tomato			Cucumber			Lettuce		
number	Δh (%)[a]	Δl (%)	Δb (%)	Δh (%)	Δl (%)	Δb (%)	Δh (%)	Δl (%)	Δb (%)
No. 1	92.7	84.3	38.4	90.7	84.1	88.9	88.4	76.9	66.7
No. 2	71.4	48.9	52.9	89.5	84.6	90.6	81.8	71.4	55.6
No. 3	84.6	81.4	46.7	95.2	95.9	93.3	75	71.4	50

[a] Δh is the increment of shoot height, Δl is the increment of leaf length, Δb is the increment of leaf width.

The maximum increment of tomato sprout height (Δh) was observed in the phytotron with PhT3 mode and amounted to 92.7%. The maximum increment of tomato leaf length (Δl) was observed in the phytotron with PhT1 mode and amounted to 84.3%. The maximum increment of tomato leaf width (Δb) was observed in the phytotron with PhT2 mode and amounted to 52.9%.

The maximum increment of cucumber shoot parameters was observed in the phytotron with PhT3 mode. The increment of height was 95.2%, increment of length was 95.9% and increment of leaf width was 93.3%.

The maximum increment of lettuce shoot parameters was observed in the phytotron with PhT1 mode. The increment of height in this PhT was 88.4%, increment of length was 76.9% and increment of leaf width was 66.7%.

The third development stage begins with the formation of leaf primordium of the main stem and ends with the differentiation of a growing point on the main stem. The formation of leaves at this stage almost terminates. The results of model plant development at this stage are presented in table 3.

Table 3. Measurements at the third stage of model plant development

Phytotron		Tomato			Cucumber			Lettuce	
number	Δh (%)[a]	Δl (%)	Δb (%)	Δh (%)	Δl (%)	Δb (%)	Δh (%)	Δl (%)	Δb (%)
No.1	21.6	41.7	52.9	20.4	48.8	67.8	32.5	71.7	74
No. 2	27.3	42.6	44.4	18.9	45.5	62.5	11.5	72.9	77.1
No. 3	38.4	40	43.3	14.6	40.6	60	0	59.5	75

[a] Δh is the increment of shoot height, Δl is the increment of leaf length, Δb is the increment of leaf width.

The maximum increment of tomato shoot height was observed in the phytotron with PhT3 mode and amounted to 38.4%. The length of tomato leaves in all three phytotrons increased equally (the increment was 41%). The maximum increment of tomato leaf width was observed in the phytotron with PhT1 mode and amounted to 52.9%.

The maximum increment of cucumber shoot height was observed in the phytotrons with PhT2 and PhT3 modes and amounted to 20.4% and 18.9%, respectively. The maximum increment of cucumber leaf length was observed in the phytotron with PhT1 mode and amounted to 48.8%, while the maximum increment of cucumber leaf width was observed in the phytotron with PhT1 mode and amounted to 67.8%. Thus, at this development stage, the best development conditions for cucumber were in the phytotron with PhT1 mode.

The maximum increment of lettuce leaf length was observed in the phytotron with PhT1 mode and amounted to 32.5%. The enhancement of lettuce length in the phytotron with PhT3 mode at this stage stopped almost completely. The maximum increment of lettuce leaf length was observed in the phytotron with PhT2 mode and amounted to 72.9%, while the maximum increment of lettuce leaf width was observed in the phytotron with PhT3 mode and amounted to 75%.

4. Conclusions

The presented results demonstrate that at the first and the second development stages all plant types (in addition to other colors) require a certain amount of blue (high-energy) radiation. Its fraction in IU radiation spectrum depends on the plant type and its growth time. Lettuce requires almost none of it, while tomato and cucumber at early growth stages need limited amount of it, since the increase of the fraction of blue radiation can lead to retarded plant development. At the third stage, the development of all plants requires mainly red radiation.

Thus, the presented results clearly testify that the energy efficacy of LED irradiation units can be increased by selecting an optimal radiation spectral content and consequently varying it depending on the plant growth stage, i.e. by creating adaptive irradiation units. Whilst the basic radiation spectra (i.e. those which will be adjusted afterwards) can be different for different types of agricultural plants.

The study also demonstrated the possibility of creating adaptive IUs on the basis of off-the-shelf industrial LEDs. Application of three LED types (blue, green and red) for greenhouse lamps is the most universal solution for adjusting the spectral content of the IU. However, white LEDs are known to have a fairly large fraction of blue spectrum component. Following the results of this work, this component can be sufficient for creating adaptive IU with optimal basic radiation spectrum in terms of color. Since white LEDs become more and more affordable, their implementation can be the most advantageous from the perspective of cost of greenhouse IUs. Therefore, the energy-efficient IU can be made on the basis of only white and red LEDs.

Acknowledgments
The work was supported by the Ministry of Education and Science of the Russian Federation in the frame of State Program in the field of scientific activities No. 8.2500.2014/K.

References

[1] Federal Law of the Russian Federation No. 261-FL as of November 23, 2009 On Energy Conservation and Increase of Energy Efficacy and on Amendments to Certain Regulations of the Russian Federation: http://www.rg.ru/2009/11/27/energo-dok.html (accessed date: February 16, 2015)

[2] NeoSol DS 520W LED Plant Grow Luminaire 48": https://www.hydrofarm.com/p/ILDS520 (accessed date: February 17, 2015)

[3] Philips Horticulture Lighting products and downloads: http://www.lighting.philips.com/main/application_areas/horticultural/overview-products.wpd

[4] Voskresenskaya N P 1965 *Photosynthesis and spectral content of light* (Moscow: Nauka) p 311

[5] Protasova N N, Wells G M, Dobrovolsky M V and Tsoglin L N 1990 Spectral characteristics of light sources and peculiarities of plant growth under conditions of artificial illumination (in Russian) *Russian Journal of Plant Physiology* **37/2** 386 – 396

[6] Tikhomirov A A, Leesovsky T M and Sidko F Ya 1991 *Spectral content of light and plant capacity factor* (Novosibirsk: Nauka (Siberian branch)) p 168

[7] Johkan M, Shoji K, Goto F, Hahida S and Yoshihara T 2012 Effect of green light wavelength and intensity on photomorphogenesis and photosynthesis in Lactuca sativa *Envirimental and Experimental Botany* **75** 128 – 133

[8] Yakovlev A N, Kozyreva I N 2013 Influence of spectral characteristics of light sources on plant growth *Proceedings of the universities. Physics. (in Russian)* **56 – 7/2** 112 – 116

[9] Yakovlev A N, Kozyreva I N, Turanov S B, Stepanov S A and Kruglikova L L 2012 Study of spectral characteristics of single leaves of plants *Proceedings of the universities. Physics. (in Russian)* **55 6/2** 158 – 162

[10] Yakovlev A N, Turanov S B, Kozyreva I N and Starodubtseva D V 2014 Sources with Different Spectra Radiation Influence on Plants Growth and Development *Advanced Materials Research* **1040** 830-834

[11] Korepanov V I, Kozyreva I N 2014 Methods for creating adaptive energy conserving irradiation units for greenhouses *Proceedings of the universities. Physics. (in Russian)* **57 – 9/3** 89-93

[12] Sventitckiy I I 1965 Measurement of optical radiation for plant growing *Svetotekhnika (in Russian)* **4** 19-23

[13] Sventitckiy I I 1972 Evaluation of photosynthetical efficacy of optical radiation *Svetotekhnika (in Russian)* **4** 23–25

[14] Turanov S B, Kozyreva I N, Goncharov A D and Yakovlev A N 2014 Physical principles for creation of LED irradiators of specified spectral content *Proceedings of the universities. Physics. (in Russian)* **57 –9/3** 94-97